The Microbiology of the Terrestrial Deep Subsurface

The Microbiology of Extreme and Unusual Environments

The Microbiology of the Terrestrial Deep Subsurface

Edited by

Penny S. Amy
Dana L. Haldeman

LEWIS PUBLISHERS

Boca Raton   New York

Publisher:	Robert B. Stern
Editorial Assistant:	Carol Messing
Project Editor:	Helen Linna
Marketing Manager:	Susie Carlisle
Direct Marketing Manager:	Becky McEldowney
Cover design:	Dawn Boyd
PrePress:	Kevin Luong
Manufacturing:	Sheri Schwartz

Library of Congress Cataloging-in-Publication Data

The microbiology of the terrestrial deep subsurface / Penny S. Amy and Dana L. Haldeman
       p. cm.
     Includes bibliographical references and index — (CRC LLC series in the microbiology of
       extreme and unusual environments)
     ISBN 0-8493-8362-5
     1. Subsurface microbiology.   2. Microbiology — microbial ecology.   I. Amy, Penny S..
       II. Title.   III. Series
   QR749.H64G78 1997
   616'.0149--dc20   96-82013
                                                                    CIP

No claim to original U.S. Government works
International Standard Book Number 0-8493-8362-5
Library of Congress Card Number 96-82013
Printed in the United States of America   2 3 4 5 6 7 8 9 0
Printed on acid-free paper

Preface

This book arose from our desire to organize a comprehensive compendium of research on the microbial ecology of the deep terrestrial subsurface. It is not all-inclusive but rather represents areas that are critical for both the understanding of subsurface research and the practice of it.

Many people have been involved with support for, and the production of, this book. First, we are grateful to all of the authors for their contributions. Recognition goes also to Ms. Mona Khalil and Ms. Chiaki Brown for reformatting countless references. We would like to thank our families for the much needed and cheerfully given support that we received during the long journey from concept to fruition.

A large measure of recognition goes to Dr. Frank Wobber of the U.S. Department of Energy for his vision of the U.S. DOE Subsurface Science Program. The interdisciplinary nature of the program has been educational and inspirational to everyone who has participated in the discovery of subsurface microbiology.

We thank the U.S. DOE for ten years of continuous funding of a most worthwhile program, and the largest microbial ecology initiative in the history of the U.S.

This book is dedicated to the husbands

of all busy and ambitious women,

and specifically,

to Michael and Jess.

The Editors

Penny S. Amy, Ph.D. is Professor of Microbiology and Chair of the Department of Biological Sciences at the University of Nevada, Las Vegas.

Dr. Amy received her B.S. and M.S. degrees from Portland State University, Portland, Oregon, in 1972 and 1974, respectively. After teaching for a number of years in Oregon and Alaska, she returned to graduate school to obtain a doctorate in marine microbiology from Oregon State University in 1982. Following post-doctoral appointments at Stanford University and Oregon State University, she was hired as Assistant Professor at the University of Nevada, Las Vegas in 1985. She was promoted to Associate Professor and Professor in 1989 and 1993, respectively.

Dr. Amy is a member of the American Society for Microbiology, the Arizona Branch of that organization, Sigma Xi, and the Association for Women in Science and the southern Nevada branch of that society.

The Board of Regents of the University of Nevada System recognized Dr. Amy for Teaching in 1989 and for Research in 1990. She was given the Barrick Scholar Award for Research in 1992 at the University of Nevada, Las Vegas, and has been recognized as one of the Distinguished Women of Southern Nevada in both 1993 and 1995.

Dana L. Haldeman, Ph.D. is currently a Research Associate in the Department of Biological Sciences at the University of Nevada, Las Vegas.

Dr. Haldeman graduated from the University of Idaho with a B.S. degree in Bacteriology in 1986. Additionally, she obtained certification for teaching chemistry and biology at the secondary level. She taught physical science and biology in the Clark County School District for three years before continuing her graduate education. Her doctoral research focused on the microbial ecology of the subsurface, and she was granted a Ph.D. in biology from the University of Nevada, Las Vegas in 1994.

Dr. Haldeman is a member of the American Society for Microbiology, the Association for Women in Science, and the Arizona-Nevada Academy of Sciences. Among other awards, she was recognized for the "Outstanding Dissertation" from the Department of Biological Sciences, College of Science and Mathematics, and the University of Nevada, Las Vegas for 1994.

Dr. Haldeman has published over 30 abstracts and peer-reviewed articles on her research. Her research interests include distribution and diversity of subsurface microbiota; microbiota associated with nuclear waste disposal; and resuscitation, dormancy and survival of subsurface bacteria.

Contributors

Penny S. Amy, Ph.D.   Department of Biological Sciences, University of Nevada, Las Vegas, Nevada

David L. Balkwill, Ph.D.   Applied Geology and Geochemistry Group, Department of Biological Sciences, Florida State University, Tallahassee, Florida

Bruce N. Bjornstad, M.S.   Pacific Northwest National Laboratory, Richland, Washington

David R. Boone, Ph.D.   Department of Biological Sciences, *and* Department of Chemistry, Biochemistry, and Molecular Biology, Oregon Graduate Institute of Science and Technology, Portland, Oregon

Fred J. Brockman, Ph.D.   Pacific Northwest National Laboratory, Richland, Washington

Nick Christoffi, Ph.D.   Pollution Research Unit, Department of Biological Sciences, Napier University, Edinburgh, Scotland

Frederick S. Colwell, Ph.D.   Department of Biotechnology, Idaho National Engineering Laboratory, Idaho Falls, Idaho

J. William Costerton, Ph.D.   Center for Biofilm Engineering, Montana State University, Bozeman, Montana

Alfred B. Cunningham, Ph.D.   Center for Biofilm Engineering, Montana State University, Bozeman, Montana

J.K. Fredrickson, Ph.D.   Pacific Northwest National Laboratory, Richland, Washington

William Timothy Griffin, Ph.D.   Golder Federal Services, Oak Ridge, Tennessee

Cheryl D. Gullett, Ph.D.   Engineering Department, Clark Atlanta University, Atlanta, Georgia

Dana L. Haldeman, Ph.D.   Department of Biological Sciences, University of Nevada, Las Vegas, Nevada

Terry C. Hazen, Ph.D.   Savannah River Technology Center, Westinghouse Savannah River Company, Aiken, South Carolina

Larry E. Hersman, Ph.D.   Life Sciences Division, Los Alamos National Laboratory, Los Alamos, New Mexico

Randy Hiebert   MSE Inc., Butte, Montana

Kevin Horrigan   Center for Biofilm Engineering, Montana State University, Bozeman, Montana

Garth James, Ph.D.   Center for Biofilm Engineering, Montana State University, Bozeman, Montana

Thomas L. Kieft, Ph.D.   Department of Biology, New Mexico Institute of Mining and Technology, Socorro, New Mexico

Philip E. Long, Ph.D.   Environmental Technology Division, Pacific Northwest National Laboratory, Richland, Washington

Aaron L. Mills, Ph.D.   Department of Environmental Sciences, University of Virginia, Charlottesville, Virginia

Christopher J. Murray, Ph.D.   Pacific Northwest National Laboratory, Richland, Washington

Tommy J. Phelps, Ph.D.   Environmental Science Division, Oak Ridge National Laboratory, Oak Ridge, Tennessee

J.C. Philip, Ph.D.   Pollution Research Unit, Department of Biological Sciences, Napier University, Edinburgh, Scotland

Robert H. Reeves, Ph.D.   Department of Biological Science, Florida State University, Tallahassee, Florida

David B. Ringelberg, M.S.   Dyntel Corporation, USACE Waterways Experiment Station, Vicksburg, Mississippi

Charles E. Russell, M.S.   Water Resources Center, Desert Research Institute, Las Vegas, Nevada

Todd O. Stevens, Ph.D.   Pacific Northwest National Laboratory, Richland, Washington

Paul J. Sturman, Ph.D.   Center for Biofilm Engineering, Montana State University, Bozeman, Montana

Bryan K. Warwood   Center for Biofilm Engineering, Montana State University, Bozeman, Montana

David C. White, Ph.D.   Center for Environmental Biotechnology, University of Tennessee, Knoxville, Tennessee, *and* Environmental Sciences Division, Oak Ridge National Laboratory, Oak Ridge, Tennessee

Contents

Section III   Applications

1

Denizens of the Deep

Penny S. Amy and Dana L. Haldeman

Subsurface microbial presence and involvement in geochemical transformations was reported decades ago (Farrell and Turner, 1931; Kuznetsov et al., 1963; Lipman, 1931), and an acceleration in subsurface research has occurred over the last decade. Much of this research has been collaborative and interdisciplinary, involving the efforts of microbiologists, geochemists, hydrologists, drilling and mining experts, as well as program support, and has thus facilitated rapid progress. A primary incentive has been the potential for microbiota to clean up contaminated underground environments for the protection or purification of water supplies through bioremediation (Ghiorse and Wilson, 1988; Madsen and Ghiorse, 1993; Balkwill et al., 1994). Not long ago most scientists firmly believed that life was absent below some fairly shallow threshold depth (Ghiorse and Wilson, 1988). Microbiota have now been recovered at depths greater than 9000 ft below the surface (Boone et al., 1995). Additionally, investigations of deep ocean sediments indicate that they are teeming with microbial life (Parks et al., 1994). Microorganisms have been investigated in both shallow and deep aquifer systems and associated sediments (Balkwill and Ghiorse, 1985; Hirsch, 1992). Sampling techniques have included both drilling/coring procedures, and mining in rock, ore, and salt deposits (see Chapters 3 and 4). European, Canadian, and American scientists have also studied the microbiology of the terrestrial subsurface for the practical reason that radioactive and other waste repositories are often built underground and microbes inhabiting those environments may impact the integrity of the waste storage facilities (see Chapters 15 and 16, and Proceedings of the 7th Annual International High Level Radioactive Waste Management Meetings, Las Vegas, 1996).

Perhaps among the most intriguing of the findings in the amalgamated effort to understand the subsurface is the fact that no one could have predicted the enormity of biomass (Gould, 1996), diversity (see Chapters 7 and 10), and potential activities (see Chapter 9) present there. For example, some microbiota have been recovered that are adapted to deep, anaerobic, and geothermally heated environments (Boone et al., 1995; Colwell et al., 1996), some with associated bioremediation capabilities (Brockman et al., 1989; Fredrickson et al., 1991) and some with other capabilities of economic importance (see Chapters 14 and 17). There is also an enormous potential to discover novel organisms (see Chapter 10); some with yet unknown talents, e.g., the ability to produce substances of medical or other economic importance. Another new and fascinating aspect of the discovery of microbes in the deep subsurface has been the recent description of a nonphotosynthetically influenced microbial community which lives under anaerobic conditions and is based on energy derived from hydrogen gas (Stevens and McKinley, 1995; and see Chapter 12).

This book, divided into three sections, brings together some of the work that has led to the current understanding of the terrestrial subsurface. Section I, *Considerations for*

Sampling, deals primarily with practical information necessary to geologically characterize a site, sample, and transport geological material for microbiological analysis. Chapter 2 was written by geologists for microbiologists interested in the geochemical and hydrological concerns in site selection and characterization. Chapters 3 and 4 deal with methods of obtaining subsurface material for microbial analysis by drilling and mining techniques, respectively. Chapter 5 describes the importance of rapid analysis of microbiological samples and the associated storage-related phenomenon, while Chapter 6 discusses the importance of the spatial heterogeneity of microbes in the subsurface, scales of heterogeneity, and geostatistical methods that can be used to enhance sampling schemes.

Section II, *Microbial Ecology and Related Materials*, provides a basic understanding of methods and techniques that have been used in other environments and have been successfully adapted to the deep subsurface. Chapter 7 gives an overall description of the microbial diversity found in the subsurface with an emphasis on chemoheterotrophic microbes. Chapter 8 describes techniques for analyzing both whole samples and individual isolates by lipid analysis. The use of lipids as a measure of biomass, and signature lipids as signals of specific microbial identity and/or community physiological status, are described for subsurface environments. Chapter 9 presents analytical methods to determine activities of bacterial functional groups, and describes results of subsurface investigations. Also presented in Chapter 9 are physical and chemical factors believed to limit microbial activities as well as problems associated with estimating *in situ* microbial activity. The phylogenetic analysis of microbial communities, using 16s ribosomal DNA sequences to relate microbes to one another and to common surface microbes, is addressed in Chapter 10. The survival, dormancy, and resuscitation of subsurface microbiota with implications for microbial ecology are topics of Chapter 11. Another chapter in this section addresses the natural cycle of lithic environments, the possibility of the terrestrial subsurface origin of life, and perhaps the colonization of the earth by exobiology and life on other planets (Chapter 12). And finally, Chapter 13 deals with the transport of microorganisms from the surface to depth, and from location to location, within the subsurface, including the presentation of models of bacterial transport. Questions of the age of deep subsurface microbes are addressed throughout this section.

Section III, *Applications*, is concerned with specific topics associated with subsurface microbiology and microbial ecology. These chapters represent areas where the concerns of people intersect with the abilities of microbes. Chapter 14 specifically deals with concerns of bioremediation of deep subsurface environments. Chapter 15 reviews knowledge of, and concerns about the role of microbes in subsurface radioactive waste repositories, focusing on work conducted in Europe. Another chapter addresses the potential of microbes as carriers or barriers to specific metals transport — a concern of the nuclear industry (Chapter 16). And lastly, Chapter 17 addresses the ability of microbes to prevent movement of subsurface compounds through the purposeful production of biobarriers.

Many exciting challenges remain in subsurface research, a few of which include: knowing about the origin and transport of subsurface microorganisms, characterizing the nature of deep subsurface viruses, searching the untapped potential of microbiota for novel products and processes, and discovering the roles of novel (from an anthropogenic viewpoint) metabolic processes and reactions in the earth's biogeochemical cycles.

References

Balkwill, D.L., D.R. Boone, F.S. Colwell, T. Griffin, T.L. Kieft, R.M. Lehman, J.P. McKinley, S. Nierzwicki-Bauer, T.C. Onstott, H.Y. Tseng, G. Gao, T.J. Phelps, D. Ringelberg, B. Russell, T.O. Stevens, D.C. White, and F. J. Wobber. 1994. D.O.E. seeks origin of deep subsurface bacteria. *EOS*, 75:395-396.

Balkwill, D.L. and W.C. Ghiorse. 1985. Characterization of subsurface bacteria associated with two shallow aquifers in Oklahoma. *Appl. Environ. Microbiol.*, 50:580-588.

Boone, D.R., Y. Lui, Z. Zhao, D.L. Balkwill, G.R. Drake, T.O. Stevens, and H.C. Aldrich. 1995. *Bacillus infernus* sp. Nov., an Fe(III)- and Mn(IV)-reducing anaerobe from the deep subsurface. *Int. J. Syst. Bacteriol.*, 45:441-448.

Brockman, F.J., B.A. Denovan, R.J. Hicks, and J.K. Fredrickson. 1989. Isolation and characterization of quinoline-degrading bacteria from subsurface sediments. *Appl. Environ. Microbiol.*, 55:1029-1032.

Colwell, F., M. Delwiche, T.C. Onstsott, Q.-J. Yao, R. Griffiths, D. Ringelberg, D. White, J.K. Fredrickson, and R. Lehman. 1996. Microbial communities from deep natural gas-bearing rocks. Am. Soc. Microbiol. Annu. Meet. Abstr., American Society for Microbiology, New Orleans, LA. N58.

Farrell, M.A. and H.G. Turner. 1931. Bacteria in anthracite coal. *J. Bacteriol.*, 23:155.

Fredrickson, J.K., F.J. Brockman, D.J. Workman, S.W. Li, and T.O. Stevens. 1991. Isolation and characterization of a subsurface bacterium capable of growth on toluene, naphthalene and other aromatic compounds. *Appl. Environ. Microbiol.*, 57:796-803.

Ghiorse, W.C. and J.T. Wilson. 1988. Microbial ecology of the terrestrial subsurface. *Adv. Appl. Microbiol.*, 33:107-172.

Gould, S.J. 1996. Microcosmos. *Natural History Magazine*, 3:21-68.

Hirsch, P. 1992. Microbiology, in G. Matthess, F. Frimmel, P. Hirsch, H.D. Schulz, and H.E. Usdowski, Eds., *Progress in Hydrogeochemistry*. Springer-Verlag, New York. pp. 308-412.

Kuznetsov, A.I., M.V. Ivanov, and N.K. Lyalikova. 1963. *Introduction to Geological Microbiology*. Engl. ed. McGraw-Hill, New York, NY.

Lipman, C.B. 1931. Living microorganisms in ancient rocks. *J. Bacteriol.*, 22:183.

Madsen, E.L. and W.C. Ghiorse. 1993. Groundwater microbiology: subsurface ecosystem's processes. in *Aquatic Microbiology: an Ecological Approach*. T. Ford, Ed., Blackwell Scientific, Cambridge, MA.

Parks, R.J., B.A. Crag, S.J. Vale, J.N. Getliff, K. Goodman, P.A. Rochelle, J.C. Fry, A.J. Weightman, and S.M. Harvey. 1994. Deep bacterial biosphere in Pacific Ocean sediments. *Nature*, 371:410-413.

Stevens, T.O. and J.P. McKinley. 1995. Lithoautotrophic microbial ecosystems in deep basaltic aquifers. *Science*, 270:450-454.

Section I

Considerations for Sampling

2

Geohydrologic and Geochemical Characterization

Bruce N. Bjornstad, Cheryl D. Gullett, and Phillip E. Long

CONTENTS

KEY WORDS: *geohydrology, geochemistry, microbiology, vadose zone, saturated zone, characterization methods, origins.*

2.1   Importance of Geohydrology and Geochemistry

The geology, hydrology, and geochemistry of the subsurface play a dominant role in impacting the subsurface microbiology. Specifically, the combined physical, hydraulic, and chemical properties of subsurface porous and fractured media control the movement of groundwater and nutrient flux that may support microbial growth and productivity. Several hypotheses proposed for the origins of microorganisms in the subsurface include: (1) they are the progeny of the original population incorporated into the sediment or rock during deposition (Balkwill et al., 1994; Chapelle and Lovely, 1990; DOE, 1992; DOE, 1994; Fredrickson et al., 1995); (2) they were transported into place more recently via migrating groundwater (Bjornstad et al., 1996; Murphy et al., 1992); or (3) a combination of these two phenomena. Subsurface microbial communities are reported for both the saturated (Amy et al., 1992; Balkwill, 1989; Balkwill and Ghiorse, 1985; Fredrickson et al., 1995; Kieft et al., 1995; Pedersen and Ekendahl, 1990; Stevens et al., 1993; West et al., 1985, 1992) as well as the unsaturated zone above the water table (Brockman et al., 1992; Colwell, 1989; Fredrickson et al., 1993; Haldeman and Amy, 1993; Kieft et al., 1993). Water availability in the unsaturated zone has been hypothesized to be the single most important environmental factor influencing microbial activity (Kieft et al., 1993).

Microbial activity and types of reactions that occur are controlled by the physical, hydraulic, and chemical properties of the host rock. Physical properties such as pore spaces and fracture openings may affect, or be modified by, microbial activity and distribution. Natural openings in rock or sediment also affect the rate and direction of groundwater flow, which control microbial transport and transfer of nutrients through the subsurface (Fredrickson et al., 1995). Permeability (i.e., hydraulic conductivity) appears to be an important parameter interacting with microbial populations. In some instances, sediments of low hydraulic conductivity contribute to the long-term maintenance of both bacteria and organic carbon by limiting the supply of soluble electron acceptors for microbial respiration (Fredrickson et al., 1995; Jacobsen and Postma, 1994). In other instances, the greatest microbial activities may occur in highly permeable sands, provided sufficient nutrient sources are available to sustain microbial activity (McMahon et al., 1991; Murphy et al., 1992). Geologic environments that contain an abundance of organic matter in the form of decayed plant or animal debris have the potential to maintain microbial populations over a long period of time, especially in zones of low hydraulic conductivity where the movement of groundwater is slow (Fredrickson et al., 1995). In the unsaturated zone, there appears to be a link between microbial activity and the amount of moisture present (Fredrickson et al., 1993; Kieft et al., 1993). The chemical nature of the solid media in the subsurface, and the groundwater itself, are likely important factors impacting the types of microbes present, as well as their productivity and distribution. In the solid phase, controlling parameters include clay content and mineralogy, as well as secondary alteration products and mineral coatings of iron and manganese oxide, and availability of electron donors and acceptors. Specific electron acceptors of interest may include O_2, Fe(III), NO_3^-, SO_4^{2-}, and HCO_3^-; common electron donors in the deep subsurface include organic carbon and H_2 (Chapelle, 1993). In addition to the electron acceptors and electron donors present in groundwater, other parameters such as pH, redox conditions, and temperature may also affect the viability and reproduction of microbes in the subsurface. Postdepositional alteration in the subsurface (i.e., diagenesis), which includes precipitation, dissolution, and recrystallization, in many cases appears to be a microbially mediated process (Ferris et al., 1994; Lovely et al., 1990; McMahon et al., 1991; Monger et al., 1991). In practice the microscopic details that control microbial survival and activity in the subsurface have not been elucidated although progress is being made. For example, work by Krumholtz and Suflita (1996) appears to demonstrate the validity of the microsite model proposed by Murphy et al. (1992). This model suggests anaerobic bacteria can exist juxtaposed to aerobic bacteria in locally reducing areas where there are concentrations of organic matter (e.g., lignite).

2.2 Geohydrologic and Geochemical Characterization Methods

Methods used to characterize the geologic, hydrologic, and geochemical conditions of the subsurface can be divided into three categories: (1) direct observations and measurements made in the field on outcrop or drill samples; (2) laboratory measurements performed on outcrop, drill, or groundwater samples extracted from boreholes; and (3) downhole measurements obtained using geophysical logging tools (Figure 2.1). Typically, a predrilling geohydrologic conceptual model is formulated that serves as a testable model of subsurface conditions. Once drilling is complete and the various data sets are integrated, then a revised geohydrologic model is developed that forms the framework for the interpretation of microbiological and geochemical results. It is beyond the scope of this chapter to discuss all of the standard petrophysical and chemical methods used to characterize the subsurface environment; parameters and methods are summarized in Tables 2.1 and 2.2. Characterization parameters listed in Tables 2.1 and 2.2 are divided into four groups: (1) physical, (2) hydraulic,

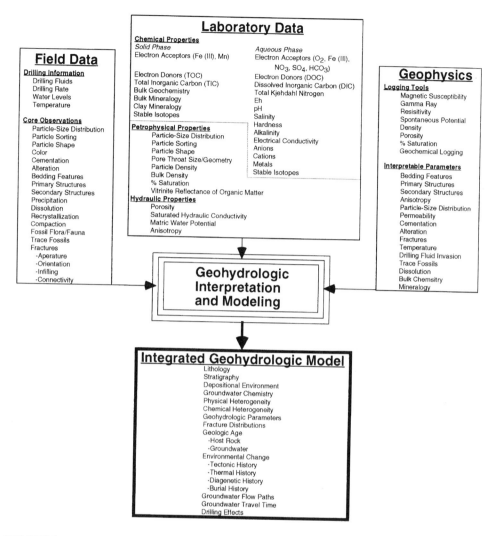

FIGURE 2.1

Possible geohydrologic and geochemical characterization data needed to construct geohydrologic model.

(3) chemical, and (4) other parameters. A comprehensive analysis of the subsurface environment encompasses both the solid and aqueous (i.e., groundwater) fractions. Most of the characterization methods are designed for measuring the properties of clastic sediments or sedimentary rocks and groundwater, but many of the methods will apply to other geohydrologic materials as well (e.g., volcanic rock, brines).

Petrophysical properties which may most influence nutrient and microbial transport in the subsurface are particle size distribution, effective porosity and permeability, and pore throat size (Fredrickson et al., 1989; Ghiorse and Wilson, 1988). The particle size distribution controls the total porosity and pore geometry of the porous media, thereby influencing the local advective flux within pores. To characterize the pore size distribution of the effective porosity, samples may be evaluated by petrographic image analysis of thin sections via petrographic microscopy (Ehrlich et al., 1991). Pore throat size directly affects local advective flux through the medium and may be a limiting factor in microbial transport between pores. Permeability will control the relative importance of advective and diffusive transport of nutrients within the sediments. Advective transport is expected to dominate in highly permeable

TABLE 2.1

Characterization Methods Used to Determine Physical and Hydraulic Parameters

Parameter	Analytical Method	Qualitative Measurement Field	Quantitative Measurement Field	Quantitative Measurement Lab	Ref. (Examples)
Physical					
Particle-size distribution	Visual	x			AGI, 1982; ASTM, 1984; Tucker, 1982
	Hydrometer			x	Lewis and McConchie, 1994
	Sieve analysis (>0.0625 mm)			x	Lewis and McConchie, 1994
	Settling tube (<0.0625 mm)			x	Lewis and McConchie, 1994
Particle sorting	Visual	x			Tucker, 1982
	Moment statistics			x	Lewis and McConchie, 1994
	Digital image analysis			x	Murphy et al., 1977
Particle shape	Visual	x			AGI, 1982; Lewis and McConchie, 1994
	Digital image analysis			x	Tucker, 1982
Pore-throat size	Mercury porosimetry			x	Lawrence, 1978
Color	Visual				
	Soil		x		Munsell, 1975
	Rock		x		GSA, 1991
Bulk density	Volume displacement			x	Blake and Hartage, 1986
Grain density				x	Head, 1992
Moisture	Desiccation			x	ASTM, 1986
Temperature	Temperature meter		x		
	Downhole temperature probe		x		
Cementation	Visual	x			
	Petrography			x	Milner et al., 1962
	Scanning electron microscope (SEM)			x	
Alteration (i.e., diagenesis)	Visual	x			
	Petrography			x	Milner et al., 1962
	Cathodoluminescence			x	Dudley, 1976; Nickel, 1978
Bedding features	Visual	x			Compton, 1962; Reinek and Singh, 1980
Primary structures	Visual	x			Compton, 1962; Reinek and Singh, 1980
	X-ray radiography			x	Bouma, 1969
	Resistivity probe		x		Schlumberger, 1991
Secondary structures	Visual	x			Compton, 1962; Reinek and Singh, 1980
	X-ray radiography			x	Bouma, 1969
	SP/resistivity probe		x		Schlumberger, 1991

TABLE 2.1 (continued)

Characterization Methods Used to Determine Physical and Hydraulic Parameters

Parameter	Analytical Method	Qualitative Measurement Field	Quantitative Measurement Field	Quantitative Measurement Lab	Ref. (Examples)
Fossil					
flora/fauna	Visual	x			
macrofossils	Petrography			x	Milner et al., 1962
microfossils	Palynology			x	Jansen and Birks, 1994
Trace fossils	Visual	x			Ekdale et al., 1984
	Petrography			x	Ekdale et al., 1984
Fractures	Visual		x		Compton, 1962
	Geophysical probes		x		Schlumberger, 1991
Hydraulic					
Porosity	Geophysical probes		x		Schlumberger, 1991
	Volume displacement			x	
	Mercury porosimetry			x	Lawrence, 1978
	Petrography			x	Murphy et al., 1977
Saturated hydraulic conductivity	Aquifer tests		x		Freeze and Cherry, 1979 Klute and Dirksen, 1986
	Permeameter			x	
Unsaturated hydraulic conductivity	Ultracentrifugation			x	Conca and Wright, 1992
	Geophysical probes		x		Schlumberger, 1991
Matric water potential	Thermocouple psychrometry			x	Kieft et al., 1993; Klute, 1986

gravels and sands ($K_{sat} \leq 10^{-2}$ to 10^1 cm/s). In less permeable sediments ($K_{sat} \leq 10^{-3}$ to 10^{-5} cm/s), nutrient flux is dominated by diffusion transport (see Chapters 13 and 16).

Primary depositional structures incorporated into the sediment at deposition impart a preferred orientation in the host rock which affects the movement of groundwater and associated chemical constituents through the geologic formation. Such a nonuniform, preferred flow distribution is referred to, hydrologically, as anisotropy. Groundwater movement and nutrient flux below the water table is a function of the saturated hydraulic conductivity, and above the water table, the matric water potential. Other features worth noting include degree of saturation (i.e., moisture), temperature, fossil flora and fauna, and trace fossils.

A number of methods are available to characterize the geochemistry of the mineral solids. Key geochemical parameters include the bulk chemistry and mineralogy of the solids. Also important are secondary alteration products such as mineral coatings (e.g., iron and manganese) that act as electron acceptors available to sustain microbial activity (Lovely et al., 1990). Other lines of evidence for alteration (i.e., diagenesis) include burial compaction, secondary porosity, precipitation of secondary mineral phases, and the maturation state of sedimentary organic matter. The chemical composition and structure of diagenetic mineral precipitates provide important clues to interpret the physical and chemical changes that have taken place over time. Past biogeochemical conditions and availability of nutrients for microbial activity may be inferred from the textural relationships of detrital constituents and diagenetic phases

TABLE 2.2

Characterization Methods Used to Determine Chemical and Other Parameters

Parameter	Analytical Method	Qualitative Measurement Field	Quantitative Measurement Field	Lab	Ref. (Examples)
Chemical (Aqueous)					
Eh	Platinum electrode		x	x	Lewis and McConchie, 1994
pH	pH meter		x	x	Lewis and McConchie, 1994
Dissolved oxygen	DO meter		x		Lewis and McConchie, 1994
	Microtitration		x		Lewis and McConchie, 1994
	Colorimetric tubes		x		
Biochemical oxygen demand	Microtitration		x		Lewis and McConchie, 1994
Salinity	Salinometer		x		Lewis and McConchie, 1994
	Conductivity probe		x		Lewis and McConchie, 1994
	Desiccation			x	Lewis and McConchie, 1994
Water hardness	Microtitration		x		Lewis and McConchie, 1994
Alkalinity	Microtitration		x		Lewis and McConchie, 1994
DOC					
Electrical conductivity					
Chemical (Solids)					
Bulk chemistry	X-ray fluorescence			x	Milner et al., 1962
	Atomic absorption			x	Kuwana, 1978
	ICP (inductively coupled plasma)			x	Golightly et al., 1978
	Natural gamma log		x		Schlumberger, 1991
Bulk mineralogy	Visual	x			Hurlburt, 1971
	X-ray diffraction			x	Zussman, 1977
	Petrography			x	Milner et al., 1962
	Grain mounts			x	Lewis and McConchie, 1994
	Heavy mineral analysis			x	Lewis and McConchie, 1994
	Electromagnetic separation			x	Lewis and McConchie, 1994
	Differential thermal analysis			x	Lewis and McConchie, 1994
	Thermogravimetric analysis			x	Lewis and McConchie, 1994
	Infrared spectrophotometry			x	Lewis and McConchie, 1994
	Mossbauer spectroscopy			x	Lewis and McConchie, 1994
	Electron microprobe			x	Brown, 1977
Clay mineralogy	X-ray diffraction			x	Zussman, 1977
TOC	Oxidative Combustion			x	Lewis and McConchie, 1994
	Wet Chemical Oxidation			x	Lewis and McConchie, 1994
Fluid inclusions	Petrography (heating/freezing stage)			x	Milner et al., 1962
	Electron Microscopy			x	Milner et al., 1962
	Raman Spectroscopy			x	Rosacco and Roedder, 1975

TABLE 2.2 (continued)

Characterization Methods Used to Determine Chemical and Other Parameters

Parameter	Analytical Method	Qualitative Measurement Field	Quantitative Measurement Field	Lab	Ref. (Examples)
Chemical (Solids and/or Liquids)					
Electron acceptors					
Electron donors (DOC)					
Metals	Voltammetry	x			Lewis and McConchie, 1994
	X-ray fluorescence	x		x	Lewis and McConchie, 1994
	Ion chromatography			x	Lewis and McConchie, 1994
	Anodic-stripping voltammetry			x	Lewis and McConchie, 1994
	Flame photometry			x	Lewis and McConchie, 1994
	Atomic absorption			x	Kuwana, 1978
	ICP (inductively coupled plasma)			x	Golightly et al., 1978
Stable isotopes	Mass spectrometer			x	Arthur et al., 1983
Total sulfur	Oxidative combustion			x	Lewis and McConchie, 1994
Oxidizable sulfur	Sulfate production			x	Lewis and McConchie, 1994
Total kjeldahl nitrogen	Digestion, distillation, titration			x	Lewis and McConchie, 1994
Anions	Ion chromatography			x	Tabatabai and Basta, 1991
	Wet chemical analysis			x	
	Atomic absorption			x	Kuwana, 1978
Cations	Ion chromatography			x	Tabatabai and Basta, 1991
Other Parameters					
Age dates	Fossil record	x			
	Radiometric			x	Mahaney, 1984
	Fission track			x	Mahaney, 1984
	Thermoluminescence			x	Mahaney, 1984
Paleotemperature	Stable isotopes			x	
	Vitrinite reflectance			x	Bostick and Freeman, 1984
	Fluid inclusions			x	Touret-Jacques, 1987
	Clay mineral ratios			x	McCarty, 1990
	Biomarkers			x	

as revealed by microscopy. Carbon, which may serve as a nutrient source for bacteria, may be a limiting factor in microbial distribution and survival.

Microorganisms appear to play an important role in the evolution of groundwater chemistry, which varies between the area of recharge and downgradient along the subsurface ground water flow path (Murphy et al., 1992). Measurements on the groundwater (Eh, pH, salinity, alkalinity, hardness, temperature, dissolved organic carbon [DOC], dissolved oxygen, anions, cations, etc.) which flows through or is trapped within geologic formations, provide data on the existing hydrochemistry, the quantity of electron donors and acceptors available to support microbial growth, and the biogeochemical evolution of the groundwater.

Other measurements are performed on both the solid and aqueous fractions of sub-surface samples. Total organic carbon (TOC) within the solid fraction provides information on the amount of organic matter that is potentially available. Concentrations of particulate organic matter (e.g., lignite, coalified plant debris, kerogen) may provide microsites for anaerobic bacteria, the fermentation products of which may provide nutrients to support populations of aerobic heterotrophs (Murphy et al., 1992). In contrast, DOC is an indication of the concentration of electron donors dissolved in ground water and directly available for the respiration of aerobic bacteria. Along the same lines, an analysis of the iron and manganese mineral coatings on particle grains or surfaces provides clues to the types and quantities of these electron acceptors that are present, while levels of electron acceptors dissolved in ground water are an indication of those immediately available to microbes.

Stable isotope analysis is another important tool used to characterize both solid and aqueous fractions. Fractionation of stable isotopes of carbon, sulfur, hydrogen, and nitrogen often reflects biogeochemical processes (Balabane et al., 1987; Fritz and Fontes, 1980; Hoefs, 1987; Hubner, 1986). In addition, stable isotopes of carbon, oxygen, and hydrogen have been used in the interpretation of paleosalinity, paleotemperature, and diagenesis (Arthur et al., 1983; Buchardt and Fritz, 1980; Hays and Grossman, 1991; Rosenfeld and Silverman, 1959). Stable isotope signatures of precipitates and products of recrystallization in the solid fraction reveal information on past climatic environments (Bowen, 1991). Stable isotope signatures of dissolved constituents in groundwater have been used to derive information on biogeochemical cycles and respiration rates (Chapelle and Lovely, 1990; Chapelle et al., 1988; Murphy et al., 1992; Phelps et al., 1994).

High-magnification scanning electron microscope (SEM) images of sediment and rock have been utilized to identify biofilms and other signs of microbial activity (Gullett and Ehrlich, 1992; Monger et al., 1991). The color of the host rock can provide an indication of the conditions of pH and Eh and the existing diagenetic environment. For example, sediments that are dark shades of gray or black are usually associated with reducing environments, while those altered yellow, red, orange, or brown are typically a sign of oxidation. Color, then, may be a useful tool for predicting microbial community structure.

Other useful information in geohydrologic and geochemical characterization include geologic age dating and paleotemperatures of the host rock and ground water. These data assist to test the hypotheses on the origins of subsurface microorganisms. For example, what is the potential that extant microbes are the progeny of those living when the rocks and sediments were laid down, and that the originally implaced microbes somehow evolved and survived the stresses associated with changes in temperature, pressure, and chemical conditions over time? Absolute age dates of the host rock may be determined radiometrically by measuring the ratios of decaying radioisotopes (U-Th, K-Ar, ^{14}C) remaining in the rock, while relative ages are determined from the fossil record. Radiometric and fossil age-dating techniques combined provide estimates of reasonable geologic age, especially in older rocks and sediments millions to hundreds of millions of years old. Other radiogenic dating methods used for more recent geological formations include thermoluminescence (TL) and fission track dating (Mahaney, 1984). TL measures the amount of light trapped in the crystal lattice as electron displacement, which is a function of sediment age since burial. In fission track dating the number of fission tracks left behind by radioactive decay in certain mineral grains (e.g., apatite, zircon) is proportional to the ^{238}U content and the age of the mineral grain (Bowen, 1978). Because ground water moves through geologic formations from areas of recharge to areas of discharge, it is ordinarily younger than the host rock. Radioactive isotopes typically used to date ground water include tritium, ^{14}C, ^{36}Cl, and ^{129}I (Davis and Murphy, 1987). Each of these radioactive isotopes has a different decay half-life and, therefore, the range of years over which these isotopes can be applied to determine groundwater ages varies from decades to millions of years.

Paleotemperature is an important parameter for deciphering the long-term changes in temperature that may have occurred in a geologic formation (Barker and Halley, 1986; Pollastro and Barker, 1986). Temperature changes can affect the types and levels of microbial activity and evolution over geologic time (Long et al., 1994). Several potential methods for measuring paleotemperature include stable isotopes, vitrinite reflectance, fluid inclusions, clay mineral ratios, and biomarkers (Table 2.3). Pressure and temperature changes in sedimentary units over time are displayed using a burial history curve (Molenaar, 1973, 1983; Molenaar and Baird, 1992).

Basic qualitative and quantitative geologic information on lithology is obtained by observing and logging cuttings or cores extracted from borings of the subsurface. Examples of the types of information collected during drilling are shown in Table 2.1. Information on the petrophysical and sedimentological parameters are obtained on core samples recovered from borings, including grain size distribution, porosity, hydraulic conductivity, and pore throat size distribution; these are commonly measured with a petrographic microscope, scanning electron microscope (SEM), or analytically in the laboratory. Paleontological evidence and sedimentary structures in combination with textural data form the basis to classify clastic materials into an appropriate lithologic category and depositional environment. Geophysical logs obtained downhole provide additional quantitative measurements of particle size, porosity, permeability, percent saturation, density, and geochemistry. Geophysical logs have an advantage over coring in that measurements can be obtained continuously or at closely spaced intervals within the borehole at a fraction of the price of drilling. A multitude of geophysical techniques have been developed to measure both cased as well as unlined boreholes (Asquith and Gibson, 1982; Schlumberger, 1989, 1991). An example summary log that graphically summarizes the results of recent field and laboratory analyses in New Mexico is presented in Figure 2.2.

Figure 2.1 shows the location of core and groundwater samples collected within a vertical borehole drilled to 230 m depth in Cretaceous- to Jurassic-age sedimentary rocks (i.e., shale, siltstone, and sandstone) in west-central New Mexico (Long et al., 1994). In the middle portion of Figure 2.1 are laboratory analytical results obtained on the core samples for pyritic sulfur, total organic carbon (TOC), and pore throat size — factors that likely influence microbial activity. To the right are two geophysical logs (resistivity and gamma ray) gathered from continuous measurements obtained with downhole probes lowered within the boring; the deflections in the resistivity and gamma ray curves are inversely proportional and provide a reliable means for estimating variations in grain size and associated permeability. The natural variability (i.e., heterogeneity) observed in most sedimentary deposits, like that in Figure 2.1, is typical of sedimentary environments. For example, the particle size distributions and hydraulic conductivity of the geologic material control the rate of groundwater movement and therefore substrate and nutrient availability to microorganisms (Barbaro et al., 1994). This creates a special challenge when trying to compare the microbiology of sediments with different physical and chemical properties (see Chapter 6).

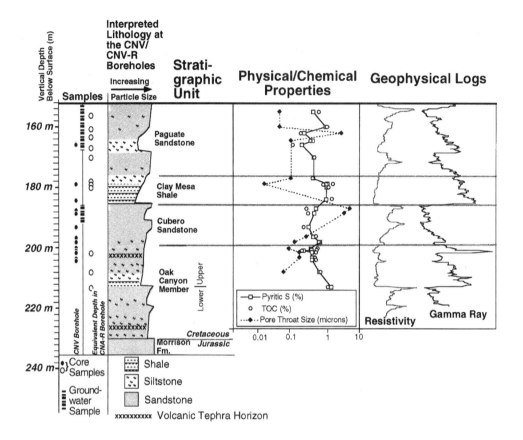

FIGURE 2.2
A comparison of some of the physical, chemical, and geophysical parameters measured in two microbial characterization boreholes drilled in west-central New Mexico. See text for description. (From Long, P.E., Fredrickson, J. K., Bjornstad, B. N., et al., *GSA Abstr. Prog.*, 26, 509, 1994. With permission.)

2.3   Acquisition of Subsurface Samples

A number of different geologic and hydrologic settings have been investigated to determine microbiological characteristics of different subsurface environments. Samples for physical, chemical, and microbiological analysis are obtained in a number of ways. Each sample type has its own set of sampling requirements: samples for physical analysis require undisturbed core samples, chemical samples require careful handling and packaging to prevent chemical cross contamination, and microbiological samples require aseptic sampling techniques (see Chapters 3, 4, and 5). Collection of samples is conducted via a number of methods ranging from sampling cleaned-off faces of near-surface outcroppings (Brockman et al., 1992; Haldeman et al., 1994; Murray et al., 1994, 1995) to the drilling of coreholes (see Chapter 4 and Balkwill et al., 1994; Balkwill and Ghiorse, 1985; Brockman et al., 1992; Fredrickson et al., 1991, 1993, 1995; Kieft et al., 1995; Lovely et al., 1990; Murphy et al., 1992; Parkes et al., 1994; Stevens et al., 1993; West et al., 1992;) and mining into walls of deep underground tunnels (see Chapter 4 and Brown et al., 1994; Haldeman and Amy, 1993; Russell et al., 1994).

Factors to consider when selecting a site for microbiological analysis include: (1) the geologic environment of deposition; (2) the age of the geologic deposits; (3) burial and exhumation history over geologic time; and (4) any associated diagenesis (i.e., alteration) that has taken place over geologic time due to precipitation, dissolution, or recrystallization by either groundwater or intense heat and pressure (i.e., metamorphism).

Depending on the purpose of the proposed research, sites selected for microbial characterization have ranged from geologically young, unconsolidated, near-surface sedimentary deposits (Barbaro et al., 1994; Brockman et al., 1991, 1992; Colwell, 1989; Fredrickson et al., 1993; Murray et al., 1994, 1995) to deeply buried lithified host rocks up to hundreds of millions of years in age (Amy et al., 1992; Balkwill et al., 1994; Brown et al., 1994; Chapelle and Lovely, 1990; Fredrickson et al., 1989, 1991; Haldeman and Amy, 1993; Jacobsen and Postma, 1994; Long et al., 1994; Lovely et al., 1990; McKinley et al., 1996; Murphy et al., 1992; Parkes et al., 1994; Pedersen and Ekendahl, 1990; Russell et al., 1994; Stevens et al., 1993). Ideally, it is best to select sites with relatively simple geologic histories to reduce complications related to dealing with a large number of changing variables introduced in geologically complex environments. Deposits in complex geologic terranes experience more environmental changes (e.g., transition from marine to fresh-water conditions, tectonic uplift and subsidence, shifts in groundwater flow patterns, and fluctuations in groundwater chemistry); unraveling the effects of these changes on the existing microbial population may be difficult or impossible.

To evaluate the origins and transport of microorganisms in the subsurface (see Chapter 13), studies have been conducted in regions with demonstrated geohydrologic isolation (Fredrickson et al., 1995; Kieft et al., 1995) to minimize the potential for contamination of the subsurface microbial community with those from the near surface. Over time, transport of microorganisms may occur naturally either via movement of ground water through pores and voids within clastic sediments or along fractures developed in fused or tightly cemented host rocks (Bjornstad et al., 1996), or a combination of the two. Studies on the transport and origins of microorganisms in fractured rocks require extensive data on fracture networks to be able to interpret movement of groundwater through fractured media. For this reason, studies on the origins of microorganisms have focused principally on porous, rather than fractured, geologic media.

Another consideration in site selection is the presence or absence of anthropogenic contamination (e.g., chemical spills, groundwater contamination plumes, etc.). These may stimulate microbial growth by providing an influx of electron donors and/or acceptors. Anthropogenic sources of contamination can be a hindrance for studying the natural background microbial population, while at the same time, beneficial toward understanding the processes involved in bioremediation (see Chapter 14).

References

AGI. 1982. AGI data sheets for geology in the field, laboratory, and office: American Geological Institute, Falls Church, VA.

Amy, P. S., D. L. Haldeman, D. Ringelberg, D. H. Hall, and C. Russell. 1992. Comparison of identification systems for classification of bacteria isolated from water and endolithic habitats within the deep subsurface, *Appl. Environ. Microbiol.*, 50:3367-3373.

Arthur, M. A., T. F. Anderson, I. R. Kaplan, J. Veizer, and L. S. Land. 1983. *Stable Isotopes in Sedimentary Geology*, SEPM Short Course No. 10. Society of Economic Paleontologists and Mineralogists, Tulsa, OK.

Asquith, B. B. and C. R. Gibson. 1982. *Basic Well Log Analysis for Geologists*. American Association of Petroleum Geologists, Tulsa, OK, 216 p.

ASTM. 1984. *Standard Practice for Description and Identification of Soils (Visual-Manual Procedure)*. Procedure D 2488-84. American Society for Testing and Materials, Philadelphia, PA, pp. 293-302.

ASTM. 1986. Laboratory determination of water (moisture) content in soil, rock, and soil-aggregate mixtures. *Annual Book of ASTM Standards*. v. 4.08, ASTM D2216, American Society of Testing and Materials, Philadelphia, PA.

Balabane, M., E. Galimov, M. Hermann, and R. Letolle. 1987. Hydrogen and carbon isotope fractionation during experimental production of bacterial methane. *Org. Geochem.*, 11:115-119.

Balkwill, D. L. 1989. Numbers, diversity, and morphological characteristics of aerobic, chemoheterotrophic bacteria in deep subsurface sediments from a site in South Carolina, *J. Geomicrobiol.*, 7:33-52.

Balkwill, D. L., D. R. Boone, F. S. Colwell, T. Griffin, T. L. Kieft, R.M. Lehman, J. P. McKinley, S. Nierzwicki-Bauer, T. C. Onstott, H.-Y. Tseng, G. Gao, T. J. Phelps, D. Ringelberg, B. F. Russell, T. O. Stevens, D. C. White, and F. J. Wobber. 1994. D.O.E. seeks origin of deep subsurface bacteria. *EOS (Trans. Am. Geophys. Union)*, 75:390-396.

Balkwill, D. L. and W. C. Ghiorse. 1985. Characterization of subsurface bacteria associated with two shallow aquifers in Oklahoma, *Appl. Environ. Microbiol.*, 50:580-588.

Barbaro, S. E., H.-J. Albrechtsen, B. K. Jensen, C. I. Mayfield, and J. F. Barker. 1994. Relationships between aquifer properties and microbial populations in the Borden Aquifer. *Geomicrob. J.*, 12:203-219.

Barker, C. E., and R. B. Halley. 1986. Fluid inclusion, stable isotope, and vitrinite reflectance evidence for the thermal history of the Bone Spring Limestone, southern Guadalupe Mountains, Texas. Society of Economic Paleontologists and Mineralogists, Tulsa, OK. p. 189-203.

Bjornstad, B. N., P. E. Long, and J. C. Lorenz. 1996. Fractures adjacent to a Miocene volcanic neck, Cerro Negro, Cebolleta Land Grant, New Mexico. *GSA Abstr. Prog.*, 28:49-50.

Blake, G. R. and K. H. Hartage. 1986. Bulk density, in A. Klute, Ed., *Methods of Soil Analysis*, Part 1. American Society of Agronomy. Madison, WI. pp. 383-409.

Bostick, N. H. and V. L. Freeman. 1984. Tests of vitrinite reflectance and paleotemperature models at the Multiwell Experimental site, Piceance Creek basin, Colorado. In C. W. Spencer and C. W. Keighin, Eds. Geologic Studies in Support of the U.S. Department of Energy Multiwell Experiment, Garfield County, Colorado. Open-File Report. U.S. Geological Survey, Washington, D.C., p. 110-120.

Bouma, A. H. 1969. *Methods for the Study of Sedimentary Structures*. Wiley-Interscience. New York. 458 p.

Bowen, D. Q. 1978. *Quaternary Geology*. Pergamon Press. New York. 221 p.

Bowen, R. 1991. *Isotopes and Climate*. Elsevier. New York. 483 p.

Brockman, F. J., T. L. Kieft, J. K. Fredrickson, B. N. Bjornstad, S. W. Li, W. Spangenburg, and P. E. Long. 1992. Microbiology of vadose zone paleosols in south-central Washington State. *Microb. Ecol.*, 23:279-301.

Brockman, F. J., T. L. Kieft, J. K. Fredrickson, B. N. Bjornstad, and P. E. Long. 1991. Moisture recharge and organic carbon as factors controlling vadose zone microbial communities at the Hanford Site. *AGU Abstr. Prog.* p. 11.

Brown, R. W. 1977. A sample fusion technique for whole rock analysis with the electron microprobe. *Geochim. Cosmochim. Acta.*, 41:435-438.

Brown, D. A., D. Choudari Kamineni, J. A. Sawicki, and T. L. Beveridge. 1994. Minerals associated with biofilms occurring in exposed rock in a granitic underground research laboratory. *Appl. Environ. Microbiol.*, 60:3182-3191.

Buchardt, B. and P. Fritz. 1980. Environmental isotopes as environmental and climatological indicators. In: *Handbook of Environmental Isotope Geochemistry*, Vol. 1, P. Fritz and J. Ch. Fontes, Eds., Elsevier, New York, p. 473-504.

Chapelle, F. H. 1993. *Ground-Water Microbiology and Geochemistry.* John Wiley & Sons, New York. 424 p.

Chapelle, F. H. and D. R. Lovely. 1990. Rates of microbial metabolism in deep coastal plain aquifers. *Appl. Environ. Microbiol.,* 56:1865-1874.

Chapelle, F. H., J. T. Morris, P. B. McMahon, and J. L. Zelibor. 1988. Bacterial metabolism and the $\delta^{13}C$ composition of ground water, Floridan aquifer system, South Carolina. *Geology,* 16:117-121.

Colwell, F. S. 1989. Microbiological comparison of surface soil and unsaturated subsurface soil from a semiarid high desert. *Appl. Environ. Microbiol.,* 55:2420-2423.

Compton, R. R. 1962. *Manual of Field Geology.* John Wiley & Sons, New York, 378 p.

Conca, J. L. and J. V. Wright. 1992. Diffusion and flow in gravel, soil, and whole rock. *Appl. Hydrogeol.,* 1:5-24.

Davis, S. N. and E. M. Murphy. 1987. Dating Ground Water and the Evaluation of Repositories for Radioactive Waste. NUREG/CR-4912, U.S. Nuclear Regulatory Commission, Washington, D.C., 181 pp.

DOE. 1992. Origins of Microorganisms in Deep Subsurface Environments. U.S. Department of Energy Subsurface Science Program. Deep Microbiology Subprogram, Phase II Preliminary Plan. DOE/ER-0556T. U.S. Department of Energy, Washington, D.C.

DOE. 1994. Origins of Microorganisms in Deep Subsurface Environments: Research Implementation Plan. U.S. Department of Energy Subsurface Science Program. October 1994. U.S. Department of Energy, Washington, D.C.

Dudley, R. J. 1976. The use of cathodoluminescence in identification of soil minerals. *J. Soil Sci.,* 27:487-494.

Ehrlich, R., S. J. Crabtree, K. O. Horkowitz, and J. P. Horkowitz. 1991. Petrography and reservoir physics. III. Physical models for permeability and formation factor. *AAPG Bull.,* 75:1547-1592.

Ekdale, A. A., R. G. Bromley, and S. G. Pemberton. 1984. *Ichnology: The Use of Trace Fossils in Sedimentology and Stratigraphy.* SEPM Short Course No. 15. Society of Economic Paleontologists and Mineralogists. Tulsa, OK.

Ferris F. G., R. G. Wiese, and W. S. Fyfe. 1994. Precipitation of carbonate minerals by microorganisms: implications for silicate weathering and the global carbon dioxide budget. *Geomicrobiol. J.,* 12:1-13.

Fredrickson, J. K., T. R. Garland, R. J. Hicks, J. M. Thomas, S. W. Li, and K. M. McFadden. 1989. Lithotrophic and heterotrophic bacteria in deep subsurface sediments and their relation to sediment properties, *Geomicrobiol. J.,* 7:53-66.

Fredrickson, J. K., D. L. Balkwill, J. M. Zachara, S. W. Li, F. J. Brockman, and M. A. Simmons. 1991. Physiological diversity and distribution of heterotrophic bacteria in deep Cretaceous sediments of the Atlantic Coastal Plain. *Appl. Environ. Microbiol.,* 57:402-411.

Fredrickson, J. K., F. J. Brockman, B. N. Bjornstad, P. E. Long, S. W. Li, J. P. McKinley, J. V. Wright, J. L. Conca, T. L. Kieft, and D. L. Balkwill. 1993. Microbiological characteristics of pristine and contaminated deep vadose sediments from an arid region. *Geomicrobiol. J.* 11:95-107.

Fredrickson, J. K., J. P. McKinley, S. A. Nierzwicki-Bauer, D. C. White, D. B. Ringelberg, S. A. Rawson, S. W. Li, F. J. Brockman, and B. N. Bjornstad. 1995. Microbial community structure and biogeochemistry of Miocene subsurface sediments: implications for long-term microbial survival. *Mol. Ecol.,* 4:619-626.

Freeze, R. A. and J. A. Cherry. 1979. *Groundwater.* Prentice-Hall. Englewood Cliffs, NJ. 604 p.

Fritz, P. and J. Ch. Fontes, Eds., 1980. *Handbook of Environmental Isotope Geochemistry*, Vol. 1. Elsevier, New York, 545 p.

Ghiorse, W. C. and J. T. Wilson. 1988. Microbial ecology of the terrestrial subsurface, *Adv. Appl. Microbiol.,* 33:107-172.

Golightly, D. W., J. L. Seeley, and P. J. Lamouthe. 1978. Argon inductively coupled plasma in the analysis of geological materials by atomic emission spectrometry. U.S. Geological Survey Professional Paper 1100. U.S. Geological Survey. Denver, CO. 246 p.

GSA. 1991. *Rock Color Chart.* Geological Society of America. Boulder, CO.

Gullett, C. D. and R. Ehrlich. 1992. Mineral biofouling: textural recognition of exopolysaccharides in deeply buried sands of the Atlantic coastal plain, South Carolina. *GSA Abstr. Prog.,* 24:232.

Haldeman, D. L., B. J. Pitonzo, S. P. Story, and P. S. Amy. 1994. Comparison of microbiota recovered from surface and deep subsurface rock, water, and soil along an elevational gradient. *Geomicrobiol. J.,* 12:99-111.

Haldeman, D. L. and P. S. Amy. 1993. Bacterial heterogeneity in deep subsurface tunnels at Rainier Mesa, Nevada Test Site, *Microb. Ecol.,* 25:183-194.

Hays, P. D. and E. L. Grossman. 1991. Oxygen isotopes in meteoric calcite cements as indicators of continental paleoclimate. *Geology,* 19: 441-444.

Head, K.H. 1992. *Manual of Soil Laboratory Testing,* John Wiley & Sons, New York, 388 p.

Hoefs, J. 1987. *Stable Isotope Geochemistry.* Springer-Verlag, New York, 241 p.

Hubner, H. 1986. Isotope effects of nitrogen in the soil and biosphere, in: *Handbook of Environmental Isotope Geochemistry,* Vol. 2. B. P. Fritz and J. Ch. Fontes, Eds., Elsevier, New York. p. 361-425.

Hurlburt, C. S. 1971. *Dana's Manual of Mineralogy,* John Wiley & Sons. New York. 579 p.

Jacobsen, R. and D. Postma. 1994. In situ rates of sulphate reduction in an aquifer (Romo, Denmark) and implications for the reactivity of organic matter. *Geology,* 22:1103-1106.

Jansen, C. R. and H. J. B. Birks. 1994. *Recurrent Groups of Pollen Types in Time.* Elsevier. Amsterdam.

Kieft, T. L., P. S. Amy, F. J. Brockman, J. K. Fredrickson, B. N. Bjornstad, and L. L. Rosacker. 1993. Microbial abundances and activities in relation to water potential in the vadose zones of arid and semiarid sites. *Microb. Ecol.,* 26:59-78.

Kieft, T. L., J. K. Fredrickson, J. P. McKinley, B. N. Bjornstad, S. A. Rawson, T. J. Phelps, F. J. Brockman, and S. M. Pfiffner. 1995. Microbiological comparisons within and across contiguous lacustrine, paleosol, and fluvial subsurface sediments. *Appl. Environ. Microbiol.,* 61:749-757.

Klute, A. 1986. Water retention: laboratory methods, in A. Klute, Ed., *Methods of Soil Analysis,* Part 1, American Society of Agronomy. Madison, WI. pp. 635-660.

Klute, A. and C. Dirksen. 1986. Hydraulic conductivity and diffusivity: laboratory methods. In A. Klute, Ed., *Methods of Soil Analysis,* Part 1, American Society of Agronomy. Madison, WI. pp. 687-732.

Kuwana, T. 1978. *Physical Methods in Modern Chemical Analysis.* Academic Press. New York. 320 p.

Krumholz, L. R. and J. M. Suflita. 1996. Ecological interactions among sulfate-reducing and other bacteria living in subsurface Cretaceous rocks. American Society for Microbiology Meetings, New Orleans, LA, N-56.

Lawrence, G. P. 1978. Stability of soil pores during mercury intrusion porosimetry. *J. Soil Sci.,* 29:299-304.

Lewis, D. W. and D. M. McConchie. 1994. *Analytical Sedimentology.* Chapman and Hall, New York.

Long, P. E., J. K. Fredrickson, B. N. Bjornstad, S. A. Rawson, T. C. Onstott, and T. Griffin. 1994. Use of thermal effects of a volcanic neck to test the origin of subsurface microorganisms. *GSA Abstr. Prog.,* 26:509.

Lovely, D. R., F. H. Chapelle, and E. J. P. Phillips. 1990. Fe(III)-reducing bacteria in deeply buried sediments of the Atlantic Coastal Plain, *Geology,* 18:954-957.

Mahaney, W. C. 1984. *Quaternary Dating Methods.* Elsevier, New York. 428 p.

McCarty, D. K. 1990. Burial Diagenesis in Two Nonmarine Tertiary Basins, Southwestern Montana. University of Montana, Masters Thesis. 73 p.

McKinley, J. P., T. O. Stevens, J. K. Fredrickson, J. M. Zachara, F. C. Colwell, K. B. Wagnon, S. A. Rawson, and B. N. Bjornstad. 1996. The biogeochemistry of anaerobic lacustrine and paleosol sediments within an aerobic unconfined aquifer. *Geomicrobiology.*

McMahon, P. B., F. H. Chapelle, W. F. Falls, and P. M. Bradley. 1991. Role of microbial processes in linking sandstone diagenesis with organic-rich clays. *J. Sediment. Petrol.*, 62:1-10.

Milner, H. B., A. M. Ward, and F. Higham, Eds., 1962. *Sedimentary Petrology*: Vol. 1, Macmillan. New York. 643 p.

Molenaar, C. M. 1973. Sedimentary facies and correlation of the Gallup sandstone and associated formations, Northwestern New Mexico. *Four Corners Geol. Soc. Mem.*, 85-110.

Molenaar, C. M. 1983. *Major Depositional Cycles and Regional Correlations of the Upper Cretaceous Rocks, Southern Colorado Plateau and Adjacent Areas.* SEPM Rocky Mountain Section. Society of Economic Paleontologists and Mineralogists. Tulsa, OK. pp. 201-224.

Molenaar, C. M. and J. K. Baird. 1992. Regional stratigraphic cross sections of upper Cretaceous rocks across the San Juan Basin, northwestern New Mexico and southwestern Colorado. U.S. Geological Survey, Open File Report 92-257. Washington, D.C.

Monger, H. C., L. A. Daugherty, W. C. Lindemann, and C. M. Lidell. 1991. Microbial precipitation of pedogenic calcite. *Geology*, 19:997-1000.

Munsell. 1975. Munsell Soil Color Charts. Macbeth, a division of Kollmorgen Co., Baltimore, MD.

Murphy, C. P., P. Bullock, and K. J. Biswell. 1977. The measurement and characterization of voids in soil thin sections by image analysis. *J. Soil Sci.*, 28:509-518.

Murphy, E. M., J. A. Schramke, J. K. Fredrickson, H. W. Bledsoe, A. J. Francis, D. S Sklarew, and J. C. Linehan. 1992. The influence of microbial activity and sedimentary organic carbon on isotope geochemistry of the Middendorf aquifer. *Water Resour. Res.*, 8:723-740.

Murray, C. J., F. J. Brockman, and B. N. Bjornstad. 1994. Spatial distribution of microorganisms in lacustrine Ringold Formation sediments, south-central Washington. *GSA Abstr. Prog.*, 26:103.

Murray, C. J., F. J. Brockman, and B. N. Bjornstad. 1995. Spatial continuity of bacterial properties in subsurface sediments from south-central Washington. *GSA Abstr. Prog.*, 27:191.

Nickel, E. 1978. The present status of cathode luminescence as a tool in sedimentology. *Miner. Sci. Eng.*, 10:73-100.

Parkes, R. J., B. A. Cragg, S. J. Bale, J. M. Getliff, K. Goodman, P. A. Rochelle, J. C. Fry, A. J. Weightman, and S. M. Harvey. 1994. Deep bacterial biosphere in Pacific Ocean sediments, *Nature*, 371:410-413.

Pedersen, K. and S. Ekendahl. 1990. Distribution and activity of bacteria in deep granitic groundwaters of southeastern Sweden. *Microb. Ecol.*, 20:37-52.

Phelps, T. J., E. M. Murphy, S. M. Pfiffner, and D. C. White. 1994. Comparison between geochemical and biological estimates of subsurface microbial activities. *Microb. Ecol.*, 28:335-349.

Pollastro, R. M. and C. E. Barker. 1986. *Application of Clay-Mineral, Vitrinite Reflectance, and Fluid Inclusion Studies to the Thermal and Burial History of the Pinedale Anticline, Green River Basin, Wyoming,* Society of Economic Paleontologists and Mineralogists, Tulsa, OK. p. 73-83.

Reinek, H.-E. and I. B. Singh. 1980. *Depositional Sedimentary Environments.* Springer-Verlag, New York, 549 p.

Rosacco, G. J. and E. Roedder. 1975. Laser Raman spectroscopy for nondestructive partial analysis of individual phases in fluid inclusions in minerals. *EOS (Am. Geophys. Union Trans.)*, 56:460.

Rosenfeld, W. D. and S. R. Silverman. 1959. Carbon isotope fractionation in bacterial production of methane. *Science*, ??:1658-1659.

Russell, C. E., R. Jacobson, D. L. Haldeman, and P. S. Amy. 1994. Heterogeneity of deep subsurface microorganisms and correlations to hydrogeological and geochemical parameters. *Geomicrobiol. J.*, 12:37-51.

Schlumberger. 1989. *Cased Hole Log Interpretation Principles/Applications,* Schlumberger Educational Services, Houston, TX.

Schlumberger. 1991. *Log Interpretation Principles/Applications,* Schlumberger Educational Services, Houston, TX.

Stevens, T. O., J. P. McKinley, and J. K. Fredrickson. 1993. Bacteria associated with deep, alkaline, anaerobic groundwaters in southeast Washington. *Microb. Ecol.*, 25:35-50.

Tabatabai, M. A. and N. T. Basta. 1991. *Ion Chromatography.* Marcel Dekker. New York.

Touret-Jacques, L. R. 1987. Fluid inclusions and pressure-temperature estimates in deep-seated rocks.

Tucker, M. E. 1982. *The Field Description of Sedimentary Rocks.* John Wiley & Sons, New York, 112 p.

West, J. M., N. Christofi, and I. G. McKinley. 1985. An overview of recent microbiological research relevant to the geological disposal of nuclear waste, *Radioact. Waste Manage.,* 6:79-95.

West, J. M., I. G. McKinley, and A. Vialta. 1992. Microbiological analysis at the Pocos de Caldas natural analogue study sites. *J. Geochem. Explor.,* 45:439-449.

Zussman, J. 1977. *X-Ray Diffraction.* Academic Press. London.

3

Methods for Obtaining Deep Subsurface Microbiological Samples by Drilling

W. T. Griffin, T. J. Phelps, F. S. Colwell, and J. K. Fredrickson

CONTENTS

KEY WORDS: *drilling, rotary, auger, cable-tool, split-spoon, vibra-core, sonic, Shelby tube, pitcher barrel, drive point, piston sampler, wireline, quality control, tracer, control, blank, replicate.*

3.1   Introduction

Collecting representative microbiological samples from the subsurface is a challenging and often costly process that typically yields small sample volumes from a limited
number of sample horizons. The very nature of subsurface sampling usually precludes opportunities to return to a sample horizon and collect additional samples.
Consequently, it is necessary to optimize the sampling opportunity by selecting from
the wide range of available technologies a drilling and sampling method that is best
suited to the environment being sampled, and by collecting and processing samples
in a manner that best preserves their integrity. Determining the most appropriate
technology requires not only knowledge of the various technologies available but
also clear sampling objectives, determination of subsurface conditions from which
the samples are to be collected, and recognition of constraints that may influence the
sampling effort.

Once a drilling and sampling technology has been selected, various quality assurance and quality control (QA/QC) procedures should be identified that can be
employed during field sampling. The goal of QA/QC is to reduce potential sample
contamination that can occur, provide measures of sample integrity, and assure that
the samples are representative of the subsurface environment from which they were
collected (Colwell et al., 1992). A number of these fundamental techniques are discussed, including recommendations on handling and processing samples, use of tracers, site decontamination procedures, and appropriate levels of documentation.

This chapter is written for the microbiologist who may need assistance in planning
a subsurface field sampling program. It describes various aspects to be considered
when selecting appropriate drilling and sampling technologies, and discusses sample
processing and QA/QC techniques that, when employed, help ensure sample
integrity.

3.2   Selecting a Sampling Method

3.2.1   Geological Assessment

Prior to selecting a specific drilling technology, it is necessary to assess the subsurface
geological environment from which the samples are to be collected. This assessment,
which should include input from microbiologists, geologists/geochemists, and
hydrologists (Russell et al., 1993a), is important for two primary reasons. First, a geological environment must be identified from which samples can be used to address
the hypotheses of the research program. For example, if the research is focused on the
presence of abundant sulfate-reducing bacteria, it may be necessary to target anaerobic, potentially sulfate-rich geological deposits such as oxbow lakes, paludal (swamp)
deposits, or coal seams. Coarse-textured sediments low in organic matter are generally oligotrophic, but may have significant concentrations of entrained nutrients in
their recharge waters. Likewise, uncompacted porous sediments may initially have
had significant volumes of ground water moving through them, but have since experienced a significant reduction in permeability due to diagenesis and are effectively
sealed from the influx of nutrients associated with ground water. In deeper formations, temperature, pressure, and the potential presence of hydrocarbons are among

the factors that may influence the presence, diversity, and activity of microorganisms. Once prospective geological environments have been identified, a sampling site can be selected. Microbiologists should coordinate this effort with geologists who understand the stratigraphy of the target environments, as well as the microbiological research hypotheses and sampling objectives of the program (Griffin, 1991; Griffin et al., 1994).

The second reason for conducting a geological assessment is that the physical characteristics of the target formations will dictate the type of drilling and sampling technology employed, and will also have considerable bearing on the QA/QC protocols used. Selection of the drilling and sampling technologies and the appropriate QA/QC protocols requires the perspectives of multiple disciplines including microbiology, geology, and those familiar with both the drilling requirements in the region and the sampling objectives of the program. If samples are to be collected from shallow, semiconsolidated to consolidated sediments, then a relatively simple and cost-effective drilling technology can usually be used. However, if the target sample intervals are in deeper geologic formations (i.e., hundreds to thousands of meters), or if drilling problems such as flowing or heaving sands, swelling clays, lost circulation zones, or abnormally pressured zones are anticipated, then the drilling technology required will be more sophisticated and costly. The strengths and limitations of each candidate technology should be identified and considered.

Once appropriate technologies have been selected, it is worthwhile to have a drilling company representative explain the various rate structures available for a drilling program, including mobilization costs, footage vs. daily rates, and standby rates. Typically, if the samples require sensitive treatment or processing, or if delays are likely to occur as a result of sample processing requirements, then a daily drilling rate may be preferable, as opposed to footage and standby rates.

3.2.2 Planning a Sampling Program

The principal objective of scientific drilling and sampling is the collection of subsurface materials that will provide the samples necessary for addressing the research hypotheses. When planning a sampling program, it is worth noting that the sampling objectives must be consistent with the available project resources.

While the geology is being evaluated and a site is being selected, specific sampling requirements can be determined. These requirements include the total number of samples to be collected, the volume or mass required, and the anticipated depths from which each sample will be collected. In determining the number of samples to be collected, it is important to consider the time required for processing and analyzing each sample. Should the number and frequency of samples exceed the capacity to process and analyze them, costly delays in drilling and sampling might be required in order to provide fresh samples. Conversely, extended drilling time between sample depths could result in inefficient utilization of laboratory personnel. Both situations can potentially impact project costs.

While having little direct impact on the type of drilling and sampling technology employed, the extent and type of sample processing to be conducted in the field can have other significant impacts on a field sampling program. Field personnel must be prepared to address all sampling requirements and conditions in the field without inflicting costly delays in the drilling and coring. Extensive processing of samples, particularly anaerobic processing, requires considerable logistical planning and support (Colwell et al., 1992; Russell et al., 1992; Fredrickson and Phelps, 1996). If cores are to be processed anaerobically, then a glove bag (Figure 3.1) and its components,

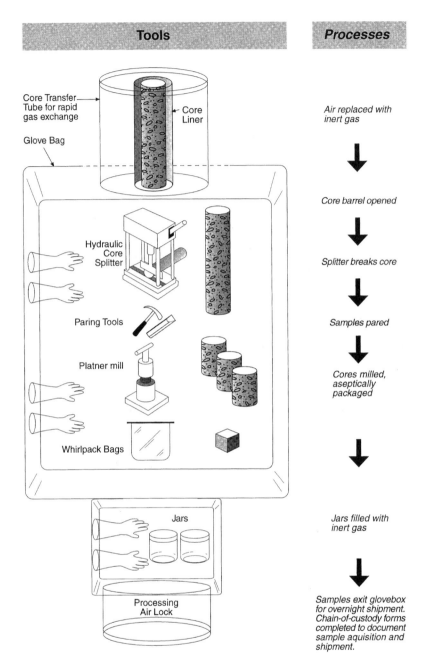

Tools

Processes

Core Transfer Tube for rapid gas exchange

Core Liner

Glove Bag

Hydraulic Core Splitter

Paring Tools

Platner mill

Whirlpack Bags

Jars

Processing Air Lock

Air replaced with inert gas

Core barrel opened

Splitter breaks core

Samples pared

Cores milled, aseptically packaged

Jars filled with inert gas

Samples exit glovebox for overnight shipment. Chain-of-custody forms completed to document sample aquisition and shipment.

FIGURE 3.1
Schematic of the core processing chamber and processing sequence. Core liners enter the core transfer tube where the air is displaced with inert gas. The core is then transferred to the anaerobic glove bag where it is split, pared, milled, and packaged in Whirlpack bags. Samples are then transferred to the smaller anaerobic chamber where they are placed in sample jars. The sample jars exit the processing chamber via the processing air lock.

including spare parts, bottled gases, and a transfer chamber that can accommodate the anticipated length of core, as well as sufficient numbers of qualified personnel, may be required on-site. Sufficient sample processing tools such as spatulas, knives, spoons, pans, and sterile sample jars or bags must be on hand, along with sterilization equipment. If samples are to be held on-site for even brief periods prior to disbursement, sample storage requirements such as refrigeration or liquid nitrogen sources need to be identified. Shipping containers, ice or ice packs, and dry ice may also need to be stocked if samples are to be shipped. Shipping arrangements should be made with the appropriate shipping agent prior to the initiation of field operations.

3.2.2.1 Defining Constraints

During development of a field sampling program it is necessary to identify constraints that may impact the selection of the sampling technology, project scheduling and implementation, and the level of effort required. The foremost constraints include funding and the availability of trained personnel. Budgeting must ensure that each phase (planning, field, and analytical) of the research program is adequately funded, and should include contingency funds to support delays or to address unforeseen drilling problems. If the budget for the drilling program is limited, it may be possible to "piggyback" onto another drilling program in order to share expenses. For example, microbiological samples were collected by the U.S. Department of Energy's Subsurface Science Program (SSP) from a petroleum exploration well drilled by Texaco in northeastern Virginia (Griffin et al., 1993).

Regulatory requirements may place additional scheduling and financial constraints on a sampling program. Some drilling sites may have access restrictions, and the driller will likely be required to have state licenses and a drilling permit. If the research project is funded by a federal agency, it may be necessary to prepare National Environmental Policy Act (NEPA) documentation for approval by that agency prior to mobilization. NEPA requirements may include preparation of environmental assessments, cultural resource surveys, and/or endangered or threatened species surveys. NEPA requirements vary between federal agencies, so specific requirements should be confirmed.

A major constraint that is impacted by all of the factors discussed previously is scheduling. Project planning, geological assessments, identifying and addressing regulatory requirements, completing the drilling and sampling operations (including delays that inevitably arise), and completing sample analyses all contribute to scheduling complexities and the time required to complete a project. Severe weather can delay starting dates and prolong the field operations. Scheduling is also an issue if the project relies on students or visiting scientists.

3.2.2.2 Documentation

It is recommended that the resolutions to the various planning issues be recorded for reference purposes. Thorough planning and documentation minimizes the potential for confusion during field operations and helps ensure that sample collection and processing is consistent, efficient, and technically sound. Field operations, including personnel requirements and responsibilities, should be described and frequently referenced during the field program. Written procedures for specific activities such as sample processing and the operation and maintenance of equipment should also be available. Inventory lists help ensure that the necessary equipment is available and operable in the field. Safety concerns and guidelines should also be documented, and Material Safety Data Sheets (MSDSs) for all chemicals and solutions on-site must be readily available. If regulatory documents such as NEPA approvals or permits have been issued, these documents should be on-site as well.

3.2.3   Drilling and Sampling Technologies

Drilling and sampling technologies vary significantly, and certain technologies are better suited than others to particular drilling conditions (Colwell et al., 1992; Russell et al., 1992; Fredrickson and Phelps, 1996; Barrow, 1994; Phelps et al., 1989; Leach, 1991). After evaluating the geology and anticipated drilling conditions, the drilling technology best suited for the site and sampling objectives can be selected. It is highly recommended that geologists and drillers who are familiar with the range of drilling technologies applicable and available in the region be consulted.

Drilling methods, or in the broader context, methods by which a borehole can be advanced, can be divided into two general categories. The first includes methods that advance the borehole without the use of circulation fluids, including direct push, auger, percussive, and vibrating technologies. The second category requires circulation media to lubricate the cutting bit, stabilize the borehole, and circulate the lithologic fragments or "cuttings" from the bottom of the borehole as the drilling bit advances. This second category includes mud and air rotary drilling. Each method has advantages and disadvantages.

Sampling tools must be compatible with the drilling method, and some tools are better suited for particular lithologies and drilling conditions than others. Whenever possible, it is recommended that alternative tools or approaches be identified and be accessible should the initial choice prove inadequate.

3.2.3.1   *Drilling Without Circulation Media*

The primary advantage of utilizing drilling technologies that do not require circulating fluids is that the potential for introducing microbial or chemical contaminants into the samples is significantly reduced. Additionally, these technologies are often the simplest and the least expensive. The major disadvantages of the noncirculation drilling technologies are their depth limitations, and in most cases, their inability to advance the borehole through indurated rock.

The simplest method of advancing a borehole is direct push. Direct push technologies vary from handheld tools to small truck-mounted rigs. They convey a sampler (i.e., Shelby tubes, soil probes) at the end of a solid or hollow direct push rod, all of which is retracted once the sample has been collected. Direct push technology is becoming more widely available and is relatively inexpensive when compared to more standard drilling technologies. The disadvantages of direct push technology are that it is ineffective in consolidated or gravelly lithologies, is limited in sample volume (3 cm diameter samples are common), and its depth limitation is typically in the realm of 10 m, depending on the lithology though depths of 100 m can be reached.

Another simple, noncirculating drilling technology is the auger drilling method. Augers advance the borehole by rotating screw-like sections of pipe, or auger "flights", into the ground. "Hollow-stem" auger flights are open in the center to allow passage of sampling equipment (Fredrickson and Phelps, 1996; Zapico et al., 1987). The smallest augers are rotated by hand, though the typical auger rig is truck or trailer mounted. Augers have the advantage of maintaining the borehole wall and keeping the borehole open while samples are collected. Depth control is precise, and under ideal conditions an auger rig may approach depths of 100 m beneath ground surface. Auger holes of <30 m depth are often completed in one day, and multiple shallow holes in noncompacted sand and silt lithologies can be completed in one day. Besides being mobile and widely available, auger drilling is relatively inexpensive, and can be accomplished by a crew of two persons. Different types of samplers can be used with auger drilling, and sample recovery and quality are generally good, particularly in the vadose zone. Disadvantages of auger drilling include its depth limitation (generally 100 m or less) and its inability to penetrate consolidated formations. In saturated unconsolidated lithologies sample recovery can be poor, especially when the formation material flows into the borehole. In these situations, drilling fluid must be added to the borehole to maintain hole stability and reduce torque. The high torque required to auger at greater depths or in finer grained sediments such as silts or clays can exceed the rig's ability to turn the augers.

One of the oldest and most common methods for advancing a borehole is percussive cable-tool drilling. Developed centuries ago, cable-tool rigs are still common in certain parts of the U.S. Relatively inexpensive to operate, cable-tool rigs are typically small- to moderate-sized truck-mounted rigs that advance the borehole by means of a downhole hammer or "jar" attached to a cable. The jar typically weighs >50 kg, and is repeatedly raised and dropped onto either a cutting tool ("hard tool") or a sampling drive barrel at the bottom of the hole. The hard tool is used to fracture consolidated material, while the drive barrel is used either to "bail" the rock fragments or cuttings from the bottom of the borehole, or to advance the borehole through unconsolidated or semiconsolidated formations. In unconsolidated formations, casing is driven immediately behind the advancing borehole to maintain borehole integrity. Cable-tool drilling can be used at depths in excess of 1000 m, and in unconsolidated fine-grained sediments can advance at a rate of meters per day. In unconsolidated lithologies where the drive barrel or other downhole sampling devices such as a split-spoon sampler can be used (Fredrickson et al., 1993) sample quality can be excellent, while sample fragments bailed from the borehole in consolidated formations may be of poor quality. Cable-tool rigs can be used in unconsolidated gravels and cobbles, and are desirable when no drilling fluids are to be used (Fredrickson et al., 1993). The biggest disadvantage of this method is that it can be painstakingly slow, particularly in consolidated formations, where advancement rates of <3 m/day are common. For deeper drilling in consolidated formations, other drilling methods are preferable.

Drilling methods that advance the borehole by vibration include some of the most recent advancements in drilling technology. Vibra-coring tools have been used for a number of years to advance boreholes and collect shallow (10 to 15 m) continuous core in unconsolidated sediments. Vibra-core units are generally handheld and operated by one or two persons. Coring is continuous, and cores are extracted with either a tripod-mounted block and tackle system or "come-along". These systems are inexpensive to operate, and sample quality is excellent. Disadvantages include their depth limitations and inability to collect consolidated material.

Sonic and roto-sonic drilling systems are significantly more advanced than vibra-coring. Typically of moderate size and truck mounted, these rigs are comprised of rapidly oscillating bearings in a top-head drive unit called a "tub". The oscillating bearings generate a mechanically induced pressure wave that is transmitted down the hole through the drillpipe, delivering a force that can exceed 120,000 kg at rates approaching 150 pulses per second (Barrow, 1994). This drilling method can be used in a wide variety of lithologies, including cobbly formations, and can have high core recovery rates in unconsolidated, saturated sands. Coring can be continuous or selective, and under ideal conditions drilling rates can approach 0.3 m/s. These systems have reached depths exceeding 200 m, but the typical limit is 120 m (Barrow, 1994). A major disadvantage of this technology is the limited availability of rigs, which is also problematic if serious mechanical breakdowns occur. The limited number of rigs also means high mobilization and operational costs, which can equal or exceed costs for standard drilling technologies.

3.2.3.2 Sampling Without Circulation Media

Noncirculation sampling tools include push, drive, and vibrating tools. These sampling tools are advantageous in that, in the absence of circulation fluids, they provide high-quality samples at relatively low cost. Their limitations, however, are generally the same limitations as the drilling method with which they are employed. Most of these sampling tools can only sample relatively shallow formations comprised of unconsolidated or "soft" sediments such as sands, silts, or clays.

Push sampling tools include Shelby tubes, pitcher barrels, and drive points. Each tool is conveyed on the end of a rod and is pushed into the formation by downward pressure. In most cases, direct push samplers will only advance through fine-grained silt and clay sediments. Sample size for Shelby tubes and pitcher barrels is typically on the order of 0.6 m in length by a minimum of 5 cm in diameter. Drive point samples typically do not exceed 3 cm in diameter. Direct push samplers can be used to advance the borehole, but are usually employed through hollow-stem augers or casing, which stabilizes the borehole wall and prevents "sloughing" of material from higher in the borehole. Direct push samplers can be sealed after they are retrieved. They are readily available and are inexpensive to use. In situations where the target sample intervals are shallow (<100 m for Shelby tubes and pitcher barrels, <10 m for drive points) and are very fine-grained, direct push samplers are a desirable option.

Drive samplers, including split-spoons, drive points (Leach, 1991; Starr and Ingelton, 1992), and piston samplers (Zapico et al., 1987), are physically driven into the sampling horizon by blows from a rig-powered hammer or weight. Drive samplers have similar limitations and are similar in attributes and application to push samplers, but have the added advantage of being able to collect samples in coarse-grained sediments and gravels. While split-spoons may, in some instances, be ineffective at retrieving unconsolidated saturated sands, these sediments can often be retrieved with piston samplers. Piston samplers create suction on the sample as the sample barrel is driven past an internal stationary piston which prevents the sample from falling out of the barrel during retrieval.

The most sophisticated of the noncirculation sampling tools is the sonic drill core barrel. Sonic core barrels are available in a number of configurations, including 1.5- to 3-m solid barrels and 1.5-m split barrels. Cores are typically 7.6 cm in diameter, and can be collected in Lexan® or aluminum liners. Sample recovery is typically high, even in saturated and unconsolidated sands, and samples appear to be minimally disturbed. Sonic sampling has had only limited application in microbiological research to date, but may prove to be a highly effective, although currently somewhat costly, method for collecting subsurface microbiological samples.

3.2.3.3 Drilling With Circulation Media

Standard rotary drilling methods that rotate a downhole bit in order to advance the borehole generally require a circulating medium that is pumped down the hole and back to the surface during drilling. Depending on the nature of the lithologies being drilled, this medium can be either air, water, foam, or bentonite-based "mud", though in rare circumstances inert gas (i.e., nitrogen or argon) has been used in microbiological sampling to minimize exposure of the core to oxygen (Russell et al., 1993b). Air and bentonite mud are the most common circulation media. Circulating media serve multiple purposes. They cool and lubricate the drill bit, and remove rock fragments or "cuttings" from the bottom of the borehole. In the case of water or bentonite mud they stabilize flowing sands and decrease the influx of formation fluids or gases by maintaining hydrostatic pressure on the borehole wall. Rotary drilling methods are among the most common and versatile available, and can vary significantly in cost depending on drilling depth, circulation medium, and whether or not drilling problems such as expanding clays, flowing sands, or lost circulation zones are encountered. The largest of the rotary rigs are usually employed in petroleum exploration and production, as well as scientific programs targeting depths of several kilometers (Proceedings of the VIIth International Symposium on the Observation of the Continental Crust Through Drilling, 1994). Moderately sized truck-mounted rigs can drill to depths of 1000 m or more. Under ideal drilling conditions, rotary rigs can drill and core >100 m/day, and may thus be the most economical, if not the only method for reaching deeper sampling horizons. A variety of sampling tools can be used with rotary/fluid circulation drilling systems.

The major disadvantage of utilizing circulation drilling methods for subsurface microbiological research (other than cost) is the potential for introducing chemical and microbiological contaminants into the sample via the circulation medium. Air can be effectively HEPA filtered at the surface prior to being pumped down the borehole, but filtration to remove microorganisms is impossible for water or mud circulation systems. Circulation media are pumped through long expanses of either open borehole or drillpipe, and thus microorganisms can become entrained in the circulation media during circulation. Water, and mud in particular, maintain warmer temperatures they acquire due to both friction and deeper borehole depths, and some drilling additives can serve as nutrient and energy sources for bacteria. Therefore, drilling fluids usually have high populations of bacteria, often exceeding 107 cells per cubic centimeter (Haldeman et al., 1995). These fluids are thus a major source of microbial contamination. Contaminated circulation media can also impact the target sample horizon ahead of the bit in permeable or fractured geological units where fluid pressures can force fluids into the formation far in advance of the bit. Air forced into the sampling horizon can cause drying and can alter subsurface chemistries (oxidation, etc.), which can subsequently impact microbial viability and activity. Fortunately, many of these contamination issues can be addressed with a rigorous quality control regimen, as discussed later.

3.2.3.4   *Sampling With Circulation Media*

A number of sampling technologies have been developed for use with circulation drilling systems. These technologies range from simple sampling devices that can be used with small, truck-mounted drilling rigs, to more sophisticated tools that require large rigs to be employed. These sampling devices have the advantage over noncirculation sampling devices in that they can be used to collect samples at depths to several thousand meters.

The simplest of these sampling devices is the core barrel conveyed on the end of the drillstring (Phelps et al., 1989). Core barrels range in length from 1.5 up to 18 m. Core diameters also vary, but typically range from 8 to 13 cm. A standard core barrel configuration, even on smaller truck-mounted rigs, includes an inner core barrel inside an outer barrel. In most cases, circulation fluids are diverted through the annulus between the two barrels by a valve at the top of the inner barrel, and continue out the coring bit and up the borehole annulus while coring is underway. Sample recovery is usually quite good, but these systems have the disadvantage of exposing the core to drilling mud as the drillstring is retrieved from the hole to recover the core barrel. More sophisticated and costly applications employed in the petroleum exploration industry include systems that seal cores in rubber sleeves or viscous gels, and that retrieve core while maintaining its original formation pressure. The latter systems are especially effective at collecting high-quality core from depths of up to several thousand meters.

Wireline core retrieval systems are cost-effective sampling technologies that are effective for collecting samples at depths to several hundred meters (Phelps et al., 1989; Clark, 1988). These systems are comprised of an inner core barrel that is rapidly lowered through the core pipe to the bottom of the borehole and then retrieved on the end of a cable. When lowered to the bottom, the inner barrel locks into an outer barrel with a spring-loaded latch; plastic or Lexan® core liners can be placed inside the inner barrel to protect the core. As with standard core barrels, drilling fluid is diverted between the inner and outer barrels. Once coring is complete, an "overshot" assembly is lowered by cable into the borehole. The overshot attaches to the latch at the top of the inner core barrel and retrieves the inner barrel to the surface (Phelps et al., 1989; Clark, 1988). Wireline systems are cost effective in that they can be used with standard truck-mounted rigs, and because the delivery and retrieval of the inner barrel is rapid relative to standard core barrels. Sample recovery in unconsolidated sediments can be poor, but in semiconsolidated to consolidated rock, sample recovery is generally good. Sample quality is typically good, although in porous materials there is potential for contamination from drilling mud. For sample targets that are several hundred meters below land surface, a wireline coring system may be the sampling system of choice.

A somewhat more sophisticated pair of wireline-conveyed sampling tools include the rotary and percussion sidewall coring tools. Rotary and percussion sidewall coring tools have been used for several years in the petroleum exploration industry, and have had limited application in microbiological sampling programs (Proceedings of the VIIth International Symposium on the Observation of the Continental Crust Through Drilling, 1994; Griffin et al., 1993). Rotary sidewall coring tools, which are effective in highly consolidated formations, consist of a small (approximately 7 cm in length, 2.5 cm in diameter) rotary coring bit at the top of the tool which hydraulically rotates out from the tool toward the borehole wall, cores a small core plug from the rock face, then rotates back into the core tool where the plug is pushed from the bit into a storage container. Core plug dimensions are approximately 6 cm × 2 cm. Percussion sidewall coring tools, which work better in less indurated formations, are comprised of a series of explosively charged hollow "bullets" connected to the tool by a steel cable. At each sample point, a hollow bullet is fired into the formation wall and then recovered as the tool is pulled up the hole. Sample plugs are approximately 7 cm × 1.5 cm.

Though sidewall coring tools are not specifically designed for use with rotary/circulation drilling systems, the tool size requires larger rigs that typically employ rotary/circulation. The advantages of these tools is that they can collect numerous samples from multiple sampling targets at depths of several thousand meters. Their disadvantages often outweigh their advantages in that they require a large rig and large-diameter borehole (>15 cm) to be employed, and their cost is relatively expensive, though in some cases less than conventional coring methods in deeper formations. Sample size is small, and in the case of the percussion sidewall cores, samples can be highly fractured and contaminated. Both tools expose the samples to the drilling mud for extended periods during sampling and retrieval. However, in cases where there are multiple deep target intervals in an existing borehole and conventional coring is restricted, sidewall coring may be an option.

3.3 Quality Assurance and Quality Control

Quality assurance and quality control (QA/QC) are critical aspects of any field sampling program, and are particularly important to subsurface sampling because biomass concentrations are typically low. QA/QC programs are implemented to minimize sample contamination, and ultimately, for measuring and evaluating sample quality and integrity. An effective QA/QC program consists of three primary components: (1) procedural controls, (2) physical controls, and (3) sample quality assessment.

3.3.1 Procedural Controls

Procedural controls include the development and use of written protocols and guidelines (i.e., sample processing and tracer injection) to ensure that the field operations are conducted in a manner that is consistent and technically sound. Procedural controls also include the documentation discussed previously in Section 2.2 of this chapter, along with additional documentation and records that should be maintained during the field program. Such documentation includes detailed logbooks maintained by microbiologists, geologists, and drillers. Sample logbooks should be maintained that record sample numbers, depths, dates, times collected, investigators for whom they were collected, special processing procedures, and other pertinent information. Additional logs, such as tracer or photographic logs, should also be maintained when applicable. Chain-of-custody (COC) forms should be considered for each sample that leaves the field site. COC forms provide field personnel with a record of samples collected, processed, and shipped, and provide investigators receiving samples a means by which to confirm receipt of the samples.

Properly maintained field documentation serves several important functions. It provides assurance that the field sampling operations were performed consistently and in accordance with detailed procedures. Field records also provide information regarding the integrity and quality of the samples collected, and serve as references for site activities.

3.3.2 Physical Controls

Physical controls include procedures implemented in the field to minimize sample contamination by exogenous microorganisms, colloids, or solutes. These procedures include decontamination of drilling, sampling, and sample processing equipment, and recommendations for site layout that reduce potential contamination pathways.

3.3.2.1 Site Layout

Sample contamination can occur by many pathways, including the circulation medium, equipment, personnel, and airborne transport. Sites can be arranged such that sample exposure to these pathways is minimized. To the degree possible, drilling rigs should be oriented such that exhaust is downwind of the drilling derrick, and all sampling tools, tool assembly areas, site access points, and sample processing facilities are upwind of the drilling rig. The borehole should have protective surface casing that is sealed or bermed, and the ground under the rig should slope away from the borehole to prevent extraneous fuel, fluids, and other contaminants from entering the borehole. Tool decontamination should be performed downwind of all site activities.

Incompatible activities such as tracer mixing and sample processing should be physically separated; if not, they should be performed at different times. Tracer storage, particularly volatile chemical tracers, should be isolated from the sample processing and storage areas, and if tracer analysis is being conducted in the field, it too should be isolated. Personnel who prepare tracer solutions should not immediately process samples nor handle sampling tools, and drillers and other personnel exposed to the tracers should minimize their access to the sample processing area.

3.3.2.2   Contamination Reduction

Sterile conditions are impossible to achieve in a subsurface sampling program. However, steps can be taken to minimize contamination. Drilling equipment is a significant source of contamination. Prior to drilling or coring, the rig, working platforms, drill rods, and sample handling equipment should be steam cleaned to remove fluids or particulates which could compromise sample integrity. Under the most strenuous conditions, steam cleaning of equipment should be repeated each time drilling tools are removed from the hole or when drilling fluids are changed. The addition of 5 mg/l chlorine residual to water used for decontamination and steam cleaning provides an effective disinfectant. Field personnel should wear clean or sterile gloves when assembling sampling tools after they have been cleaned.

All sample processing tools should be cleaned and flame sterilized or autoclaved, wrapped in sterile foil, and stored in sterile bags between each sample processing event. Glove bags, laminar-flow hoods, and other sample processing equipment should also be cleaned and disinfected between processing events. Sample containers can be autoclaved in the laboratory and sealed prior to being sent to the field, or can be autoclaved in the field. Core liners should be autoclaved and wrapped prior to field operations, due to their size. The use of sterile core liners inside sampling tools reduces sample exposure to contaminated drilling fluids, and can provide a means of transporting the sample from the rig to the processing area or laboratory with minimum sample exposure to air. Air drilling with HEPA filters can minimize the introduction of particulate contamination to the borehole. When possible, circulation systems should be single-pass or should circulate in contained, closed loops to reduce exposure of the circulation medium to contaminants. An enclosed circulation system also helps in maintaining cleanliness on the rig and site, reduces the contamination of sampling equipment and lab facilities, and reduces the loss of volatile tracers in the circulation medium.

3.3.3   Sample Quality Assessment

Sample integrity is the cornerstone of any microbiological field sampling program (Russell et al., 1992). Numerous opportunities exist for sample contamination, particularly when drilling is involved. As a result, samples are often contaminated to varying degrees by exogenous microorganisms or solutes. The evaluation of sample quality is often difficult in that it is often a subjective, qualitative judgment made from multiple lines of evidence. Accurate sample quality assessment requires two primary components in a field sampling program: (1) the implementation of a tracer program (Phelps et al., 1989; Chapelle and Lovley, 1990; Chapelle, 1992), and (2) the use of controls, blanks, and replicate samples. The use of microbiological and chemical tracers provides information on the extent and type of contamination that may have been introduced during sample collection and processing. Controls, blanks, and replicates provide evidence of contamination introduced during sample handling. Together, these two components address the contaminant pathways between the sample's native environment and its ultimate destination in the laboratory.

3.3.3.1   Controls, Blanks, and Replicates

Control, blank, and replicate samples are used to assess sample integrity and to determine if sample processing and transportation have potentially impacted the samples (Russell et al., 1992; Phelps et al., 1989; Fredrickson et al., 1993; Lehman et al., 1995). Controls include samples spiked with known microorganisms or chemical contaminants to assess the detection sensitivities and limits of analytical techniques, as well as potential physical impacts that may have occurred on samples during processing and shipping. Blank samples consist of sterile samples, preferably similar to those being sampled in the field. Blanks are also processed in the field in the same manner as authentic samples to determine if contamination has been introduced by processing or during shipping. Replicates, or sample "splits", are collected periodically for independent laboratory analyses to assess the repeatability of analytical results. Comparison of analytical results for controls, blanks, and replicates with those from actual samples can indicate whether or not artifacts were introduced during coring and core processing, or later by laboratory procedures.

3.3.3.2   *Tracers*

Tracers are introduced to the circulation medium or at the borehole/sample interface to identify potential sources of sample contaminants. The most effective tracer programs use multiple tracers that mimic the physical and chemical processes by which exogenous microorganisms and chemicals can contaminate a sample. Multiple tracers also enable distinction of contamination that arises from various sources (i.e., drilling fluids, cuttings, sidewall slough, surface materials, or contamination from adjacent boreholes). Tracers often provide the best means to assess sample integrity, and can prove unequivocally that a sample has been contaminated. They can also give some indication as to the source and degree of contamination.

The U.S. Department of Energy's Subsurface Science Program (SSP) has used tracers extensively in its field programs (Colwell et al., 1992; Phelps et al., 1989; Russell et al., 1992; Phelps and Russell, 1991). SSP investigators strive for a millionfold reduction in tracer concentration between the point of contact where the tracer was introduced and the sample itself. Reduction by six orders of magnitude after on-site paring and subcoring provides evidence that contamination represents less than 1 mg/kg of sediment. Table 3.1 identifies several tracers than can be used to evaluate chemical contamination of subsurface sediments, including their method of introduction, recommended initial concentrations, detection limits, and log protection factors. Table 3.2 describes additional techniques that can be used to help evaluate the degree of contamination by exogenous microorganisms. If feasible, it is recommended that independent tracer analyses be performed by alternative methods for increased defensibility of sample quality.

Dyes — Rhodamine WT (Crompton and Knowles, Green Hills, PA), a fluorescent dye, has been used as a tracer in aqueous drilling fluids at >20 mg/l (Russell et al., 1992; Phelps and Russell, 1991) (Table 3.1). The dye's red color allows visual detection of contaminated sediments at concentrations >0.2 mg/kg, enabling rapid discrimination of severely contaminated materials. Rhodamine concentrations can be quantified by mixing equal volumes of water and sample, centrifuging the particulates, and measuring the fluorescence of the supernatant (excitation wavelength: 546 nm, fluorescence: 590 nm) (Russell et al., 1992; Phelps and Russell, 1991). The detection limit of rhodamine by fluorimetry is approximately 2 µg/l, or approximately 4 orders of magnitude below the drilling fluid concentration (Table 3.1). Other investigations have used fluorescein dye as a tracer with limited success (Pederson and Ekendahl, 1990; Behrens, 1988). Potential problems of fluorescent dyes include their sensitivity to pH and oxidizing agents such as sodium hypochlorite. Rhodamine dyes have been used for decades in water flow tests but questions regarding their toxicity persist.

Perfluorocarbons — Perfluorocarbon tracers (PFTs) are nontoxic and inert, and are used in atmospheric testing (Senum and Dietz, 1991) and in clinical applications. Several different PFTs have been used as tracers in subsurface investigations (Russell

TABLE 3.1

Tracers Used to Evaluate Chemical Contamination of Subsurface Sediments

Tracer	Method of Introduction	Initial Concentration (mg/l)	Detection Limit in Samples (μg/l)	Protection Factor (log)
Rhodamine	Soluble in aqueous drilling fluids	20	0.5–2.0	4
Bromide	Soluble in aqueous drilling fluids	>500	10	5
(K or Na salt)	Concentrated solution directly to borehole	>500	10	5
Potassium (as Cl or Br salt)	Soluble in aqueous drilling fluids	>500	10	5
Sulfate	Present in bentonite fluids or soluble	>150	10	4
ClO_4 (as $NaClO_4$)	Sterile solution in core barrels or liners	200	2000	3
Perfluorocarbons	Add as methanol emulsion for 1 mg/l solubility	1	0.00001	8
	Meter and vaporize solution	1	0.001	6

TABLE 3.2

Techniques Used to Evaluate Potential Microbiological Contamination of Cored Subsurface Sediments

Technique	Method of Introduction	Initial Concentration	Detection Limit in Samples	Protection Factor (log)
Latex fluorescent microspheres, 1.0 μm carboxylated	Dry mix with glass beads, attach to core barrel	$5 \times 10^9/g; \times >200$ g	$10^3/g$	>5
	In bag in core barrel or liner that breaks upon coring	$3 \times 10^{10}/ml \times 10$ ml	$10^3/g$	>5
Community-level physiological profiles[a]	Comparative analysis	NA	NA	3
Bacterial membrane lipid profiles[a]	Comparative analysis	NA	NA	2
Exogenous microorganisms[b]	Add to circulating fluids	$<10^6/g$	$10^0/g$	5
Serendipitous indigenous microbes	Native, no introduction	$10^3–10^7/ml$	$10^3/ml$	>3
Bacterial plate counts[a]	Comparative analysis	NA	NA	2
Biochemical profiles of bacterial isolates[a]	Comparative analysis	NA	NA	2

Note: NA = not applicable.
[a] Analyses based upon pattern recognition.
[b] Regulated.

et al., 1992; Senum and Dietz, 1991; and McKinley and Colwell, 1996). The PFTs are typically emulsified in a 1% solution and added at 1 mg/l to drilling fluids (Table 3.1). Pyrolysis gas chromatography (GC) using capillary columns and electron capture detection (ECD) provides sensitive (picogram per gram) detection of PFTs. PFTs can also be measured using GC-ECD, but the sensitivity is lower. PFTs can be detected at eight orders of magnitude below their concentration in circulation fluids. PFTs have also been used as gaseous tracers by metering a PFT tracer solution into a heating unit which delivered volatilized tracer at milligram per liter levels into the air or inert gaseous drilling fluids as they were pumped downhole (Colwell et al., 1992) (Table 3.1). PFTs are versatile in that they can be used to trace gaseous or aqueous contaminants, and different PFTs provide multiple means of detecting drilling fluid intrusion into subsurface samples. However, the extremely sensitive detection of PFTs (6 to 8 orders of magnitude detection range) requires rigorous adherence to strict tracer handling protocols to prevent nondrilling-associated contamination, such as can result from PFTs on clothing, work surfaces, and lab equipment. Subsamples for off-site QA/QC must be collected immediately during sample processing, sealed in gas-tight vessels, or placed in a solvent in which PFTs are highly soluble (i.e., methanol), and refrigerated to prevent off-gassing of PFTs.

Water-soluble ionic tracers — Chemicals indigenous to bentonite drilling fluids, such as barium, sulfate, or ammonium, can be detected at 10^4 dilution and can be used as tracers for drilling fluid intrusion into samples (Russell et al., 1992; Chapelle and Lovley, 1990; and Phelps and Russell, 1991) (Table 3.1). Unfortunately, sulfate and ammonium can be subjected to chemical and biological transformations such that they are not conservative. Potassium or sodium bromide can be added to aqueous drilling fluids as conservative solute tracers at concentrations <1000 mg/l (Table 3.1). Ionic tracers such as bromide can be measured on-site using ion-specific probes, while more sensitive off-site analyses, such as ion chromatography, can provide increased sensitivity (i.e., detection at a 10^5 dilution of drilling fluids). Pressure generated from the drilling fluid column in the borehole (drilling muds are typically 20% denser than water) may facilitate the infiltration of water and soluble ions such as potassium, bromide, or sulfate into uncored formations in advance of the drill bit, or during a hiatus in drilling. It is worth noting that contamination by solutes in this manner may not represent microbial contamination, since the accumulation of drilling mud, or "wall-cake", on the borehole wall and bottom may retard bacterial transport while allowing solute tracers to permeate into uncored sediments.

Particulate tracers — Fluorescent bacteria-sized latex beads (0.4 to 1.2 μm diameter, Polysciences Inc., Bay Shore, NY) have been widely used as tracers of particulate migration (Colwell et al., 1992; Chapelle, 1992; Harvey et al., 1989). As tracers for subsurface sampling, microspheres are typically placed in Whirlpak bags and fastened to the inside of the core barrels or liners prior to sampling (Table 3.2). Upon coring the bag breaks, allowing microspheres to perfuse through fluids at the bottom of the borehole as coring advances. Microscopic analysis of core samples is used to assess further migration of the microspheres into the sample. Laboratory analysis using fluorescence microscopy and 10× objectives can detect microspheres at a concentration of approximately $10^4/g$, a 4 to 5 order of magnitude reduction from initial concentration (Table 3.2). In vadose zone investigations, particulates such as glass beads can be used as a tracer carrier (Fredrickson et al., 1993).

Community-level bioassays — Community-level physiological profiles (CLPPs) can be used to compare qualitative characteristics of the microbial community in the drilling fluids with communities in sampled materials (Lehman et al., 1995) (Table 3.2). The CLPP involves inoculation of sole carbon source-bearing wells in microtiter plates (BIOLOG, Inc., Hayward, CA) with sample particles, incubating the sample, and measuring the ability of the community to use a specific carbon source based on reduction of a redox-sensitive dye. This procedure provides a metabolic "fingerprint" of a community (Garland and Mills, 1991). Phospholipid and storage polymers of microbial cells can also be utilized in a QA/QC program. Phospholipid fatty acid (PLFA) profiles of microbial membrane lipids provide indicators of biomass (White et al., 1991; Haldeman et al., 1994), whereas analysis of individual PLFA components can be used to compare the community structure of the drilling fluids to those in the subsurface samples. Dramatic differences in PLFA profiles suggest that microbial communities from the drilling fluids differ from those of the sediments, providing additional evidence of sample integrity (Lehman et al., 1995; White et al., 1991; Haldeman et al., 1994).

Microbial tracers — The addition of microorganisms with distinguishing characteristics such as morphology or pigmentation (Wood and Ehrlich, 1978; Fournelle et al., 1957; Colwell et al., 1994) to drilling fluids and the sample/borehole interface is an excellent tracer technique, but regulatory constraints have limited this approach. In some instances, indigenous, nonpathogenic species are being approved for reintroduction to the subsurface. Harvey et al. (1984, 1989) reintroduced fluorescein-tagged groundwater microorganisms (Table 3.2) to study bacterial transport in the subsurface. Serendipitous microbial tracers may occur as a natural enrichment in drilling fluids. Beeman and Suflita (1989) observed that nonfecal coliforms commonly colonize drilling fluids. After 3 days the concentration of coliform bacteria in the drilling fluids varied between 10^3 to 10^6 colony forming units (CFU) per milliliter, whereas coliforms were rarely detected in subsurface samples (Beeman and Suflita, 1989). When hydrogen gas accumulated in an open borehole in Washington State, aerobic hydrogen-utilizing microorganisms were present in the borehole fluid at densities exceeding 10^5/ml. This density of organisms far exceeded typical *in situ* biomass density, and thus served as an indicator of contamination when subsurface sample biomass reflected the borehole fluid community (Bjornstad et al., 1994).

Even if microbiological tracers are not added to the drilling fluids, microorganisms that are naturally occurring in the fluids can be used in quality assurance (Table 3.2). Dilution-plate counts for aerobic heterotrophic bacteria performed on subsurface samples and drilling fluids can be used as a general indicator of potential contamination by comparing colony morphology (Pedersen and Ekendahl, 1990; Balkwill, 1989; Balkwill et al., 1989). Also, isolates from distinct sample environments can be assayed for biochemical, genetic, and/or physiological characteristics. As a general rule, isolates from one sample environment are rarely observed in others. Unusually high numbers of organisms in clays may suggest contamination from drilling fluids, since clays typically exhibit $<10^2$ CFU/g, while coarser sediments may exhibit a range of viable aerobic heterotrophic bacteria between 10^1 to 10^7 CFU/g. Enumerations of physiological groups, along with comparisons of physiological and metabolic traits of isolates and communities in core samples and drilling fluids, are effective methods for assessing potential contamination of subsurface samples.

Other types of analyses of microbial isolates that can be used for QA/QC include comparisons of total lipid profiles by MIDI analysis (White et al., 1991), comparison of nucleic acid sequences (Vescio and Nierzwicki-Bauer, 1994), and DNA homology (Jiminez, 1990). These analyses can also be used to distinguish microbial communities separated either vertically by variations in lithology, or horizontally (Lehman et al., 1995; Balkwill et al., 1989; Jiminez, 1990). If the microbial community structure and/or characteristics of the isolates from the core samples are distinct from each other and from the drilling mud, then it is unlikely that the subsurface samples have been severely contaminated.

3.4 Field Sample Processing

Sample processing in the field should be done according to clearly defined procedures that have been tailored to the lithologies being sampled, and in particular, to the microorganisms those lithologies are thought to host. Field sample processing can be rigorous and time-consuming, and requires planning to be properly implemented.

The following sections describe the major considerations for field sample processing, including descriptions of processing procedures for both unconsolidated and consolidated samples.

3.4.1 Sample Processing Chamber

When sampling aerobic subsurface environments, and only aerobic processes are to be examined, sample processing can be done in a portable laminar-flow hood (Harvey et al., 1984, 1989). However, if the redox status of the environment is unknown, or is known to be anaerobic, samples should be processed in a glove bag filled with an inert gas such as argon or nitrogen (Figure 3.1). Processing in an inert atmosphere will minimize oxidation of samples and facilitate survival of strict anaerobic microorganisms, though it should be noted that in the absence of an oxygen-scrubbing system, the atmosphere in the glove bag may contain trace quantities of oxygen. The addition of hydrogen gas and palladium catalysts to remove oxygen will aid in maintaining anaerobic conditions in the processing chamber, though the long-term exposure of samples to hydrogen may result in growth and/or enhancement of metabolic activities of microorganisms (Cragg et al., 1992).

A diagram of a sample processing chamber used in the U.S. Department of Energy's Subsurface Science Program is shown in Figure 3.1. In this chamber the core transfer tube is a 2-m-long pipe, 30 cm in diameter, with latched doors and vacuum/flushing ports at both ends. Geologic logging and sample paring is accomplished in a modified 2.5-m-long × 1.0-m-wide bag through pairs of gloved sleeves (Coy Laboratory Products, Ann Arbor, MI). An air lock connects the paring chamber with a sample disbursement chamber which is 0.6 m wide × 1.3 m long and has gloved sleeves and a 30-cm diameter air lock. In small-scale field operations where only unconsolidated materials are encountered, smaller configurations can be used. Materials placed in the entrance port of the processing chamber are subjected to more than three flush and evacuation cycles with an inert gas (typically argon) to displace most of the air. The interior air lock is opened and materials are passed into the inert atmosphere of the processing chamber (Figure 3.1).

3.4.2 Processing Unconsolidated Materials

Geological descriptions and photographic documentation of the core should occur before processing while the core is in the glove bag. All sample processing manipulations should be performed using sterile latex gloves worn over the chamber sleeve black rubber gloves; frequent changes of the sterile latex gloves helps prevent sample contamination. Paring of samples is accomplished with flame-sterilized or autoclaved paring tools. As a general guideline, one half of the outermost portions of the samples are pared away in addition to all of the portions which appear contaminated with drilling fluids or other extraneous materials (Figure 3.2). Paring requirements will vary, however, depending on such factors as sample permeability, core diameter, and fracture density. Drilling fluids will have had the greatest sample contact at the ends and outermost surfaces of the core. Samples of drilling fluid should be collected from the circulation system before, during, and after tracers have been introduced and stored in screw-cap tubes for both microbial and tracer analyses (QA/QC). Samples of surficial drilling fluid, outer core surfaces, parings, and interior portions should also be collected for QA/QC analyses. These samples should be clearly labeled as to their location and physical condition (i.e., highly fractured, coated in drilling fluid, etc.) to aid in their identification and in the interpretation of sample

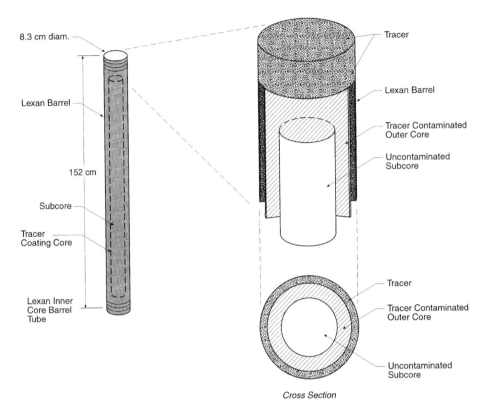

FIGURE 3.2

Paring of drilling fluid and tracer-contaminated outer material from subsurface core sample to obtain an uncontaminated and tracer-free subcore for microbiological and geochemical analyses.

integrity. On-site tracer analyses, if performed, should be initiated while samples are being processed, and subsamples for off-site tracer analyses should be stabilized, packaged, and refrigerated.

After paring and subcoring is completed, samples are placed into sterile pans and passed to the sample disbursement chamber (Figure 3.1). Here the samples are transferred into sterile Whirlpak bags. The bags and samples are weighed, labeled, described, logged in the sample logbook, and placed into canning jars. To minimize exposure to oxygen, the jars may be further flushed with an inert gas via a cannula and then sealed. Sealed jars are placed into the exit air lock and removed from the chamber. Between each sample processing event, the bag and its equipment contents should be thoroughly cleaned and disinfected.

3.4.3 Processing Consolidated Materials

When consolidated materials or rocks are sampled, hydraulic core splitters (Sepor Inc., Wilmington, CA) can be positioned in the paring chamber to "slice" the materials into subcores or "cookies" that can be carefully pared and fractured into small fragments (Figure 3.2). Hydraulic core splitters are recommended to limit scatter of the parings. The chisel blades of the hydraulic core splitters require frequent cleaning with dilute bleach (~0.5% sodium hypochlorite) to remove particulates and potential contaminants. Core slices can be further pared with additional hydraulic core splitters or manually with hammers and chisels. Upon paring to an innermost ~1 cm^3, the sample fragment can be packaged or further disaggregated with a Platner mill (Figure 3.1), in which rocks are crushed into fragments weighing 0.05 to 10 g. If smaller sizes are required by investigators, samples can be further reduced in a ball mill; however, it is recommended that further reduction be conducted in the controlled environment of the laboratory, since further reduction can result in drying and oxidation of the sample. As with unconsolidated materials, the pared sample fragments are transferred through an air lock to the disbursement chamber. Samples are weighed, logged, segregated, and placed into sterile Whirlpak bags. It is important for consolidated materials to be double bagged before placing them into jars, to cushion the impact of rocks hitting the glass sides during sample transport.

The procedures described above pertain to field projects requiring homogenization and dispersal of samples to multiple investigators. If homogenization of samples is not a priority, another option is to ship intact core to investigators, which places the burden of sample paring and tracer sample collection on the receiving investigator. Shipping intact core is only practical with consolidated cores that have low permeability, thus preventing disaggregation and inner core contamination from the outer core. However, intact cores may be preferable to prevent changes in the microbial community (see Chapter 4).

3.4.4 Sample Disbursement

Processing procedures are typically completed within 2 to 3 h of sample recovery, though consolidated samples may take somewhat longer. Samples acquired at the surface by early afternoon are frequently shipped the same day, arriving at their destination the following morning. Sealed jars containing the samples can be shipped overnight in protective Styrofoam® boxes with cold packs, or in the case of frozen samples, dry ice. It is recommended that investigators receiving samples be notified of impending sample shipments, and care should be taken to avoid shipments over the weekend which can delay sample delivery and compromise sample preservation, particularly during warm weather. Upon receipt of the samples, investigators should inspect the shipment and notify field personnel if samples were compromised in transit. Investigators should initiate their laboratory experiments as soon as possible after receiving the samples (see Chapter 4).

3.5 Summary

Considerable emphasis should be placed on the task of selecting the appropriate drilling and sampling technologies. After the scientific objectives have been defined, an appropriate geological regime for sampling should be identified. Additional factors such as budget, personnel requirements, and scheduling must also be considered when planning a field program. Geologists or drillers should be consulted to determine the most technically sound and cost effective drilling and sampling method for collecting samples from the target strata.

The very nature of coring, processing, and disbursing samples impacts the physical, chemical, and biological properties of subsurface materials. Once the appropriate drilling and sampling technologies have been selected and field operations have begun, a number of QA/QC procedures can be employed in the field to ensure that samples are the highest quality possible, and are representative of the environment from which they were collected. These procedures include the development of protocols and guidelines for implementation of the field program, adequate documentation of field activities, and optimization of the site layout. Rigorous decontamination of drilling, sampling, and sample processing equipment helps reduce the number of potential contaminant pathways. Critical components of a QA/QC program include the use of tracers during sampling to allow evaluation of the degree of sample contamination resulting from drilling and sample processing. Samples of drilling fluid, parings, outer core, and inner core, as well as sample controls, blanks, and replicates should be collected and processed to help identify potential sources of contamination during sample collection and processing. Sample processing can be performed on-site for both consolidated and unconsolidated materials in an inert atmosphere, thereby minimizing sample exposure to oxygen and contaminants.

Results from scientific drilling programs at numerous sites with varied lithologies have demonstrated that the guidelines described herein are broadly applicable for obtaining subsurface samples for microbiological and geochemical scientific investigations.

Acknowledgments

The authors wish to express their appreciation to Dr. F. J. Wobber and the Subsurface Science Program (SSP) of the U.S. Department of Energy's Office of Health and Environmental Research for supporting field research in which considerable expertise has developed in collecting subsurface microbiological samples. The authors also wish to acknowledge the many SSP participants who have contributed to the field sampling programs from which this expertise evolved.

References

Balkwill, D.L., 1989, Numbers, diversity, and morphological characteristics of aerobic, chemoheterotrophic bacteria in deep subsurface sediments from a site in South Carolina, *Geomicrobiol. J.*, 7, 33.

Balkwill, D.L., Fredrickson, J.K., and Thomas, J.M., 1989, Vertical and horizontal variations in the physiological diversity of aerobic chemoheterotrophic bacterial microflora in deep southeast coastal plain sediments, *Appl. Environ. Microbiol.*, 55, 1058.

Barrow, J.C., 1994, The resonant sonic drilling method: an innovative technology for environmental restoration programs, *Ground Water Monitoring Remediation*, reprint, spring.

Beeman, R.E. and Suflita, J.M., 1989, Evaluation of deep subsurface sampling procedures using serendipitous microbial contaminants as tracer organisms, *Geomicrobiol. J.*, 7, 223.

Behrens, H., 1988, Quantitative bestimmung von uranin, eosin and pyranin in gemischen mittels fluoreszenzmessung bei definierten pH-Werten, *Steir. Beitr. Z. Hydrogeol.*, 39, 117.

Bjornstad, B.N., McKinley, J.P., Stevens, T.O., Rawson, S.A., Fredrickson, J.K., and Long, P.E., 1994, Generation of hydrogen gas as a result of drilling within the saturated zone, *Ground Water Monitoring Remediation*, 14, 140.

Chapelle, F.H., 1992, *Ground-Water Microbiology and Geochemistry*, John Wiley & Sons, New York, chap.9, pp. 208-231.

Chapelle, F.H. and Lovley, D.R., 1990, Rates of microbial activity in deep coastal plain aquifers, *Appl. Environ. Microbiol.*, 56, 1865.

Clark, R.R., 1988, A new continuous sampling wireline system for acquisition of uncontaminated minimally disturbed soil samples, *Groundwater Monitoring Rev.*, 8, 66.

Colwell, F.S., Stormberg, G.J., Phelps, T.J., Birnbaum, S.A., McKinley, J., Rawson, S.A., Veverka, C., Goodwin, S., Long, P.E., Russell, B.F., Garland, T., Thompson, D., Skinner, P., and Grover, S., 1992, Innovative techniques for collection of saturated and unsaturated subsurface basalts and sediments for microbiological characterization, *J. Microbiol. Methods*, 15, 279.

Colwell, F.S., Pryfogle, P.A., Lee, B.D., and Bishop, C.W., 1994, Use of a cyanobacterium as a particulate tracer for terrestrial subsurface applications, *J. Microbiol. Methods*, 20, 93.

Cragg, B.A., Bale, S.J., and Parkes, R.J., 1992, A novel method for the transport and long-term storage of cultures and samples in an anaerobic atmosphere, *Lett. Appl. Microbiol.*, 15, 125.

Fournelle, H.J., Day, E.K., and Page, W.B., 1957, Experimental ground water pollution at Anchorage, Alaska, *Public Health Rep.*, 73, 203.

Fredrickson, J.K., Brockman, F.J., Bjornstad, B.N., Long, P.E., Li, S.W., McKinley, J.P., Wright, J.V., Conca, J.L., Kieft, T.L., and Balkwill, D.L., 1993, Microbiological characteristics of pristine and contaminated deep vadose sediments from an arid region, *Geomicrobiol. J.*, 11, 95.

Fredrickson, J.K. and Phelps, T.J., 1996, Subsurface sampling and drilling, in *Manual of Environmental Microbiology*, C.J. Hurst, ed., ASM Press, Washington, D.C., pp. 526–540.

Garland, J. and Mills, A., 1991, Classification and characterization of heterotrophic microbial communities on the basis of patterns of community-level sole-carbon-source utilization, *Appl. Environ. Microbiol.*, 57, 2351.

Griffin, W.T., 1991, Role of the site geologist in a subsurface microbiological sampling program, in *Proc. First Int. Symp. Microbiology Deep Subsurface*, Fliermans, C.B. and Hazen, T.C., Eds., Westinghouse Savannah River Company, Information Services Section Publications Group, Aiken, SC, 5-61.

Griffin, W.T., Phelps, T.J., Colwell, F.S., Lehman, R.M., Russell, B.F., McKinley, J.P., and Stevens, T.O., 1993, Subsurface microbial ecology of the Taylorsville Triassic Basin, King George County, Virginia, presented at Am. Assoc. Petroleum Geologists Annu. Convention, April 25 to 28, 110.

Griffin, W.T., Russell, B.F., and Colwell, F.S., 1994, Microbiological sampling procedures for the deep subsurface, Proc. VIIth Int. Symp. Observation of the Continental Crust Through Drilling — A Decade of Drilling Discoveries, Santa Fe, April 25 to 30, 297.

Haldeman, D.L., Amy, P.S., White, D.C., and Ringelberg, D.B., 1994, Changes in bacteria recoverable from subsurface volcanic rock samples during storage at 4°C, *Appl. Environ. Microbiol.*, 60, 2697.

Haldeman, D.L., Amy, P.S., Russell, C.E., and Jacobsen, R., 1995, Comparison of drilling and mining as methods for obtaining microbiological samples from the deep subsurface, *J. Microbiol. Methods*, 21, 305.

Harvey, R.W., Smith, R.L., and George, L., 1984, Effect of organic contamination upon microbial distributions and heterotrophic uptake in a Cape Cod, Mass. aquifer, *Appl. Environ. Microbiol.*, 48, 1197.

Harvey, R.W., George, L.H., Smith, R.L., and LeBlanc, D.R., 1989, Transport of microspheres and indigenous bacteria through a sandy aquifer. Results of natural- and forced-gradient tracer experiments, *Environ. Sci. Technol.*, 23, 51.

Jiminez, L., 1990, Molecular analysis of deep-subsurface bacteria, *Appl. Environ. Microbiol.*, 56, 2108.

Leach, L., 1991, An aseptic procedure for soil sampling in heaving sands using special hollow-stem auger coring, in *Proc. First Int. Symp. on Microbiology Deep Subsurface*, Fliermans, C.B. and Hazen, T.C., Eds., Westinghouse Savannah River Company, Information Service Section Publications Group, Aiken, SC, 2-3.

Lehman, R.M., Colwell, F.S., Ringelberg, D.B., and White, D.C., 1995, Combined microbial community-level analyses for quality assurance of terrestrial subsurface cores, *J. Microbiol. Methods*, 22, 263.

McKinley, J.P. and Colwell, F.S., 1996, Application of perfluorocarbon tracers to microbial sampling in subsurface environments using mud-rotary and air-rotary drilling techniques, *J. Microbiol. Methods*, 26, 1–9.

Pedersen, K. and Ekendahl, S., 1990, Distribution and activity of bacteria in deep granitic groundwaters of southeastern Sweden, *Microb. Ecol.*, 20, 37.

Phelps, T.J., Fliermans, C.B., Garland, T.R., Pfiffner, S.M., and White, D.C., 1989, Methods for recovery of deep terrestrial subsurface sediments for microbiological studies, *J. Microbiol. Methods*, 9, 267.

Phelps, T.J. and Russell, B., 1991, Drilling and coring deep subsurface sediments for microbiological investigations, in *Proc. First Int. Symp. Microbiology Deep Subsurface*, Fliermans, C.B. and Hazen, T.C., Eds., Westinghouse Savannah River Company, Information Service Section Publications Group, Aiken, SC, 2-35.

Proc. VIIth International Symposium on the Observation of the Continental Crust Through Drilling — A Decade of Drilling Discoveries, 1994, Sante Fe, April 25 to 30, 326 pp.

Russell, B.F., Phelps, T.J., Griffin, W.T., and Sargent, K.A., 1992, Procedures for sampling deep subsurface microbial communities in unconsolidated sediments, *Groundwater Monitoring Rev.*, 12(1), 96.

Russell, B.F., Griffin, W.T., and Phelps, T.J., 1993a, Microbiological sampling strategies — fundamentals for the collection of representative materials, presented at Int. Symp. Subsurface Microbiology, Bath, U.K., September 19–24, A-08.

Russell, B.F., Colwell, F.S., Stormberg, G., McKinley, J., and Innis, P., 1993b, Use of argon gas with wireline rotary coring systems for the collection of subsurface materials: a case study, presented at Int. Symp. Subsurface Microbiology, Bath, U.K., September 19-24, A-08.

Senum, G.I. and Dietz, R.N., 1991, Perfluorocarbon tracer tagging of drilling muds for the assessment of sample contamination, in *Proc. First Int. Symp. Microbiology Deep Subsurface*, Fliermans, C.B. and Hazen, T.C., Eds., Westinghouse Savannah River Company, Information Service Section Publications Group, Aiken, SC, 7-145.

Starr, R.C. and Ingelton, R.A., 1992, A new method for collecting core samples without a drilling rig, *Groundwater Monitoring Rev.*, 12(1), 91.

Vescio, P.A. and Nierzwicki-Bauer, S.A., 1994, Extraction and purification of PCR amplifiable DNA from lacustrine subsurface sediments, *J. Microbiol. Methods*, 21, 225.

White, D.C., Ringelberg, D.B., Guckert, J.B., and Phelps, T.J., 1991, Biochemical markers for *in situ* microbial community structure, in *Proc. First Int. Symp. Microbiology Deep Subsurface*, Fliermans, C.B. and Hazen, T.C., Eds., Westinghouse Savannah River Company, Information Service Section Publications Group, Aiken, SC, 4-45.

Wood, W.W. and Ehrlich, G.G., 1978, Use of baker's yeast to trace microbial movement in ground water, *Groundwater*, 16, 398.

Zapico, M.M., Vales, S., and Cherry, J.A., 1987, A wireline piston core barrel for sampling cohesionless sand and gravel below the water table, *Groundwater Monitoring Rev.*, 7, 74.

4

The Collection of Subsurface Samples by Mining

Charles E. Russell

CONTENTS

KEY WORDS: *mining, deep subsurface microbiology, indigenous, endolithic microorganisms, anabiosis.*

4.1   Historical Utilization of Mines for the Acquisition of Subterranean Microorganisms

Mines frequently have been used by twentieth century microbiologists as a source of samples in the study of the microbial ecology of the geologic formations that the mines penetrate. These investigations can be classified, based upon the goals of the original researchers, into three categories. The earliest studies utilized mines to acquire samples of microorganisms thought to be responsible for the formation of, or for detrimentally affecting, economic minerals and/or oil and gas deposits. Those

0-8493-8362-5/97/$0.00+$.50
© 1997 by CRC Press LLC

studies, related to mineral deposits, often focused on the role of microorganisms in acid mine drainage. The second type of investigation focused on the potential affects of allochthonous (foreign or introduced) and autochthonous (indigenous) microorganisms on the reliability of potential geologic repositories for the storage of high-level nuclear waste (see Chapter 15). The third category of research utilized mines to obtain samples of indigenous endolithic microorganisms to assess the possibility of anabiosis or dormancy, and more recently, to characterize the presence, abundance, and metabolic capabilities of deep subsurface microorganisms for the purpose of *in situ* bioremediation.

The following is a review of representative publications that document the utilization of mines to acquire subterranean microorganisms. Potential factors that may detrimentally affect sampling are considered, and recommendations for future sampling efforts are made.

4.1.1 Sampling of Microorganisms Associated with Minerals of Economic Importance

The majority of mines that have been sampled in microbiological investigations were excavated for the purpose of exploiting mineral deposits of economic importance. Casual observations of acidic drainage or measurements of high concentrations of dissolved metals in mine effluent prompted serious inquiries into the role of microorganisms in the oxidation of sulfide minerals (Atlas and Bartha, 1987; Kuznetsov et al., 1963). The following is a review of some of the locations where investigations utilized mines to acquire microorganisms associated with the oxidation of metal sulfides, sulfur, and fuel deposits. However, sampling methodologies were often not recorded in these early studies.

4.1.1.1 *Sampling of Microorganisms Involved in the Oxidation of Ore Bodies or Fuel Deposits*

Sulfide and oxide ores — The role of *Thiobacillus ferrooxidans* and other chemolithoautotrophs in the oxidation of sulfide minerals has been well documented (see Table 4.1). Ecological samples acquired by the investigators listed consisted of water draining from drifts within the mine, the mine portal, or from leach pads. Solid samples were obtained from leach pads or from freshly mined ores. Sampling techniques, although not well described, often consisted of aseptic sampling of the mine water, sediments, and ores. In other cases, aseptic sampling may not have been a concern due to the high acidity of the environment (a pH often approaching 1) or the desire of the investigator to sample allochthonous organisms.

Hydrocarbon and sulfur deposits — Epigenetic sulfur deposits have been postulated to form as a result of both abiogenic and biogenic processes (Ivanov, 1957). A number of investigations were conducted in the 1950s to determine the role of microorganisms in the formation and destruction of sulfur deposits. A review of relevant literature is presented in Table 4.2. Unfortunately, sampling protocols were not described, denying the opportunity to evaluate the relevance of the methods used to sample microorganisms. Oxidation of hydrocarbon deposits via microorganisms has also been the subject of several investigations. One of the earliest was conducted in the Borislavskoye mine, where oil was obtained by mining methods (Kuznetsov et al., 1963). A few unidentified microorganisms were isolated by L. D. Shturm and G. I. Rozanova from oil within the mine (reported in Kuznetsov et al., 1963). Sampling techniques were not cited.

TABLE 4.1

A Review of Investigations Utilizing Mines for the Purpose of Studying the Role
of Microorganisms in the Oxidation of Sulfide Ores

Investigators, Location of Mine, Type of Mine, and Organism Isolated		Description of Sampling and Results
Investigators: Location of mine: Type of mine: Organisms isolated:	Lyalikova (1961) Kola Peninsula, Russia Copper/nickel sulfide *Thiobacillus ferrooxidans*	Out of 23 mine water samples taken from the Nittis-kumuzhye deposit, 20 contained *Thiobacillus ferrooxidans*, however, numbers are low due to extreme cold and lack of pyrite in the deposit; 7 out of 14 mine water samples taken from the Kaula deposit contained *Thiobacillus ferrooxidans*. Numbers were low for similar reasons.
Investigators: Location of mine: Type of mine: Organisms isolated:	Ivanov et al. (1958) Armenia Polymetallic sulfide *Thiobacillus ferrooxidans*	Microbiological samples consisted of water taken from the mine portal and the drift walls. All samples contained *Thiobacillus ferrooxidans*.
Investigators: Location of mine: Type of mine: Organisms isolated	Bryner et al. (1954) Bingham County, Utah Copper sulfide Autotrophic iron and sulfide oxidizers	Aqueous microbiological samples taken from leaching streams discharging from exposed ore piles. All samples contained bacteria.
Investigators: Location of mine: Type of mine: Organisms isolated:	Johnson et al. (1979) Conway Valley, North Wales Iron sulfide *Thiobacillus ferrooxidans, Thiobacillus thiooxidans,* and heterotrophic bacteria found in "streamers"	Drainage water from the mine was sampled in sterile polypropylene screw-cap bottles. The bottles were rinsed in the effluent and filled by submerging with the mouths directed against the water current. Samples were filled to capacity to prevent oxidation. Bacterial "streamers" in mine pools and walls were also sampled.
Investigators: Location of mine: Type of mine: Organisms isolated:	Lyalikova (1960) Middle Urals, Russia Copper and iron sulfide *Thiobacillus ferrooxidans*	Water samples were taken from fracture seeps in and around the ore body. Samples of freshly mined ore were also taken. The number of cells of *Thiobacillus ferrooxidans* in 1 ml of water or 1 g of rock ranged from 0 to 10,000.
Investigators: Location of mine: Type of mine: Organisms isolated:	Yenikyev and Sindarovsky (1947) Red Guard Deposit, Ural Mountains, Russia Copper/iron sulfide *Thiobacillus ferrooxidans*	Microbiological samples consisted of water ponded in areas where highly sulfurous ore spontaneously combusted, where there were water and rusty encrustations on exposed ore, and from water from unburned ore. The number of *Thiobacillus ferrooxidans* in 1 ml of water or 1 g of ore ranged from <100 to 100,000.
Investigators: Location of mine: Type of mine: Organisms isolated:	Golacheva and Karavaiko (1977) East Kasakhstan Polymetallic sulfide Previously unidentified species of thermophilic *Thiobacillus*	Samples of this organism acquired in places of spontaneous ignition of ore deposits.
Investigators: Location of mine: Type of mine: Organisms isolated:	Groudev et al. (1978) Vlaikov Vrah, Bulgaria Low-grade copper sulfide and oxide ores A variety of ferrobacilli are present	The majority of microbiological samples came from water and ore samples in an area used for natural and bacterial leaching.

TABLE 4.1 (continued)

A Review of Investigations Utilizing Mines for the Purpose of Studying the Role
of Microorganisms in the Oxidation of Sulfide Ores

Investigators, Location of Mine, Type of Mine, and Organism Isolated		Description of Sampling and Results
Investigators: Location of mine: Type of mine: Organisms isolated:	Gupta et al. (1977) Variety of locations in India Iron sulfide Mixed flora	Microbial samples were obtained from water discharging from the mines.
Investigators: Location of mine: Type of mine: Organisms isolated:	Brierley (1978) Chino Mine, Hurley, NM Copper/iron sulfide Moderately thermophilic, iron-oxidizing bacteria	Samples used to isolate the organism were taken from an actively leached area that was subjected to temperatures of up to 63°C.
Investigators: Location of mine: Type of mine: Organisms isolated:	Bryner and Jameson (1958) Bingham Canyon, Utah Copper/iron sulfide Chemosynthetic, autotrophic bacteria	Microbiological samples were acquired from a leaching stream discharging from exposed ore bodies.
Investigators: Location of mine: Type of mine: Organisms isolated:	Lyalikova and Lebedeva (1984) Northern Caucuses, Russia Molybdenum tungsten deposit Molybdenum oxidizing bacteria	Bacteria isolated were from mine water and mined ore samples.
Investigators: Location of mine: Type of mine: Organisms isolated:	Rusin et al. (1993) Saguochi County, Colorado Silver-bearing manganese oxide ore 300 Isolates were found. The most efficient manganese reducer was *Bacillus polymyxa* strain D1	A total of 11 locations, which consisted of ponds, inlets to ponds, ore heaps, and outcrops, were aseptically sampled. Samples taken from the ore heap, and from sediments in the inlets to the ponds were obtained using a hollow copper coring device.
Investigators: Location of mine: Type of mine: Organisms isolated:	McCready et al. (1986) Unknown Uranium oxide *Thiobacillus ferrooxidans*	*Thiobacillus ferrooxidans* were isolated from six different mine water samples.

4.1.1.2 *Sampling of Microorganisms Associated with Acid Mine Drainage*

An extensive amount of research has been conducted to determine the role of micro-organisms in the production of acid mine drainage. Acid mine drainage is a serious problem that consists of a combination of autoxidation and microbially induced sulfide oxidation (H_2S or metal sulfide to H_2SO_4). The resultant highly acidic mine drainage kills aquatic life and renders waterways unsuitable for use (Atlas and Bartha, 1987). Acid mine drainage has been documented in sulfide mines and in coal mines that have relatively high concentrations of iron sulfide. More recent investigations have begun to focus on the role of nonacidophilic sulfur-oxidizing bacteria.

The majority of microbiological studies associated with acid mine drainage utilized isolated organisms from water samples that drained coal mines and waste heaps, metal sulfide mines, and ore refuse or leach piles. Sample protocols typically involved the use of sterile bottles and storage of the sample at low temperatures during transport to the laboratory (Colmer et al., 1949; Colmer and Hinkle, 1947; Ehrlich, 1963; Tuttle et al., 1968, 1969). In addition, solid samples were collected from the mine walls, floors, and waste heaps. Some samples were collected at depth via a calibrated stainless-steel hand corer (Belly and Brock, 1974).

TABLE 4.2

A Review of Investigations Utilizing Mines for the Purpose of Studying the Role
of Microorganisms in the Oxidation of Hydrocarbon and Sulfur Deposits

Investigators, Location of Mine, Type of Mine, and Organism Isolated		Description of Sampling
Investigators: Location of mine: Type of mine: Organisms isolated:	Ivanov et al. (1958) Gaurdak Deposit, Russia Sulfur-bearing limestone *Thiobacillus thiooxidans*	Groundwater and sulfur-bearing rocks were tested for acidity and the presence of *Thiobacillus thiooxidans*.
Investigators: Location of mine: Type of mine: Organisms isolated:	Ivanov (1957) Shor-su, Russia Epigenetic sulfur deposit *Vibrio desulfuricans, Thiobacillus thioparus, Thiobacillus thiooxidans*	Groundwater, sulfur, and oil from a nearby oil deposit were sampled in an investigation to determine the origin of epigenetic sulfur deposits.
Investigators: Location of mine: Type of mine: Organisms isolated:	Karavayko (1961) Russia Sulfur deposit *Thiobacillus thiooxidans*	Samples of ore were acquired to determine the role played by bacteria in the weathering of sulfur deposits. Samples used to isolate *Thiobacillus thiooxidans* consisted of freshly mined ore, sulfur-bearing limestone, and weathered ore.

The role of microorganisms in the production of alkaline mine drainage has also been investigated. Samples of alkaline mine drainage were acquired from a bituminous coal mine in the Ohio Valley and transported to the laboratory for enrichment (Tabita et al., 1970). Sampling techniques were not described; however, various types of *Thiobacillus* sp. were isolated.

4.1.2 Sampling of Microorganisms in Geologic Nuclear Waste Repositories

Several studies have focused on allochthonous and autochthonous bacteria that survive in mines. The objective of these studies was to characterize the types of microorganisms present in potential repositories for the storage of high-level nuclear waste and to ascertain if those microorganisms could affect the viability of the repository for the safe rentention of radioactive waste (see Chapters 15 and 16).

Several active and abandoned mines in Cornwall, Derbyshire, and Cumbria, England were examined for the presence, abundance, and metabolic activity of microorganisms that may have affected the structural integrity of steel and concrete within a potential repository or that may have enhanced radionuclide migration (Christofi et al., 1984). The majority of the mines were excavated in granite, with the remainder in limestone. Sampling methods consisted of collecting water samples in sterile glass bottles which were filled to the brim, sealed, and placed on ice for transport to the laboratory for microbiological and chemical analyses. Solid samples were scraped from tunnel walls into sterile polyethylene bags using sterile spatulas and were also placed on ice. Transportation to the laboratory occurred as soon as possible; however, if a delay was anticipated then the samples were placed into a refrigerator during the delay (Christofi et al., 1984).

An investigation at Rainier Mesa, located within the Nevada Test Site, Nevada, focused on the impact of diesel exhaust on the distribution and abundance of subsurface microbiota in tunnel walls and perturbed invert materials (Haldeman et al., 1995). Rainier Mesa has geological characterisitics similar to Yucca Mountain, a proposed underground nuclear repository site within the Nevada Test Site. Microbiological samples were acquired in triplicate by aseptically chipping approximately 100 g of rock into sterile plastic bags. Microbiological samples were taken from the mine wall at the surface and at depths of 1 cm and 3 cm into the wall. Contamination, due to sampling procedures, was assessed by painting adjacent sample points with 1 μm fluorescent microbeads and then replicating the sampling procedure at the painted areas. The presence of sample beads at any given depth indicated contamination (Haldeman et al., 1995). All samples bags were sealed, placed on ice, then transported to the laboratory for processing within 6 h of sampling. Three invert samples were aseptically obtained by pounding 15-cm-long presterilized core barrels (6 cm in diameter) into the tunnel floor. Cores were wrapped in sterile foil and transported to the laboratory as described. The cores, prior to analysis, were pared to remove any surfaces that were potentially contaminated during the sampling and transport process.

Water samples and frozen consolidated Tertiary clay samples were acquired from the Mol underground facility in Belgium (Christofi et al., 1985). Four 250-ml water samples were aseptically collected from the main shaft. One sample was immediately inoculated for the cultivation of aerobic heterotrophs and sulfate-reducing bacteria while the rest were transported on ice to the laboratory where identical analyses were conducted. It is interesting to note that only the sample inoculated on-site demonstrated the presence of microorganisms. The authors suspected that transportation of the microorganisms may have been deleterious and recommended future work on the entire range of microorganisms be carried out on-site (see Chapter 5 for additional information regarding storage phenomena). In addition, solid frozen clay samples were collected via a pneumatic drill. Unfrozen clay samples were also collected and transported for analysis.

In a similar study conducted within the Stripa mine, Sweden, water samples were collected from two boreholes and a stream (Pederson and Ekendahl, 1992). Stripa samples were acquired from boreholes associated with the mine using packers (inflatable rubber elements) that isolated the depth intervals of 799 to 807 m, 812 to 820 m, and 970 to 1240 m below the land surface. Sterile sampling equipment and sterile containers were used for the transport of the sample material back to the laboratory. Inoculations for total aerobic heterotrophs and sulfate-reducing bacteria were conducted on site (Christofi et al., 1985; Pederson and Ekendahl, 1992). These techniques were considered adequate as the isolation of both allochthonous and autochthonous organisms was desired (Christofi and Philp, 1991).

Allochthonous and autochthonous microorganisms were also sampled at the Underground Research Laboratory in the Archean Lac du Bonnet batholith in southeastern Manitoba, Canada (Brown et al., 1994), the South Crofty mine and Youd's level mine in the U.K. (Christofi et al., 1984; West et al., 1986), the Konrad iron mine and the Asse salt mine in France (West et al., 1986). Sample protocols were similar to those previously described.

Studies that focus on the ecology of mined repositories often disregard the origin of the microbiota (e.g., Sand and Bock, 1991; Zheregyateva et al., 1991). Both autochthonous and allochthonous organisms are considered important, and therefore sampling methods often focus on the surface or near-surface of the mine walls, backs, and inverts. For this reason, sampling protocols for these studies are often quite similar to those used to study microbially enhanced weathering of exposed rock outcrops, stone monuments, and buildings. A review of nondestructive and destructive techniques for sampling surface and near-surface colonies is given by Hirsch et al. (1995). Nondestructive techniques consist of photographic registration of microbial surface growth, swabbing or washing the microorganisms from the surface, printing of the microorganisms using a solid agar medium, demonstration of *in situ* exoenzymatic activity by microbial cleavage of fluorogenic substrate analogues, and the collection of airborne microorganisms. Destructive methods consist of detachment of surface material, rock coring, and methods of desorbing organisms from rock and inoculating solid and liquid media with rock fragments.

4.1.3 Sampling Indigenous Endolithic Microorganisms

The sampling of indigenous endolithic organisms is a complex matter. It requires that only those bacteria that were in the geologic formation prior to the creation of the mine be sampled. Many of the previously described studies were little concerned with the origin of the organisms, i.e., allochthonous or autochthonous (Christofi and Philp, 1991). The focus was on the presence and potential effects of bacteria on mineral resources or man-made structures. Many of the bacteria previously mentioned were allochthonous, therefore aseptic sampling techniques involved the use of sterile containers and aseptic techniques during the sampling and transport process to prevent contamination from sources not associated with the mine or activity occurring therein, but there was no attempt to sample exclusively native microbes.

An even more complicated situation arises when one studies anabiosis or dormancy (see Chapter 11). Bacteria sampled for this purpose must be representative of organisms that were emplaced when the geologic strata were originally laid down or that became isolated in the strata over geologic time periods. Attempts must be made to find sample locations that preclude transport of bacteria from other locations, e.g., transport of surface microbiota to depth by ground water to the sample point. Sampling procedures for these types of studies are the most rigorous and require ample documentation of protocols.

4.1.3.1 Sampling for Ecological Considerations

One of the first attempts to isolate indigenous endolithic organisms from a mine was in the Ukhta Oil Fields in Russia, where oil reserves were at relatively shallow depths and were extracted via mines. Samples of oil, water, and rock were taken under aseptic conditions immediately upon the oil-bearing stratum being broken open and under conditions that prevented contamination by drilling fluids or tools (Andreyevskiy, 1959; Kuznetsov et al., 1963).

A series of studies were conducted in zeolitized ash-fall tuffs to determine the presence, abundance, metabolic diversity, types, and spatial variability of indigenous endolithic microorganisms. These studies were conducted in a series of mines (called tunnels) constructed at depth into the side of Rainier Mesa, Nevada Test Site (Amy et al., 1992; Haldeman and Amy, 1993; Haldeman et al., 1993). Samples were taken from multiple tunnels at depths ranging from 50 to 450 m and lateral distances separated by 1 to 10 km from both preexisting and from freshly mined tunnels.

Preexisting tunnels were sampled by using sterile impact hammers to remove the outer few centimeters of exposed rock (Amy et al., 1993; Haldeman and Amy, 1993). Surface contamination was assessed by swabbing the area to be sampled with R2A or water-moistened sterile swabs before and after creation of the new rock face. R2A agar fall-out plates were opened during sampling for approximately 5 min to ascertain if the samples were being contaminated by air-borne dust. Several hundred grams of rock were acquired from each location by hand-chipping the samples with a sterile chisel and hammer into sterile plastic bags. During sampling, exposed rock surfaces were periodically swabbed to determine whether surface contamination had occurred. Fungi, which were abundant at the surface but absent in the sample rock, were used as a fortuitous surface contamination marker. All samples were sealed, placed on ice, then transported to the laboratory and workup initiated within 6 h of being sampled. Rapid initiation of analyses was determined to be important when investigating autochthonous communities (see Chapter 5). Aqueous samples from mine seeps were collected into sterile containers and placed in coolers containing ice for transport to the laboratory (Amy et al., 1992).

New tunnels were mined using an alpine miner, a mining machine consisting of a rotating drum covered with tungsten carbide "teeth". The drum was hydraulically pressed against the mine face, gouging out chunks of ash-fall tuff with minimal heat production and without the use of fluids. The previously described sample methods were used once a sample point was encountered. Dust problems were alleviated by ventilating air within the mined chamber to the outside of the mine (Haldeman and Amy, 1993; Haldeman et al., 1993).

An experiment conducted simultaneously with the one above compared the microbiology of mined samples to cored samples acquired from the same location (Haldeman et al., 1993). A 3-m horizontal core (4.7 cm diameter) was drilled utilizing mine water tagged with 26 mg/l of the conservative tracer lithium bromide. Bromide concentration within the cores of greater than 0.02 mg/l was used as an indication of contamination by drilling fluid. Five cores were extracted at 0.75-m intervals along the core barrel. The cores were extruded from the barrel onto a sterile surface, the core was aseptically pared, transferred to sterile containers, placed on ice, and then transported to the laboratory for analysis. The rock was then mined from around the core hole; mined samples, taken in the manner previously described, were collected adjacent to the locations of the cored samples. Tracer results indicated the cores to be grossly contaminated with bromide, whereas only one of the mined samples contained bromide above background levels, indicating the mined samples had a greater degree of sample integrity relative to the cored samples.

4.1.3.2 Sampling for the Study of Anabiosis or Dormancy

Some of the earliest studies of anabiosis utilized coal mines to sample for indigenous microorganisms which were thought to have been entrained in the rock since coalification (Lipman, 1931). Early experiments utilized commercially available coal, but subsequent investigations used samples collected from approximately 122 m deep in the Primrose vein of the Otto mine near Pottsville, Pennsylvania (Turner, 1932). Sampling techniques were not documented. However, a second study, conducted at the same location, stated that the sample site was accessible to surface waters (Farrell and Turner, 1931), and thus suggests the possibility of contamination with surface microbiota. Samples of anthracite were collected at a 1-ft depth below (or back into) the exposed rock face and in formations that were free from bone and pyrite. The coal samples were placed in large cans of formalin and transported to the laboratory. In addition, aqueous samples and soil samples were taken adjacent to the anthracite sampling point and at the mine entrance.

Recent research on anabiosis has included culturing of microorganisms entrapped in salt mines. One such study involved samples from the Carey salt mine in Hutchinson, Kansas (Reiser and Tasch, 1960). Rock salt samples were collected at a depth of approximately 200 m where active mining occurred; water samples from mine effluent were collected at an approximate depth of 35 m. Open petri dishes were used to collect air-borne microorganisms during sampling of the rock. Samples were transported to the laboratory where extensive sterilization and control procedures were implemented to insure contamination of the samples did not occur in the laboratory.

A second study involving microorganisms in salt was conducted on mined samples of Paleozoic Zechstein salt (Dombrowski, 1963). Samples were not collected at points near faults or the upper layers of the salt formation. Specimens showing signs of recrystallization were discarded. Only salt crystals obtained from undisturbed points in the middle of the larger automorphic to subautomorphic beds of salt were used. Samples were collected from these layers and transported to a laboratory in an unspecified manner. Extensive sterilization techniques and control procedures were used in the laboratory to prevent contamination. Subsequent testing of the *Pseudomonas* isolated in this study showed it to be identical to modern *Pseudomonas* (De Ley et al., 1966).

A third study of Archaeal halophiles was conducted at the Winsford salt mine in Cheshire, England and from the Boulby potash mine in Cleveland, England (Norton et al., 1993). Samples of brine, recrystallized salt sludges, wall efflorescences, moist surface crystals from tunnel walls, and halite samples from recently blasted working faces were collected. Surface brine pools, soils from two dry saline springs, and brine derived from solution sampling at a nearby mine were also sampled. Liquid samples were collected by pipette into sterile universal vials (Norton et al., 1993). Solid samples were acquired with sterile wooden tongue depressors and placed into sterile plastic bags. Recently exposed halite crystals (2 to 5 g) were placed into vials of absolute alcohol and transported to the laboratory within 5 h. The efficacy of the surface sterilization procedure was tested by trapping orange-pigmented halobacteria into halite crystals. The 4- to 5-mm crystals were placed into a thick suspension of pink-pigmented halobacteria for a period of 4h. The crystals were then placed into an ethanol solution for 2 h. Only orange-pigmented bacteria were subsequently isolated from the laboratory-grown crystals.

Halophiles were also isolated in another study designed to compare membrane ATPases of entrapped organisms to those of modern organisms (Stan-Lotter et al., 1993). Samples were collected from Triassic and Jurassic salt deposits mined at the Winsford salt mine in Cheshire, England and the Salzburg salt mine at Bad Ischl, Austria after blasting from newly exposed faces and kept in absolute ethanol for transport to the laboratory.

Isolation of bacteria from ancient salt deposits has also been conducted at the Waste Isolation Pilot Plant, New Mexico (Vreeland et al., 1995). This study focused on the sterile sampling of primary salt crystals in which the original chevron growth pattern had been preserved. These samples were analyzed using phenotypic and chemotaxonomic techniques, with the results being compared to samples taken from recrystallized salt, brines discharging into the mine, as well as surface samples. Results indicate a wide variety of halophilic bacteria were present, with some representing well-known groups of halophiles while others appeared to belong to a unique bacterial family composed largely of organisms isolated from crystalline samples. Since few recognized halophilic species have been isolated from materials preserved with ancient salts, there is a distinct possibility that the ancient isolates are represented in modern-day culture collections. Consequently, the evidence needed to support claims of anabiosis may rest more with geological/paleontological evidence than with biological (Vreeland et al., 1995).

4.2 Sampling Considerations Specific to Indigenous Endolithic Microorganisms Within Mines

Mines offer a cost-effective environment in which to acquire isolated or three-dimensional sample distributions. Subsurface indigenous endolithic microorganisms can be obtained for the purpose of studying subsurface ecology, anabiosis, or the effect of these organisms on anthropogenic structures. However, several factors within the mine environment may affect the quality of the samples, thus skewing analyses and resultant interpretations.

4.2.1 Factors Affecting the Sampling of Microorganisms in Mines

Several factors should be considered prior to utilizing a mine to acquire samples for the purpose of isolating indigenous endolithic microorganisms. These factors consist of the mining technology used to create the mine, hydrogeology of the mine, type of sample (aqueous vs. solid), location of the sample point with respect to fractures, method of sample acquisition, mine ventilation, mine drainage, and presence of heavy equipment operating within the mine. Variations in these factors will determine whether or not the microorganisms within a sample are representative of indigenous endolithic microorganisms, indigenous endolithic microorganisms that have been impacted by mining activities, or allochthonous microorganisms.

Mining is typically conducted by either blasting, excavation, or a combination of the two. Blasting with explosives has two potentially detrimental side effects. The propagation of the compressional wave created by the blast through the geologic medium will fracture it and may force compressed gas and blast-borne particulates into the rock. Blasting also introduces large amounts of nitrates and carbon oxides into a system that may be oligotrophic. These compounds may be utilized by bacteria as carbon and/or energy sources and may cause shifts in community composition. Excavation, on the other hand, is mechanical. Contaminants from this process are less intrusive and are less likely to impact indigenous microorganisms. However, most excavation processes require diesel-powered loaders, miners, and ore trains. Exhausts from these pieces of equipment introduce large quantities of oxidized hydrocarbons into the mine. The exhaust is unlikely to permeate deep into the rock matrix, yet microorganisms on the surfaces of the walls, ceilings (backs), and floors (inverts) of the mine may be affected (Haldeman et al., 1995). In addition, fuel spills and compaction of mine floors can be expected in mines excavated by mechanical means.

The hydrogeology of the mine must also be considered prior to sampling for microorganisms. Mined geologic formations that consist of a permeable matrix, are highly fractured, and/or are located at shallow depths, are much more likely to be continually exposed to allochthonous microorganisms transported by recharge water from the surface. Samples for indigenous microorganisms need to be derived from low permeability rocks that are isolated from surface recharge. A related issue is concerned with the type of sample being collected. Microorganisms isolated from ground water or recharge water samples have been shown to be distinctly different in both number and types than microorganisms isolated from the geologic medium (e.g., Amy et al., 1992; Pederson, 1993). The objective of the study must be considered prior to utilizing sediment, rock, or water samples.

The method of sample acquisition also affects the type of microorganisms that will be subsequently isolated. Grab samples from the walls, invert, or ceiling of the mine are likely to include allochthonous microorganisms introduced by the ventilation system, the process of mining, or routine traffic. These samples would be adequate for studying the microbial communities that survive in mines but may not be representative of indigenous or endolithic microorganisms. Samples acquired by cores would be subject to all of the problems associated with drilled and cored microbial samples (see Chapter 3). Samples acquired directly by blasting are likely to be fractured and exposed to blast-borne surface contamination, whereas samples acquired by mechanical mining are susceptible to contamination by the tool used to expose the rock. A combination of mechanical mining followed by manual removal of the outer layers of rock and the manual sampling of deeper layers will likely produce the best results (Haldeman et al., 1993; Russell et al., 1994).

Mine ventilation has two potential effects. The first consists of producing air currents within the mine capable of mobilizing air-borne contaminants throughout the mine. This is especially important during the actual sampling process. A second is the potential for desiccation of the walls of the mine by the circulated air, producing drier conditions at the air/rock interface with greater concentrations of dissolved minerals than would normally be present. These processes may have the combined effect of introducing allochthonous microorganisms to potential sample points as well as changing the ambient conditions of a sample point via desiccation. Sampling methods and the objective of the study must consider these factors. For example, Kieft et al. (1993) studied the effect of variable water potential on microbial abundance and activity. Samples for this study were acquired via a hand-held electric jack hammer from a few centimeters into a tunnel wall in Rainier Mesa, Nevada. The results detected decreasing desiccation with depth. Other investigators sampling in nearby similar tunnels utilized diesel-powered alpine miners to acquire samples from meters into the rock to minimize the effect of desiccation and potential contamination (Haldeman et al., 1993; Russell et al., 1994).

Drainage within the mine is likely to contain both allochthonous and autochthonous microorganisms. Sample locations for studying indigenous endolithic microorganisms should avoid areas where mine drainage or fracture flow has occurred in a manner that may potentially impact the sample point.

4.2.2 Recommendations for Future Sampling Efforts

Future sampling efforts for the purpose of studying indigenous endolithic microorganisms need to address the aforementioned factors to insure the quality of the samples. The following steps can be considered as general recommendations for these future efforts. However, these recommendations are applicable to only certain types of mines in semiconsolidated geologic formations.

Mines should be chosen that penetrate relatively unfractured and impermeable rock at depth, because samples derived from these locations will be less likely to contain recently transported microorganisms. Mines should be chosen that have been mechanically excavated in the region of the sample point. Excavations by blasting are more likely to introduce contaminants into nearby unmined formations. Sample locations should be carefully chosen to minimize potential impact from water emanating from fractures, mine drainage, exhaust from heavy equipment within the mine, and desiccation or contamination from mine ventilation.

Once a sample location has been chosen, ideally, a new adit (or section of tunnel) should be mechanically excavated into previously unexposed rock. Several steps need to be taken to insure that contamination does not occur. Vacuum ventilation needs to be present to remove dust created by the mining process. Fluorescently labeled microspheres may be applied to the outer surface of the sample point, functioning as a marker for verification that surficial dust has not contaminated the sample. Additional quality controls on dust contamination include the use of open petri dishes as fall-out plates and the use of biological markers for surface contamination, e.g., fungi. Sterile swabs should be periodically wiped across the rock before, during, and after mining to assess changes in the microbial community with depth. Analysis of the microbiota may provide additional evidence of sample integrity, especially if organisms commonly associated with humans and/or a surficial environment are found.

There is insufficient experience with the mining of microorganisms to determine how much rock should be removed before sampling can begin. One previous study, conducted in ash-fall tuff, removed 0.76 m of rock prior to initial sampling (Haldeman et al., 1993), while another study demonstrated that the removal of only 1 cm of rock was sufficient to remove surface microbiological contamination (Haldeman et al., 1995). Rock properties such as permeability and hydraulic conductivity may aid in the decision of how much rock removal will be necessary.

Once a sample point has been reached, fluorescently labeled sterile microspheres should be painted on the surface, then the outer 2 to 3 cm of rock should be aseptically removed by hand. When the sample point is reached, rock may be chiseled into sterile containers and the containers may be flushed with inert gases depending on analytical purpose, i.e., if anaerobes are being investigated prolonged exposure to oxygen may preclude subsequent culture or detection. The samples should then be placed on ice and immediately transported to the laboratory for workup. Ideally, sample processing should begin immediately on-site once the sample has been obtained (Christofi et al., 1991). The presence of microspheres within the samples will be indicative of contamination during the sampling process.

Not all mines will lend themselves to these processes. Mines constructed in dense, hard rock, such as highly consolidated sandstone, do not lend themselves to mechanical mining methods or to hand chipping. Excavation by blasting is the only recourse. Care in selection of the sample location to minimize contamination via mine drainage, traffic, or ventilation should be taken and the liberal use of fluorescently labeled microspheres or other tracers during blasting will help to assess the degree of contamination. In addition, locations within the mine affected by invasive mine structures such as rock bolts should be avoided.

Acknowledgments

The compilation and writing of this chapter would have been impossible without the logistical support of the Desert Research Institute and the guidance given by Dr. Roger Jacobson.

References

Amy, P. S., C. Durham, D. Hall, and D. L. Haldeman. 1993. Starvation survival of deep subsurface isolates. *Curr. Microbiol.*, 26:345.

Amy, P. S., D. L. Haldeman, D. Ringleberg, D. H. Hall, and C. Russell. 1992. Comparison of identification systems for classification of bacteria isolated from water and endolithic habitats within the deep subsurface. *Appl. Environ. Microbiol.*, 58:3367.

Andreyevskiy, I. L. 1959. The ways of utilizing oil microbiology in the oil extraction business. *Tr. VNIGRI*, 131.

Atlas, R. M. and R. Bartha. 1987. *Microbial Ecology: Fundamentals and Applications*. 2nd ed., Benjamin/Cummings, Menlo Park, CA. 442 p.

Belly, R. T. and T. D. Brock. 1974. Ecology of Iron-oxidizing bacteria in pyritic materials associated with coal. *J. Bacteriol.*, 117:726.

Brierley, J. A. 1978. Thermophilic iron-oxidizing bacteria found in copper leaching dumps. *Appl. Environ. Microbiol.*, 36:523.

Brown, D. A., D. C. Kamineni, J. A. Sawicki, and T. J. Beveridge. 1994. Minerals associated with biofilms occurring on exposed rock in a granitic underground research laboratory. *Appl. Environ. Microbiol.*, 60:3182.

Bryner, L. C., J. V. Beck, D. B. Davis, and D. G. Wilson. 1954. Microorganisms in leaching sulfide minerals. *Ind. Eng. Chem.*, 46:2587.

Bryner, L. C. and A. K. Jameson. 1958. Microorganisms in leaching sulfide minerals. *Appl. Microbiol.*, 6:281.

Christofi, N. and J. C. Philp. 1991. Microbiology of subterranean waste sites. *Experientia*, 47:5240

Christofi, N., J. M. West, and J. C. Philp. 1985. The Geomicrobiology of European Mines Relevant to Radioactive Waste Disposal. FLPU8 85-1. British Geological Survey Report. Keyworth, U.K.

Christofi, N., J. M. West, J. C. Philp, and J. E. Robbins. 1984. The Geomicrobiology of Used and Disused Mines in Britain. FLPU 89-5. British Geological Survey Report, Keyworth, U.K.

Colmer, A. R. and M. E. Hinkle. 1947. The role of microorganisms in acid mine drainage. *Science*, 106:253.

Colmer, A. R., K. L. Temple, and M. E. Hinkle. 1949. An iron-oxidizing bacterium from the acid drainage of some bituminous coal mines. *J. Bacteriol.*, 59:317.

De Ley, J., K. Kersters, and I. W. Park. 1966. Molecular-biological and taxonomic studies on *Pseudomonas halocrenaea*, a bacterium from Permian salt deposits. *Antonie van Leeuwenhoek*, 32:315.

Dombrowski, H. 1963. Bacteria from Paleozoic salt deposits. *Ann. NY Acad. Sci.*, 108:453.

Ehrlich, H. L. 1963. Microorganisms in acid drainage from a copper mine. *J. Bacteriol.*, 25:350.

Farrell, M. A. and H. G. Turner. 1931. Bacteria in anthracite coal. *J. Bacteriol.*, 23:155.

Golovacheva, R. S. and G. I. Karavaiko. 1977. A new facultative thermophilic *Thiobacillus* isolated from sulphide ore. *Abstr. Symb. Microbial Growth on C1-compounds*. U.S.S.R. Academy of Science. Moscow, 108.

Groudev, S. N., F. N. Genchev, and S. S. Gaidarjiev. 1978. Observation on the microflora in an industrial copper dump leaching operation. in *Abstr. Symb. Microbial Growth on C1-compounds*. U.S.S.R. Academy of Science. Moscow, 108.

Gupta, R. C., M. M. Nandi, and R. R. Sant. 1977. Microbiological leaching of copper sulfide concentrate. *Indian J. Technol.*, 15:245.

Haldeman, D. L. and P. S. Amy. 1993. Bacterial heterogeneity in deep subsurface tunnels in Rainier Mesa, Nevada Test Site. *Microb. Ecol.*, 25:183.

Haldeman, D. L., P. S. Amy, D. Ringleberg, and D. C. White. 1993. Characterization of the microbiology within a 21 m^3 section of rock from the deep subsurface. *Microb. Ecol.*, 26:145.

Haldeman, D. L., P. S. Amy, C. E. Russell, and R. Jacobson. 1995. Comparison of drilling and mining as methods for obtaining microbiological samples from the deep subsurface, *J. Microbiol. Methods*, 21:305-316.

Haldeman, D. L., T. Lagadinos, L. Hersman, A. Meike, and P. S. Amy. (in Review). The effects of diesel exhaust on the microbiota within a tuffaceous system.

Haldeman, D. L., L. Hersman, A. Meike, and P. S. Amy. 1995. The use of microspheres as tracers to investigate sampling integrity in subsurface tunnels impacted by diesel exhaust. Abstr. Am. Soc. Microbiology Meet., Washington, D.C., N-142.

Hirsch, P., F. E. W. Eckhardt, and F. J. Palmer, Jr. 1995. Methods for the study of rock-inhabiting microorganisms — A mini review. *J. Microbiol. Methods*, 23:143.

Ivanov, M.V. 1957. The participation of microorganisms in the formation of the sulfur deposits at Shor-su. *Mikrobiologiya*, 26:5.

Ivanov, M. V., N. N. Lyalikova, and S. I. Kuznetzov. 1958. The role of *Thiobacteria* in the weathering of rocks and sulfide ores. *Seriya Biologicheskaya*, 2.

Johnson, D. B., W. I. Kelso, and D. A. Jenkins. 1979. Bacterial streamer growth in a disused pyrite mine. *Environ. Pollut.*, 18:107.

Karavayko, G. I. 1961. On the microzonal distribution of oxidizing processes in the sulfur ore of the Rozdol'skoye deposit. *Mikrobiolgiya*, 30:3.

Kieft, T. L., P. S. Amy, F. J. Brockman, J. K. Frederickson, B. N. Bjornstad, and L. L. Rosacker. 1993. Microbial abundance and activities in relation to water potential in the vadose zones of arid and semiarid sites. *Microb. Ecol.*, 26:59.

Kuznetsov, A. I., M. V. Ivanov, and N. K. Lyalikova. 1963. *Introduction to Geological Microbiology.* Engl. ed., McGraw-Hill, New York, pp. 124-154 and 202-203.

Lipman, C. B. 1931. Living microorganisms in ancient rocks. *J. Bacteriol.*, 22:183.

Lyalikova, N. N. 1960. The part played by *Thiobacillus ferrooxidans* in oxidizing the sulfide ores in the copper pyrite deposits of the Middle Urals. *Mikrobiologiya*, 29.

Lyalikova, N. N. 1961. Role of bacteria in oxidation of sulfide ores in copper-nickel deposits of Kola peninsula. *Mikrobiologiya*, 30:135.

Lyalikova, N. N. and E. V. Lebedeva. 1984. Bacterial oxidation of molybdenum in ore deposits. *Geomicrobiol. J.*, 3:307.

McCready, R. G. L., D. Wadden, and A. Marchbank. 1986. Nutrient requirements for the in-place leaching of uranium by *Thiobacillus ferrooxidans*. *Hydrometallurgy*, 17:61.

Norton, C. F., T. J. McGenity, and W. D. Grant. 1993. Archaeal halophiles (halobacteria) from two British salt mines. *J. Gen. Microbiol.*, 139:1077.

Pederson, K. 1993. The deep subterranean biosphere. *Earth-Sci. Rev.*, 34:243.

Pederson, K. and S. Ekendahl. 1992. Incorporation of CO_2 and introduced organic compounds by bacterial populations in groundwater from the deep crystalline bedrock of the Stripa mine. *J. Gen. Microbiol.*, 138:369.

Reiser, R. and P. Tasch. 1960. Investigation of the viability of osmophile bacteria of great geological age. *Trans. Kansas Acad. Sci.*, 63:31.

Rusin, P. A., J. E. Sharp, K. L. Oden, R. G. Arnold, and N. A. Sinclair. 1993. Isolation and physiology of a manganese-reducing *Bacillus polymyxa* from an Oligocene silver-bearing ore and sediment with reference to Precambrian biogeochemistry. *Precambrian Res.*, 61:231.

Russell, C. R., R. Jacobson, D. L. Haldeman, and P. S. Amy. 1994. Heterogeneity of deep subsurface microorganisms and correlations to hydrogeological and geochemical parameters. *Geomicrobiol. J.*, 12:37-51.

Sand, W. and E. Bock. 1991. Biodeterioration of mineral materials by microorganisms. Biogenic sulfuric and nitric acid corrosion of concrete and natural stone. *Geomicrobiol. J.*, 9:129-138.

Stan-Lotter, H., M. Sulzner, E. Egelseer, C. F. Norton, and L. I. Hochstein. 1993. Comparison of membrane ATPases from extreme halophiles isolated from ancient salt deposits. *Origins Life Evol. Biosphere*, 23:53.

Tabita, R., M. Kaplan, and D. Lundgren. 1970. Microbial ecology of mine drainage. In C. T. Holland. Ed., *Third Symp. Coal Mine Drainage Problems*. Bituminous Coal Research, Inc. Monroeville, PA. 94 p.

Turner, H. G. 1932. Bacteria in Pennsylvania anthracite. *Science*, 76:122.

Tuttle, J. H., P. R. Dugan, C. B. Macmillian, and C. I. Randles. 1969. Microbial dissimilatory sulfur cycle in acid mine water. *J. Bacteriol.*, 97:594.

Tuttle, J. H., C. I. Randles, and D. R. Dugan, 1968. Activity of microorganisms in acid mine water: I. Influence of acid water on aerobic heterotrophs of a normal stream. *J. Bacteriol.*, 95:1495.

Vreeland, R. H., A. Piselli, L. Shilling, L. Fortis, C. Pogue, and S. Angelini. 1995. Distribution of halophilic bacteria in the 250 million year old Salado Formation: recovery of cultures and comparison to isolates from surface sources. 1995 Geological Soc. Am. Annu. Meet. New Orleans, LA. Abstr. BTH-18. p. A-305.

West, J. M., N. Christofi, J. C. Philp, and S. C. Arme. 1986. Investigations on the Populations of Introduced and Resident Microorganisms in Deep Repositories and Their Effects on Containment of Radioactive Wastes. British Geological Survey. Keyworth, U.K. p. 3-13.

Yenikyev, P. B. and I. S. Sindarovskiy. 1947. Development of the fire area of the Main line in the "Red Guard" mine, *Gorn. Zh.*, 6.

Zheregyateva, T. V., E. V. Lebedeva, and G. I. Karaviako. 1991. Microbiological corrosion of concrete structures of hydraulic facilities. *Geomicrobiol. J.*, 9:119-127.

5

The Storage-Related Phenomenon: Implications for Handling and Analysis of Subsurface Samples

Dana L. Haldeman

CONTENTS

KEY WORDS: *storage-related phenomenon (SRP), sample handling, viable but nonculturable, resuscitation, growth, community composition, diversity, moisture, microbial redistribution, perturbation, temperature, succession, dormancy, oxygen, homogenization, freezing, anaerobes.*

5.1 What Is the Storage-Related Phenomenon (SRP)?

In efforts to characterize microbiology, ecologists have often attempted to obtain samples for microbiological analysis with as little disruption as possible (Deming and Colwell, 1985; Hirsch, 1992; Kinkel et al., 1992; Kruse and Iverson, 1995; Yayanos, 1995). Additionally, samples have often been stored in the dark, at temperatures believed to inhibit bacterial activity, during transport to the laboratory or until analysis could be initiated (Atlas and Bartha, 1987; Clesceri et al., 1989; Wollum, 1982). However, even when perturbation is minimized and samples are held at temperatures believed to be restrictive to microbial growth (commonly 10°C), successional changes in microbial communities can occur, causing what will be described as the storage-related phenomenon (SRP). These changes may include increased culturability, increased activity, and decreased diversity of microbiota with time, but may also result in the recovery of unique bacterial types only after the storage process (Haldeman et al., 1993, 1995; Hirsch, 1992).

Claude Zobell (1943) first described the SRP (or what he termed the "bottle effect") in water samples that exhibited increased culturable counts over time. He hypothesized that the observed increase in microbial abundance was due to the concentration of nutrients on sampling container walls and subsequent microbial growth. This fits well with more recent investigations that suggest attachment to surfaces may enhance microbial survival in oligotrophic environments (Marshall, 1988; Poindexter, 1981). There are more current and applied interests in this phenomenon as well, especially with regards to the safety of stored bottled waters (Ferreira et al., 1993, 1994), foods that are stored (Saguay, 1992), and in standard methods for the analysis and treatment of waste waters (Clesceri et al., 1989). For example, it is recommended that samples for coliform analysis be transported to the laboratory within 6 h of sample collection, and that analysis be initiated within 2 h of storage to prevent erroneous results. The SRP has also been reported to occur in aquifer samples (Hirsch and Rades-Rohkol, 1988), in soils (Harris et al., 1993; Petersen and Klug, 1994), and in deep subsurface materials (Amy et al., 1992; Brockman et al., 1992; Fredrickson et al., 1995; Haldeman et al., 1994, 1995a, 1995b).

This chapter will present information regarding specific changes that have been observed in SRP of terrestrial subsurface samples, and hypotheses that have been posed to explain them. The importance of these observations with regard to sampling strategies and bioremediation efforts will also be addressed.

5.2   SRP in Terrestrial Subsurface Samples

5.2.1   Community-Level Changes

The most consistent community-level changes that have been observed from the analysis of stored subsurface samples are increased culturable counts, increased microbial activity, and decreased culturable diversity (Amy et al., 1992; Brockman et al., 1992; Fredrickson et al., 1995; Haldeman et al., 1994). Factors that may be important in impacting the magnitude of community response, and changes in the composition of the microbial community, have also been reported (Brockman et al., 1992; Haldeman et al., 1994, 1995b), and will be discussed in following sections.

Typical findings of a storage experiment investigating culturable organisms are presented in Figure 5.1. Samples were aseptically collected at approximately 400 m deep within the tunnels at the Nevada Test Site (see Chapter 4). Analysis was initiated in less than 6 h after collection and included spread-plating on R2A agar. After a minimum of 2 weeks incubation at 24°C, culturable organisms were enumerated and the proportion of occurrence of each colony type was noted to ascertain culturable diversity (Atlas and Bartha, 1987). Figure 5.1A depicts increasing culturable counts observed in subsurface materials of various types over the 45-d storage period. Figures 5.1B and 5.1C show changes in diversity and the numbers of colony types that were recovered. The SRP did not increase uniformly in all materials. For example, culturable counts increased by four orders of magnitude in the water sample, but did not change significantly in the zeolitized tuff. Likewise, less dramatic changes

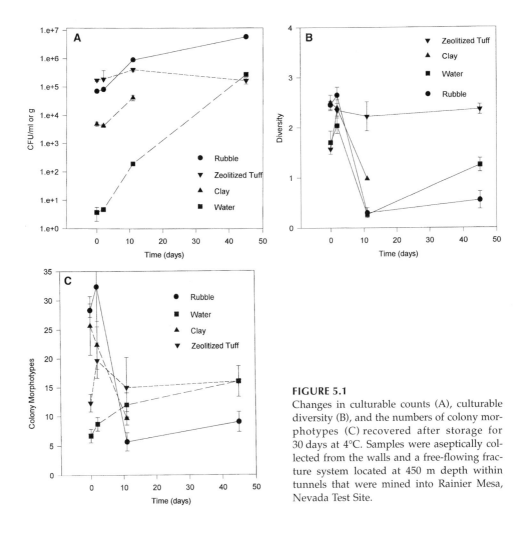

FIGURE 5.1

Changes in culturable counts (A), culturable diversity (B), and the numbers of colony morphotypes (C) recovered after storage for 30 days at 4°C. Samples were aseptically collected from the walls and a free-flowing fracture system located at 450 m depth within tunnels that were mined into Rainier Mesa, Nevada Test Site.

were observed in the diversity and the number of colony types recovered from the zeolitized tuff samples. Similar results have been reported by others, i.e., the observation of increased numbers of culturable organisms and decreased diversity, with variable responses dependent upon the sample type (Brockman et al., 1992; Fredrickson et al., 1995; Haldeman et al., 1994, 1995b).

5.2.2   Factors Influencing SRP

Moisture content may be an important factor controlling the extent of microbial change seen in SRP. In the experiment described above, the least dramatic community shifts were observed in the zeolitized rock (Figure 5.1). This rock had the lowest moisture content of the samples tested (4.7%). In other experiments, the addition of water (storage of zeolitized tuff in a slurry form; 1:10 w/v crushed tuff in 0.1% sodium pyrophosphate) resulted in a more dramatic increase in abundance as compared to unamended rock (data not shown). Brockman, et al. (1992) reported more dramatic increases in glucose mineralization from samples with higher moisture contents. Moisture may alter the distribution of nutrients and thereby increase their bioavailability. Kieft et al. (1993) have shown that lack of moisture limited activity in some

TABLE 5.1

The Proportion of the Recoverable Community That Was Common
among Three Replicate Spread Plates at Each Time Point

Sample Type	Moisture Content (%)	Time of Sampling (days)			
		0	2	11	45
Water	100	ND	68	74	87
Rubble Rock	9.4	54	51	95	93
Zeolitized Rock	4.7	34	38	60	58
Clay	35.1	39	39	59	ND

Note: Average values are presented. Three replicate plates were analyzed
for each of three subsamples taken from each sample type at each
time point.

vadose zone samples collected from the subsurface of arid and semiarid sites. This
result may be due to more negative water potentials limiting the cells' access to soluble nutrients.

The permeability of samples may also influence SRP. Table 5.1 presents the percent
of the recoverable community that was common among replicate plates throughout
the storage of different sample types. Particularly in the water and rubble rock, the
microbiota that were recovered after storage demonstrated a more even distribution
among replicate plates. In the clay and zeolitized samples, the percent of colony types
found among replicates did not increase as dramatically throughout storage. The clay
had a higher water content than the rubble rock, but due to the lack of well-connected
pores (a characteristic of clays) diffusion of water may have been limited in comparison. Zeolitized tuffs are also noted for their low permeability to water — a hydraulic
conductivity of approximately 5×10^{-12} m/s based on helium flow through cores
(Story, 1994). After storage, the zeolitized rock had a lower percent of colony types in
common among replicates as compared to water and rubble rock. These data substantiate hypotheses concerning redistribution of nutrients and water, because enhanced
SRP was seen in samples with higher permeability (see Figure 5.1). Further, the data
suggest that the redistribution of microbiota, as a result of sampling and sample handling, may be influenced by sample permeability. The mechanism of "microbial redistribution" is not known, but may be the result of physical movement of microbiota
with water due to grinding or mixing samples. Alternatively, redistribution may be a
function of successional changes that occur during storage, i.e., the growth and colonization of new microenvironments within the stored sample, or the awakening of
dormant bacterial types (see Section 3 below).

Sample perturbation may also be important with regard to SRP. In Figure 5.2A, the
increase in culturable count within the intact sample was more gradual and never
reached the peak that was observed in the replicate sample that was homogenized
prior to storage. Other investigators have observed similar trends (Fredrickson et al.,
1995). Homogenization of a sample may cause the physical redistribution of nutrients
or water, thus making substrates available for microbial growth. In soil environments,
increased decomposition of organic matter by microbiota has been attributed to
increased availability of substrates as a direct result of grinding, or tillage perturbation (Craswell and Waring, 1972; Rovira and Greacen, 1957). Work by Brockman,
et al., substantiates these findings in subsurface sediments (unpublished data;
Figures 5.2A and 5.2B). A lag in culturability and glucose mineralization was
observed in intact but not in homogenized samples (Figure 5.2, A and B). In addition,
oxygen concentration can affect SRP; the magnitude of changes were generally
greater in the samples incubated at lower oxygen concentrations (Brockman et al.,

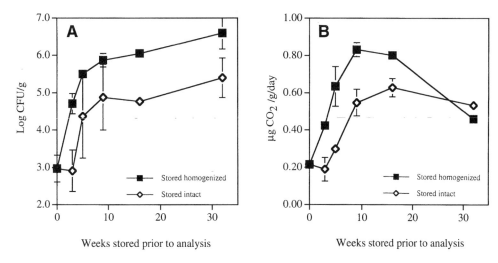

FIGURE 5.2
Changes in aerobic heterotrophic plates counts (A) and glucose mineralization (B) observed during storage of intact and homogenized samples. Samples were aseptically collected from a vadose zone in south-central Washington state at 15 m depth.

1992; data not shown). A review by Kusnetov et al. (1979) states that products of oxygen metabolism (H_2O_2) may be overly stressful to the recovery of oligotrophic microorganisms, and in part may explain the enhanced SRP observed in samples held at lower oxygen concentrations.

Temperature can strongly influence the rate at which the SRP occurs. In Figure 5.3, it can be seen that changes in culturable count began to occur relatively quickly — in less than 24 h in the case of samples stored at room temperature. Petersen and Klug (1994) have also reported more dramatic SRP occurring in soil samples that were stored at warmer temperatures. Storage of samples at –20°C may decrease changes in microbial communities, but the effect of freezing on microbial communities has not been well studied, and should not be used when analysis includes culturing or monitoring activity. Freezing may disrupt membrane integrity and impact recovery. For example, Nelson and Parkinson (1978) reported differences in survival for three bacterial isolates that were dependent upon moisture level, storage time, and thaw rate. Freezing may be the best method for sample preservation when extraction-based quantification and diversity methods are to be used, e.g., direct extraction of nucleic acids (see Chapter 10) or community-level fatty acid analysis (see Chapter 8) of the microbial communities present in samples.

5.2.3 Methods of Analysis

Some researchers have reported an insignificant change in the microbiota of samples that were stored (Madsen and Bollag, 1989; Balkwill et al., 1988). These results were probably dependent on the types of analyses that were conducted to monitor change; direct counts or other measures of total biomass may indicate nominal microbial change. Changes in total cell counts and total biomass have been negligible, even when SRP has been reported. Numbers of culturable cells often approach total cell numbers during storage (Brockman et al., 1992; Haldeman et al., 1995b). These results are confirmed when comparing the ratio of PLFA (indicative of living biomass) to DGFA (indicative of dead biomass) during storage (see Chapter 8). PLFA was

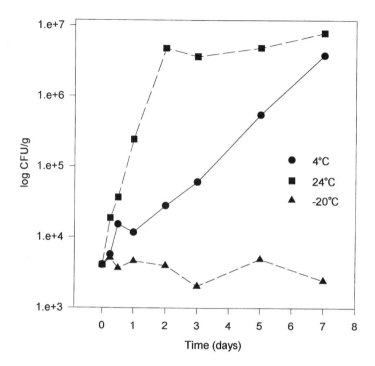

FIGURE 5.3

Effect of temperature on the recovery of aerobic heterotrophic bacteria during storage. Zeolitized tuff was aseptically collected from the walls of tunnels that have been mined into Rainier Mesa, Nevada Test Site (approximately 400 m depth).

recovered in a higher proportion after storage, and suggested that much of the biomass after storage was living (Haldeman et al., 1995b). This may provide an explanation of where nutrients for cell growth and resuscitation originate. Some substrates may be provided by the death and lysis of biomass of the once extant community.

Diversity as measured by the recovery of culturable microorganisms has its disadvantages. Of primary importance is the fact that often only a small percentage of total microbiota are recovered by plating techniques. Activity measurements (see Chapter 9), which do not rely on the culture of microbiota, may be used to assess populations of bacteria (i.e., all species capable of the utilization of a particular growth substrate), but cannot provide insight into shifts in abundance of individual microbial types. Phospholipid fatty acid (PLFA) and diglyceride fatty acid analyses (DGFA) provide other methods of analysis without reliance on culturing. Data suggest that indeed shifts did occur in microbial community composition, while total biomass remained relatively constant or decreased (Haldeman et al., 1995b; Peterson and Klug, 1994). For example, the storage of samples at 4°C caused a shift in the fatty acid patterns of the microbial community (Figure 5.4). After 30 days of storage, whole-community PLFA analysis indicated that different microbial communities had developed. Further, analysis of specific fatty acids indicated that there was an increase in the mole percent of terminally branched saturates (indicative of Gram-positive organisms) which mirrored the increases observed in the recovery of Gram-positive organisms after storage. Similar results have been obtained for samples stored at 15°C for 32 weeks (see unpublished data of Brockman et al. in Chapter 8).

Community-level PLFA analysis may also provide insight into changes that might be occurring in anaerobic microbial populations. Research to elucidate changes that occur in anaerobic microbial communities due to sampling perturbation of subsurface

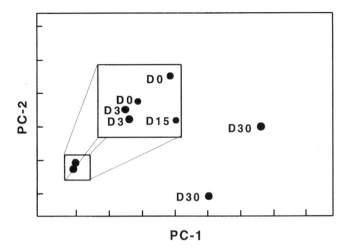

FIGURE 5.4
Principal components plot demonstrating a shift in the phospholipid fatty acid composition of microbial communities after storage. D0, D3, D15, and D30 indicate the length of time (days) that samples were stored. A blow-up is used to provide detail of the five sample points that were obscured in the plot. Zeolitized tuff was aseptically collected from the walls of tunnels that were been mined into Rainier Mesa, Nevada Test Site (approximately 400 m depth).

materials has not been initiated. However, work reported by Harris et al. (1993) states that shifts in anaerobic populations occurred within microbial communities extant in the anaerobic portion of soil stock piles.

5.2.4   Changes in Microbial Composition

When looking at individual microbial populations, dramatic shifts in community composition have been reported (Haldeman et al., 1994, 1995b). Some members of the microbial community can no longer be cultured after storage, others are culturable only after storage, while others increase or decrease in abundance throughout the process. Based on fatty acid composition, genera of culturable microbiota exhibiting the largest shifts in abundance during storage included *Arthrobacter*, *Acinetobacter*, and genera unmatched to the MIDI (Microbial ID, Inc., Newark, Delaware) database or distantly related to *Gordona* (Haldeman et al., 1994).

Perhaps one of the most interesting findings is the recovery of new microbial types only after the storage process (Hirsch, 1992; Haldeman et al., 1994, 1995b). When comparing microbiota based on MIDI analysis, genera of organisms that have been recovered only after storage included *Arthrobacter*, *Pseudomonas*, *Micrococcus*, *Acinetobacter*, and *Methylobacterium* (Haldeman et al., 1994). Questions have arisen as to the origins of these microbial populations. They were not the result of contamination, as analysis of sterile controls did not result in the recovery of bacteria.

5.3   Hypotheses Concerning Community Shift

Several hypotheses have been posed to explain the occurrence of new microbial types that are recovered only after the storage process: (1) "after" storage isolates result from the growth of organisms that were previously below the limit of detection, (2) "after" storage organisms result from the resuscitation of organisms that could not initially be cultured by standard plating techniques, or (3) both phenomena occur.

5.3.1 Proliferation Hypothesis

In support of the growth hypothesis, some microbial types that were recovered after community storage may have had a competitive advantage as compared to isolates recovered only before storage; i.e., those having high K_s and low μmax, or so-called "r"-strategists (Williams, 1984). The mean doubling times of "after" storage isolates were nearly one half of those measured for "before" storage isolates (those found only upon initial plating and absent at subsequent time points) (Haldeman et al., 1994). Additionally, the "after" storage isolates demonstrated the ability to use a wider variety of carbon substrates as sole carbon sources as compared to "before" storage isolates. Perturbations associated with sampling and sample handling may have changed the abundance and types of available substrates, and thus those organisms with the ability to use a wide variety of substrates for growth (eurytrophs) may have had a competitive advantage and an increased chance of success. As a result, they would become more dominant members of the microbial communities, as was the observed result after the storage process.

However, this still does not explain the occurrence of all of the new microbial types that are recovered from stored samples. Three "after" storage isolates that were grown under what one might consider more optimal growth conditions (i.e., in an oxygenated, low-nutrient medium as opposed to unamended rock) had extremely low doubling times at the storage temperature ($4°C$). It is unlikely that these isolates could have increased in abundance to detectable numbers during the storage time. For example, isolate R2D30-3, a Gram-positive, non-spore-forming microorganism, was not detected upon initial plating. The doubling time of this organism at $4°C$ was approximately 9 days. Assuming the isolate was present but below the detection limit (plate count = 5.18×10^4 CFU/g) at the initiation of the experiment, storage for 30 days would have allowed this microbe to increase in abundance by approximately 3.3-fold, or to a detection limit of 1.73×10^5 CFU/g. After storage, plate counts of this organism were 3.91×10^6 and the isolate was recovered in high abundance (7% of the microbial community which was cultured after storage). Thus, cell proliferation cannot adequately explain the abundance of this organism. Another possible explanation is the resuscitation and recovery of vegetative cells that were previously dormant or viable-but-nonculturable (VBNC) bacteria (see Chapter 11).

5.3.2 Resuscitation Hypothesis

In opposition to the r-strategy of rapid growth, another lifestyle strategy may include enduring extended periods of starvation. However, when nutrients again become available, starving microorganisms may not be competitive (Veldkamp et al., 1984). These starving organisms may be dormant or injured cells, and may require specific conditions to allow resuscitation. Researchers have shown that dormant, injured, or VBNC organisms may be resuscitated after exposure to an osmotic or temperature flux (Murno et al., 1989; Nilsson et al., 1991), low-nutrient media (Bissonette et al., 1975; Caldwell, et al., 1989), soil extract media (Lopez and Vela, 1981), liquid media (Kapreyelants and Kell, 1993), the addition of specific metabolites (Henis, 1987), or by

TABLE 5.2

Numbers of Culturable, Viable, Dead, Total, and Viable But
Nonculturable (VBNC) Cells Before and After Antibiotic Treatment[a]

	Cell Counts[b]				
	Culturable	Viable	Dead	Total	VBNC
Before					
R2D30-3a	1.2 e6	1.3 e9	4.0 e8	1.7 e9	1.3 e9
R2D30-3b	8.7 e6	1.1 e9	5.7 e8	1.7 e9	1.1 e9
R2D30-3c	9.7 e7	1.9 e9	4.2 e8	2.3 e9	1.8 e9
After					
R2D30-3a	1.4 e6	4.9 e7	1.9 e7	6.7 e7	4.8 e7
R2D30-3b	1.7 e5	8.2 e7	2.4 e7	1.1 e8	8.2 e7
R2D30-3c	7.5 e4	7.8 e7	3.1 e7	1.1 e8	7.8 e7

[a] Cell cultures that had been starved for over 2 months were incubated
overnight in R2 broth amended with 15 μg/ml penicillin G. After antibiotic treatment, cells were washed in R2 broth to remove residual antibiotic
before counting. BacLight was used to enumerate live, dead, and total
cells. Culturable cells were determined by plating on R2 agar and VBNC
were determined by subtraction of culturable from live cell numbers. On
average, only 8% of the viable cells were culturable before treatment, and
only 0.02% of viable cells were culturable after treatment.

[b] Strains were analyzed in triplicate. Values are the mean of triplicate
spread plates and duplicate slides.

unknown mechanisms in microcosms without added carbon (Byrd et al., 1991). Any
or all of these enriching phenomena may have contributed to the resuscitation of bacteria in subsurface samples and may have occurred as a result of the perturbations
associated with sampling and sample handling.

Other evidence supports the hypothesis of VBNC bacteria in subsurface environments. When subjected to carbon starvation for over 2 months, cultures of "after" storage isolates (described in the section above) were found to contain large numbers of
VBNC cells. Table 5.2 depicts results of experiments with isolate R2D30-3. The difference between the culturable counts and viable cell counts (determined by staining with
BacLight, Molecular Probes, Eugene, OR), represents VBNC cells. Further, the experiment demonstrated that the proportion of detectable VBNC could be increased with the
addition of penicillin, which was added to eliminate actively growing cell populations.

5.3.3   Proliferation and Resuscitation

Both the proliferation and resuscitation of bacteria probably occur in SRP. Additionally, other hypotheses can no doubt be put forth to explain SRP. However, it is clear
that examination of environmental samples by several techniques, i.e., culturing,
measuring activities, direct counting, and PLFA analysis, is necessary to best describe
changes that occur in samples over time. Analyses based on community-level DNA
extraction and genetic methods may provide additional information on the nature of
microbial community shifts and may be better able to address the origins of previously nonculturable microbiota.

5.4 Importance of SRP to Sampling Considerations

Sample handling concerns are of immediate importance in determining field sampling logistics (see Chapter 3). Standard procedures in our laboratory and by other Deep Subsurface Science Program investigators include shipping freshly collected samples to analysis laboratories via overnight express mail, or setting up portable laboratories on-site for immediate sample analysis. Samples are shipped in coolers containing ice to inhibit microbial activity. If microorganisms are not going to be cultured or activity analyses are not to be conducted, samples may be frozen or chemically fixed with a preservative before shipment. Specific experimental objectives may require sample homogenization, especially when multiple investigative teams are working together on common materials. However, leaving materials intact may decrease changes associated with sample handling, and may be required to investigate microbiological spatial heterogeneity (see Chapter 6).

Methods for handling environmental samples may include immediately enclosing them in anaerobic systems, or flushing sampling containers with inert gases to prevent or slow changes that might take place in anaerobic populations of microbiota (Parkes et al., 1994; and see Chapter 3). Work on deep ocean sediments demonstrated no overall significant differences in total anaerobic cell counts when comparing gassed samples before and after storage under anaerobic conditions at 4°C (Parkes et al., 1995). Exposure of samples to oxygen levels that are greater than *in situ* levels may cause changes in microbial communities. Work by Brockman et al. (unpublished data) has shown that incubation of samples at low oxygen concentrations may elicit a greater shift in aerobic populations and activity than incubation at atmospheric oxygen concentrations. Of additional concern are problems associated with the desiccation that may occur when preparing samples. Research has indicated that drying may decrease the numbers of culturable bacteria that one is able to recover from a sample (Kieft et al., 1993). Moisture and oxygen content, temperature of transport and storage, and additional undiscovered variables may influence SRP. The best solution to obtaining results representative of the environment is to analyze samples as soon as possible after collection.

5.5 Concluding Remarks

Understanding SRP is important for any microbial ecologist that is working with samples from the field, and it is also of interest in applied microbial ecology. For example, with regards to bioremediation, microbial communities may be stimulated simply by perturbation alone, perhaps without addition of air, water, or nutrients. Physical disturbance may stimulate microorganisms capable of degrading or cometabolizing a contaminant, serve to accelerate bioremediation, and allow more cost-effective implementation. In other instances, stimulation of microbiota due to perturbation may prove detrimental. For example, increased microbial activity may serve to accelerate biodeterioration or corrosion of barriers used in waste storage, and may thus impact long-term stability of waste drums, concrete structures, etc. Obviously, conditions within a waste repository may require designs to limit perturbation or other factors that increase SRP such as increased moisture contents.

Of interest to general microbial ecology is the understanding of successional changes that occur in microbial communities with time, after a perturbation. Botanists may be familiar with the idea of a seed bank which serves to reduce the risk of population extinction. Reservoirs of seeds in soils (sometimes hundreds to thousands per square meter) have been reported, some of recent origin, but some many years old (discussed by Levin, 1990, and McGraw et al., 1991). Depending on prevalent environmental conditions, only seeds capable of success may germinate, while others are left in the "soil bank" as a source for future years. SRP may be uncovering portions of the "microbial seed bank" that become active in response to conditions in microenvironments, i.e., specific conditions at the time of sample analysis. Some of what is measured as SRP may be due to microbial growth, while other successional changes may be due to the awakening, or "germination" so to speak, of dormant bacterial types.

Acknowledgments

Thanks to Abie Igbinovia, Melinda Duong, and Theodore Lagadinos for microbiological analyses conducted in the laboratory of Penny Amy, David Ringelberg, and D. C. White for fatty acid analyses, and Fred Brockman for critical review and data presented in Figure 5.2.

References

Amy, P. S., Haldeman, D. L., Ringelberg, D., Hall, D. H., and Russell, C. E. 1992. Comparison of identification systems for the classification of bacteria isolated from water and endolithic habitats within the deep subsurface, *Appl. Environ. Microbiol.*, 58: 3367.

Atlas, R. M. and Bartha, R. 1987. *Microbial Ecology: Fundamentals and Applications*, 2nd ed., Benjmin/Cummings, Menlo Park, CA.

Balkwill, D. L., Leach, F. R., Wilson, J. T., McNabb, J. F., and White, D. C. 1988. Equivalence of microbial biomass measures based on membrane lipid and cell wall components, adenosine triphosphate, and direct counts in subsurface aquifer sediments, *Microb. Ecol.*, 16: 73-84.

Bissonette, J. K., Jezeski, J. H., McFeters, G. A., and Stuart, D. C. 1975. Influence of environmental stress on enumeration of indicator bacteria from natural water. *Appl. Environ. Microbiol.*, 29: 186-194.

Brockman, F. J., Kieft, T. L., Fredrickson, J. K., Bjornstad, B. N., Li, S. W., Spangenburg, W., and Long, P. E. 1992. Microbiology of vadose zone paleosols in south-central Washington state. *Microb. Ecol.* 23: 279-301.

Byrd, J. J., Xu, H. S., and Colwell, R. R. 1991. Viable but non culturable bacteria in drinking water. *Appl. Environ. Microbiol.*, 57: 875-878.

Caldwell, B. A., Ye., C., Griffiths, R. P., Moyer, C. L., and Morita, R. Y. 1989. Plasmid expression and maintenance during long-term starvation-survival of bacteria in well water. *Appl. Environ. Microbiol.*, 55: 1860-1864.

Clesceri, L. S., Greenberg, A. E., and Trussell, R. R. 1989. *Standard Methods for the Examination of Water and Wastewater*. American Public Health Association, Washington, D.C., pp. 8-45, 9-35, and 9-43.

Craswell, E. T. and Waring. S. A., 1972. Effect of grinding on the decompositions of oil organic matter: Oxygen uptake and nitrogen mineralization in virgin and cultivated cracking clay soils. *Soil Biol. Biochem.*, 4: 435-442.

Deming, J. W. and Colwell, R. R., 1985. Observations of barophilic microbial activity in samples of sediment and intercepted particulates from the Demerara Abyssal Plain. *Appl. Environ. Microbiol.,* 50: 1002-1006.

Ferreira, A.-C., Morrais, P. V., and da Costa, M. S. 1993. Alterations in total bacteria, iodonitrophenyltetrazolium (INT)-positive bacteria, and heterotrophic plate counts of bottled mineral water. *Can. J. Microbiol.,* 40: 72-77.

Ferreira, A.-C. 1994. Alterations in total bacteria, iodonitrophenyltetrazolium (INT)-positive bacteria, and heterotrophic plate counts of bottled mineral water. *Can. J. Microbiol.,* 40: 72-77.

Fredrickson, J. K., Li, S. W., Brockman, F. J., Haldeman, D. L., Amy, P. S., and Balkwill, D. L. 1995. Time-dependent changes in viable numbers and activities of aerobic heterotrophic bacteria in subsurface samples. *J. Microbiol. Meth.,* 21, 253.

Haldeman, D. L., Amy, P. S., White, D. C., and Ringelberg, D. B., 1994. Changes in bacteria recoverable from subsurface volcanic rock samples during storage at 4°C. *Appl. Environ. Microbiol.,* 60, 2697.

Haldeman, D. L., Amy, P. S., Russell, C. E., and Jacobson, R. 1995a. Comparison of drilling and mining as methods for obtaining microbiological samples from the deep subsurface, *J. Microbiol. Meth.* 21: 305-316.

Haldeman, D. L., Amy, P. S., Ringelberg, D., White, D. C., Gharen, R. E., and Ghiorse, W. C. 1995b. Microbial growth and resuscitation alter community structure after perturbation. *FEMS Microbiol. Ecol.,* 17: 27-38.

Harris, J. A., Birch, P., and Short, K. C. 1993. The impact of storage of soils during opencast mining on the microbial community: A strategist theory interpretation. *Restoration Ecol.,* 88-100.

Henis, Y. 1987. Survival and dormancy of bacteria. In Y. Henis, Ed., *Survival and Dormancy of Microorganisms.* John Wiley & Sons, New York.

Hirsch, P. 1992. Microbiology. in Matthess, G., Frimmel, F., Hirsch, P., Schulz, H. D., and Usdowski, H.-E., Eds., *Progress in Hydrogeochemistry,* Springer-Verlag, New York. pp. 308-412.

Hirsch, P. and Rades-Rohkohl, E. 1988. Some special problems in the determination of viable counts of groundwater microorganisms. *Microb. Ecol.,* 16: 99.

Kapreyelants, A. S. and Kell, D. B. 1993. Dormancy in stationary-phase cultures of *Micrococcus luteus*: Flow cytometric analysis of starvation and resuscitation. *Appl. Environ. Microbiol.,* 59: 3187-3196.

Kieft, T. L., Amy, P. S., Brockman, F. J., Fredrickson, J. K., Bjornstaad, B. N., and Rosacker, L. L. 1993. Microbial abundance and activities in relation to water potential in the vadose zones of arid and semiarid sites. *Microb. Ecol.,* 26: 59-78.

Kinkel, L. L., Nordheim, E. V., and Andrews, J. H., 1992. Micorbial community analysis in incompletely or destructively samples systems. *Microb. Ecol.,* 24: 227-242.

Kruse, C. W. and Iversen, N. 1995. Effect of plant succession, ploughing, and fertilization on the microbiological oxidation of atmospheric methane in a heathland soil. *FEMS Microbiol. Ecol.,* 121-128.

Lopez, J. G. and Vela, G. R. 1981. True morphology of the Azotobacteriaceae-filterable bacteria. *Nature,* 289: 588-590.

Madsen, E. L. and Bollag, J.-M. 1989. Aerobic and anaerobic microbial activity in deep subsurface sediments from the Savannah River Plant. *Geomicrobiol. J.,* 93-101.

Marshall, K. C. 1988. Adhesion and growth of bacteria at surfaces in oligotrophic habitats. *Can. J. Microbiol.,* 34: 503-506.

Murno, P. M., Gauthier, M. J., Greittmayer, V. A., and Bongiovanni. J., 1989. Influence of osmoregulation process on starvation survival of *Escherichia coli* in sea water. *Appl. Environ. Microbiol.,* 55: 2017-2024.

Nelson, L. M. and Parkinson, D. 1978. Effect of freezing and thawing on survival of three bacterial isolates from an arctic soil. *Can. J. Microbiol.,* 24: 1468-1474.

Nilsson, L. L., Oliver, J. D., and Kjelleberg. S., 1991. Resuscitation of *Vibrio vulnificus* from the viable but non culturable state. *J. Bacteriol.,* 173: 5054-5059.

Parkes, R. J., Cragg, B. A., Bale, S. J., Getliff, J. M., Goodman, K., Rochelle, P. A., Fry, J. C., Weightman, A. J., and Harvey, S. M. Deep bacterial biosphere in Pacific ocean sediments. *Nature,* 371: 410-413.

Petersen, S. O. and Klug, M. J. 1994. Effects of sieving, storage, and incubation temperature on the phospholipid fatty acid profile of a soil microbial community. *Appl. Environ. Microbiol.,* 60: 2421-2430.

Poindexter, J. S. 1981. Oligotrophy: Fast and famine existence. *Adv. Microb. Ecol.,* 5: 63-89.

Rovira, A. D. and Greacen, E. L. 1957. The effect of aggregate disruption on the activity of microorganisms in the soil. *Aust. J. Agric. Res.,* 8: 659-673.

Saguay, I. 1992. Simulated growth of *Listeria monocytogenes* in refrigerated foods stored at variable temperatures. *Food Technol.,* 69-71.

Story, S. P. 1994. Microbial Transport in Volcanic Tuff: Rainer Mesa, Nevada Test Site. Thesis. University of Nevada, Las Vegas.

Veldkamp, H., van GemerdenH., Harder, W., and Laanbroek, H. J. 1984. Competition among bacteria: an overview. In Klug M. J. and Reddy, C. A., Eds., *Current Perspectives in Microbial Ecology.* American Society for Microbiology, Washington, D.C. pp. 279-290.

Williams, S. T. 1984. Oligotrophy in soil: Fact or fiction. In Fletcher, M. and Floodgate, G. D., Eds., *Bacteria in Their Natural Environments.* American Society for Microbiology, Washington, D.C. pp. 81-110.

Wollum, A. G. 1982. Cultural methods for soil microorganisms, *Methods for Soil Analysis, Part 2, Chemical and Microbiological Properties,* Page, A. L., Ed., American Society of Agronomy, Madison, WI, 781.

Yayanos, A. A. 1995. Microbiology to 10,500 m in the deep sea. *Annu. Rev. Microbiol.,* 49: 777-805.

Zobell, C. E. 1943. The effects of solid surfaces upon bacterial activity. *J. Bacteriol.,* 46: 39.

6

Microbiological Heterogeneity in the Terrestrial Subsurface and Approaches for Its Description

Fred J. Brockman and Christopher J. Murray

CONTENTS

KEY WORDS: *averaging scale, cross-correlogram, geostatistics, microbial distribution, microbiological spatial heterogeneity, microbiological spatial variability, multivariate analysis, variogram.*

6.1   Introduction

Microbial populations or activities can exhibit several types of spatial distributions, including randomness, clumping, and uniformity, as well as spatial patterns that possess various forms of regularity. Random patterns or uniformity can be related to homogeneity of the environment or the ability of a microorganism or a group of microorganisms to adapt to a wide range of conditions. Clumping can be caused by interactions between the microorganisms and their supporting environment, including interactions with other microorganisms. Tests to examine a data set for clumping or homogeneity can be found in Ludwig and Reynolds (1988) and other texts on mathematical ecology.

Some macroorganisms display complex spatial patterns in population density that are caused by a close association between the organism and its environment, e.g., terrestrial plant densities are often controlled by annual rainfall, soil type, and the slope and aspect of the area in which the plants grow. The underlying environmental variables that may control the distribution of a macroorganism often exhibit spatial continuity over an area since many of those variables result from the operation of continuous physical, chemical, and geological processes. Based on fundamental ecological principles, there is strong reason to believe that improved sampling designs might reveal the spatial relationships between environmental properties and microbiological properties in the subsurface, similar to those observed in geostatistical studies of macroecology (Rossi et al., 1992).

The variability in the spatial distribution of a microbiological property or process* can be important for several reasons. One of the most important is that if there is some continuity to the distribution of a microbiological property in space, then, by definition, the value of the property present at one location will not be independent of the value at nearby locations. The spatial dependence of the data will make it difficult to apply classical statistical methods to the study of the property, since most classical statistical methods (e.g., analysis of variance) assume independence of the data. In addition, any attempt to estimate the value of the property at unsampled locations should incorporate the spatial dependence between the data. If the spatial structure of a variable is known, it is possible to optimize the design of a sampling scheme to reduce the overall uncertainty in mapping the property over an area (Englund and Heravi, 1993). This can lower the overall cost of sampling, while increasing the value of the data obtained.

Quantification of spatial patterns in subsurface environments is important because it is well known that geologic, hydrologic, and geochemical properties are not constant in space; rather, they are autocorrelated or related over certain length scales. Temporal variability can also be important in subsurface systems that receive seasonal recharge. In order to better understand heterogeneous subsurface sytems, it is critical to (1) sample such that the spatial and temporal variability and patterns are adequately captured, and (2) understand what is causing the variability. Improved understanding in these two areas will increase the ability to predict, and ultimately model, the distribution of microbiological properties and the responses of microbial communities to environmental perturbations such as subsurface contaminant transport. Improved understanding of the spatial and temporal distribution of microbiological properties is also critical due to the relative difficulty and high cost of obtaining large numbers of subsurface samples.

---

* Hereafter, the word property (or properties) will stand for a measurable quantity regardless of whether it is considered a distinctive attribute of something (property) or a result of a continuous action or series of changes (process). This usage is necessary because the distinction between a property and a process changes depending on the context and the discipline.

6.2 Heterogeneity Exists at Numerous Scales

Spatial heterogeneity is inherently hierarchical with multiple scales of heterogeneity, ranging from micrometers to kilometers and greater. Parkin (1993) defines four scales of variability for describing microbial processes in soil. Microscale variability exists at the microscopic level (micrometers) to the size of soil aggregates (millimeters), and addresses the immediate environment(s) in which the cell finds itself. Variability at the plot scale is focused on land management methods and is generally measured in meters up to the size of a few acres. Landscape scale variability addresses factors such as soil type and landscape position. Variability at the regional scale includes factors such as climatic regime and vegetation type.

Similar scales of microbiological heterogeneity exist in the subsurface environment but are defined by the scales of the geologic structures. Different physiological types of microorganisms may exist in pores or pore networks that have different mineral phases (e.g., iron, sulfur, carbonate), different types of sedimentary organic matter (kerogen, bitumen, alkanes), or different redox status due to the supply rate of electron acceptor(s) and donor(s). Pore-scale heterogeneity is difficult to examine in many subsurface environments due to the relative paucity of microorganisms. For example, several thorough studies to examine microorganisms by scanning electron microscopy (SEM) in pristine sediments with up to 10^6 cells per gram have been unsuccessful or so few organisms were found that a systematic study could not be conducted (S. Birnbaum, personal communication; C. Gullett, personal communication). A notable exception is a study in eastern coastal plain sediments in the U.S. (Gullett, 1993).

Heterogeneity at the scale of individual laminae, thin lenses or beds, or inclusions (millimeters to tens of centimeters thick) appears to have been largely ignored in the literature, yet probably accounts for major differences in microbiological properties and processes where the laminae or thin lenses exhibit wide variability in permeability or geochemical properties. Studies partially addressing this scale of heterogeneity have focused primarily on variability between strata. Indeed, the majority of subsurface studies have been at the scale of geological strata. A stratum is a geologic unit of like materials that represents continuous deposition, is homogeneous relative to the composition of adjacent strata, and typically ranges in thickness from tens of centimeters to tens of meters. Geologic strata can be continuous for distances of meters to kilometers. Geologic strata are usually the focus of subsurface characterization because problems such as contaminant transport, poor recovery from oil reservoirs, and biogeochemical impacts on water quality manifest themselves at this scale.

Heterogeneity also exists at scales greater than the strata, and can be considered as ecosystem-level variability in microbial properties. Subsurface ecosystems can be defined by their relative distance from sources of transported nutrients, the relative flux of nutrients or recharge (for vadose zones), temperature, and the climatic and geologic conditions that were responsible for deposition of the sediments and which control the predominant types of electron donors/acceptors and their bioavailability. For example, eastern coastal plain sediments of the U.S. and fractured basalt systems in the arid western U.S. represent different subsurface ecosystems.

The scale that is important for a particular problem or scientific question is defined by the scale of heterogeneity that dominates the behavior of the system under investigation. Questions about global biogeochemical fluxes or the amount of subsurface biomass require data on ecosystem scale heterogeneity. If and when a contaminant plume will impact a drinking water well requires information on microbiological properties at the stratum and laminal scale. For example, continuous thin sand lenses or beds could dominate the flow of a subsurface system and result in breakthrough of contaminant much more quickly than predicted (Ronen et al., 1993). Heterogeneity may also dominate the microbiology of a subsurface system if inclusions are highly reactive or the predominant source of electron donor, electron acceptor, or nutrients (Murphy et al., 1992). Investigations of how microorganisms colonize the subsurface, intraspecies mutualistic interactions in the subsurface, or other mechanistic studies may require spatial information at the pore and pore network scale.

6.3   Microbiological Heterogeneity in Soils

Environmental microbiology has historically been focused on isolating microorganisms and defining and understanding the various microbial processes in pure culture and in the environment. Only recently have studies focused on the distribution of microorganisms in relation to microbiological properties. These studies have been conducted in surface soils and on plants because subsurface environments were of limited interest prior to the 1980s and uncontaminated subsurface samples are relatively difficult and expensive to obtain.

Environmental properties and microbiological properties in soils show high variability (often several hundred percent of the mean) and possess distributions that are typically positively skewed and approximate the log-normal. A variable is log-normally distributed if the logarithm of the variable produces a normal distribution of values. Microbial variables showing these behaviors include denitrification (Parkin, 1987; Parkin and Robinson, 1989), methane production in an aquifer 1 to 2 m below the surface (Adrian et al., 1994), and microbial populations on plant leaves and roots (Hirano et al., 1982; Loper et al., 1984). Log-normally distributed physical and chemical properties in soils include water flux (Jones and Wagenet, 1984), pore water velocity and apparent diffusion coefficient (Biggar and Nielsen, 1976), hydraulic conductivity (Nielson et al., 1973; Warrick et al., 1977), soil-water diffusivity (Nielson et al., 1973), and nitrate (Tabor et al., 1985; Parkin et al., 1987; White et al., 1987). Log-normal distributions are thought to arise because a particular property can result from numerous effects or properties that are multiplicative (Aitchinson and Brown, 1957), and because overlapping spatially nonhomogeneous distributions of different properties yield patches of optimal, suboptimal, and excluding conditions that control the distribution of species (Ugland and Gray, 1982) and specific microbial activities (Parkin, 1987; Parkin and Robinson, 1989). These findings suggest that quantitative relationships should exist between the spatial distributions of microbiological properties and their controlling environmental properties.

Studies on denitrification in aerobic soils have examined the cause of positively skewed distributions of microbial activity. In one study, denitrification was found to be localized to soil microsites or "hotspots" possessing high organic matter; 25 to 85% of the activity in a 100 g soil core was associated with 0.4 to 0.08% of the soil mass (Parkin, 1987). A computer simulation was used to test the hypothesis that a highly skewed frequency distribution could result from a patchy distribution of an underlying variable. A computer algorithm generated points that yielded a patchy distribution of points, and denitrification activity was randomly assigned to each point from a field-derived frequency distribution of maximum denitrification potential. The synthetic field was randomly sampled and a denitrification rate calculated for each sample by volume-averaging the "hotspots" and other regions contained in each sample. Sample histograms of the denitrification rate generated from the computer simulation were highly skewed and very similar to histograms of actual field data (Parkin, 1987). The patchiness of denitrification was also confirmed in a study of the influence of sample size on the measurement of denitrification (Parkin et al., 1987). In this study, the denitrification rate in 1.73-cm diameter cores was always statistically less than in 5.40-cm diameter cores, and was shown to be due to the fact that small cores tended to miss the most active "hotspots" of denitrification. Most of the variability in denitrification occurred at distances less than 10 cm, leading the authors to conclude that 10 to 15 kg of soil was required to obtain a representative denitrification rate from the soil cores being studied. Because the ability to denitrify is fairly ubiquitous in soil microorganisms and is controlled by carbon and nitrate availability and factors affecting oxygen status in microsites (moisture content, soil texture, bulk density), these results suggest that many microbial properties may display similar spatial heterogeneity.

Temporal variability is also an important component of variability in soils. For example, over 80% of the annual nitrogen loss from forest soils by denitrification occurred during 3- to 6-week periods in the spring and fall (Groffman and Tiedje, 1989). These periods were related to the activity of forest trees and the times when nutrient influx would likely be high. The spring period was after soils thawed and before tree "leaf break", whereas the fall period was coincident with leaf senescence and litterfall. In a separate study, greater than 90% of the annual methane release from salt marsh soils was found to occur in the late summer and early fall, possibly due to temperature effects on methanogens and/or other organisms affecting activity of methanogens (King and Wiebe, 1978).

While the subsurface environment differs from soil environments in several important ways, knowledge of the spatial and temporal variability of microbiological properties in soils (as discussed in this section and in Sections 5.2.1 and 5.3.1) provides an important reference point for conceptualizing heterogeneity and designing sampling strategies in subsurface environments. Such a reference point is especially useful because of the higher cost and greater difficulty of subsurface sampling.

6.4   Microbiological Heterogeneity in Subsurface Sediments

The subsurface environment can exhibit very pronounced spatial heterogeneity of physical properties as a result of the stratified nature of the deposits and the actions of geochemical processes over geologic time periods. Subsurface environments greater than 10 to 20 m deep experience mixing only from hydrologic phenomena, which are strongly influenced by the geologic properties of the sediments. As a result, subsurface environments can possess much stronger contrasts in physical and chemical properties over short distances (centimeters to meters) than in soils, which in turn control moisture flux, nutrient flux, and redox conditions. Thus, there is potential for a high degree of microbiological heterogeneity in the subsurface at these scales.

Soils and subsurface environments can differ dramatically in their rates of nutrient flux. Soils receive a great deal of nutrient input from plants and animals. Nutrient cycling in soils is very dynamic as a result of the mixing processes that operate there (e.g., mixing by plant roots, soil insects and worms, burrowing by megafauna, freezing and thawing, mixing by hydrological phenomena), and due to daily and seasonal changes in temperature and precipitation. In contrast, subsurface systems (at depth) lack direct plant and animal inputs and are generally nutrient poor. Many subsurface systems receive very low levels of nutrient flux from surface recharge, and some microbial ecosytems may persist to a large degree on nutrients that were present within the sediments when they were deposited. Nutrient limitation results in microbial populations and activities that may be many orders of magnitude lower in subsurface sediments than in soils. Differences in nutrient flux between surface soils and subsurface systems are likely to result in significant differences in microbiological spatial continuity and in the scale of microbiological heterogeneity.

The following three sections will review what is currently known about microbiological spatial heterogeneity in subsurface sediments. The application of geostatistical methods to subsurface microbiology will be discussed in Section 6.5.

6.4.1   Centimeter-Scale Heterogeneity

Several lines of evidence suggest that microbial distribution in some subsurface sediments may be highly discontinuous, even for very short distances. For example, aerobic heterotrophic plate counts (AHPC) performed in an identical manner in different laboratories using 10-g subsamples from homogenized core intervals (0.15 to 0.5 m long, and weighing hundreds to thousands of grams) yielded differences of two or more orders of magnitude (in the 5 to 10% of the core samples that were investigated in several studies (Brockman et al., unpublished data). This phenomenon was evident in samples up to 265 m deep from the southeastern coastal plain (South Carolina, U.S.) and from deep vadose sediments up to 100 m deep in the arid western U.S. AHPC varied by two to four orders of magnitude in sets of samples tens of cm apart (sampling zone, 360 to 470 m deep) in a separate borehole in South Carolina and was attributed to variations in sediment texture, hydraulic conductivity, and the presence of lignite (Fredrickson et al., 1991).

In a detailed study of a 21-m^3 volume of saturated rock from 400 m deep (Nevada, U.S.), AHPC varied by three orders of magnitude and direct counts varied by two orders of magnitude (Haldeman et al., 1993). The volume of rock used in that study was chosen because of its apparent geological homogeneity, and the microbiological variability could not be correlated to measured physical or chemical properties of the rock (Russell et al., 1994). The authors suggested that the lack of correlation at their site may be related to the inability to measure physical and chemical conditions in the microniche where the microbes exist, or that microbes may have been dormant for very long periods and thus reflect past chemical or physical conditions to a greater degree than current conditions. Thus, in some subsurface environments correlations between microbiological properties and environmental properties may be difficult or impossible to measure, or the correlations may not reflect present-day interactions between microorganisms and their environment (see also Section 6.5.3.2).

In a shallow vadose zone in Canada (3 to 5 m deep), AHPC varied several orders of magnitude in samples separated by tens of centimeters to as little as 2 mm (Severson et al., 1991). The authors suggested the variability in AHPC may have been caused by variability in the portion of the sample volume affected by fractures containing much higher microbial densities. Using 1 cm^3 samples, the coefficient of variability for cultured aerobic heterotrophs was up to 40 times greater in an unsaturated sediment (White Bluffs, WA, U.S.) than in a nearby surface soil (Stevens and Holbert, 1995). In a different study using other sediments from White Bluffs, cultured aerobic heterotrophs varied three orders of magnitude in 1-cm^3 intact samples located several centimeters apart (Brockman et al., 1992).

Some of the variability in the above studies may have been related to the time that samples were held before processing; however, several studies specifically controlled this variable and many order of magnitude differences were still observed (Brockman et al., 1992; Haldeman et al., 1993; Stevens and Holbert, 1995). Although results based on the ability to culture organisms may be criticized, these studies suggest that microbiological properties may exist in "hotspots", similar to those discovered for denitrification in soils (Parkin, 1987) and methane production in a very shallow aquifer just 1 to 2 m below the soil surface (Adrian et al., 1994). However, the frequency, size, or distribution of "hotspots" in the subsurface may be different than in soils and very shallow aquifers.

From an applied perspective, variability at the scale of centimeters may or may not have importance for a particular problem. For example, if a site is tens of meters on a side and variability in the magnitude of the measured property does not exist at that scale, then the spatial structure is not important from an applied perspective. On the other hand, if the magnitude of the property or if the frequency, size, or distribution of "hotspots" changes with location within a site, then spatial heterogeneity may be important for a particular field application (e.g., bioremediation). A major problem with subsurface systems is that an infinitesimal volume of the site is sampled. This extreme undersampling, combined with sediment samples that are often mixed or homogenized before assaying a 10-g sample, makes it difficult to determine whether microbiological heterogeneity at the centimeter scale may be an important controlling factor for addressing applied issues, and whether variations in chemical and physical properties are linked to variations in microbiological properties at that scale.

To address these issues, recent research at the White Bluffs site has focused on centimeter-scale spatial heterogeneity in microbiological, chemical, and physical properties (Brockman et al., 1995; Murray et al., 1995). The site receives 17 cm of annual precipitation and the sampling interval (8 to 11 m deep) has received approximately 15 µm of annual recharge over the last 12,000 years (E. Murphy, unpublished data). Vertical boreholes were sampled at 5-cm intervals and horizontal boreholes sampled at 5-cm and 30-cm intervals. A 10-g subcore was assayed from each sample for aerobic microbial metabolic potential. Metabolic potential was not detected in approximately two thirds of the samples after 74 d of incubation with ^{14}C-glucose (0.2 µmol/g), ^{14}C-acetate (0.2 µmol/g), and inorganic nutrients. Other experiments were conducted by dissecting cores into many 0.1-g, 1-g, 10-g, and 100-g portions. Metabolic potential after 60 d was rarely detected in the 0.1-g and 1-g samples, but was detected in one third of the 10-g samples and 70% of the 100-g samples (Brockman et al., 1995). A similiar pattern was present for subsurface methanogenic activity in another study at a different site (Jones et al., 1989). The study at White Bluffs indicated that organisms existed in locations separated by several centimeters of volume that were devoid of measurable metabolic potential. Further, samples in excess of 100 g were required to provide a mean with low variance; this sample size integrated over a sufficiently large volume to intercept the rare locations where microcolonies or individual cells existed that were capable of growth. Such a pattern may be due to the presence of highly localized nutrients or the pattern may be the result of localized extinction processes coalescing over centuries or millennia because of the extremely low recharge and nutrient flux. If additional microorganisms were not transported into the environment by a catastrophic recharge event or climatic change during that time period, only highly specialized microorganisms, able to persist in a dormant state over very long periods of time, would survive.

In contrast to the extremely low recharge site in Washington, vadose zone and saturated cores (4 to 6 m deep) were taken from a site in Virginia, U.S. (approximately

15 cm of annual recharge), were dissected into multiple 0.1- to 100-g samples, and assayed for aerobic and anaerobic metabolic potential. Results showed that activity was present in all adjacent 0.1-g samples and the variance was relatively low. These results indicated that microsites of microbial colonization were much more dense than at the White Bluffs site. However, at other depths the aquifer had much lower activities, suggesting that larger samples may be required to yield a representative mean and low variance at these locations (F. Brockman and C. Murray, unpublished data). These studies highlight the need for averaging of microbiological properties over an appropriate sampling volume.

6.4.2   Stratum-Scale Heterogeneity

The great majority of efforts to characterize subsurface microbiology have focused on variability between geologic strata, at scales ranging from meters to tens and hundreds of meters, in part because differences in chemical and physical properties are readily apparent and sampling is relatively easy. Microbiological properties between strata typically vary by many orders of magnitude (Table 6.1). Detailed studies of the variability within and between strata (Fredrickson et al., 1991; Kieft et al., 1995), or between general stratigraphic types that have been sampled at many locations (Phelps et al., 1994b) provide insight on the average differences between strata (see footnotes f,g,h in Table 6.1) and thus minimize the effect of outlier values that may severely bias interpretations and conclusions when only a small number of samples are considered.

Analysis of ground water from different strata or locations for byproducts of microbial metabolism has also indicated high variability. This approach is especially useful for contaminated sites due to the higher levels of microbial metabolism and a greater ratio of the biomass existing as unattached cells. Methane production varied two and three orders of magnitude between points <8 m apart at two proximal sites within a very shallow aquifer (0.6 to 1.5 m to groundwater) and the means at the two sites differed by two orders of magnitude (Adrian et al., 1994). Hydrogen concentrations in a benzene-, toluene-, ethylbenzene-, and xylene-contaminated shallow aquifer (4.5 m to groundwater) in South Carolina, U.S. showed iron reduction, sulfate reduction, and methanogenesis were highly dynamic in space and time, and were related to the influx (and lack of influx) of water from precipitation events and associated transport of sulfate (Vroblesky and Chapelle, 1994). Sampling of groundwater using a multilevel sampler (Ronen et al., 1987) offers fine-scale resolution of dissolved constituents and microbiological properties. For example, a multilevel sampler identified a several-meters-thick zone of high nitrate, denitrification, and unattached biomass in an aquifer contaminated with sewage (Smith et al., 1991) that went undetected in a previous groundwater sampling effort.

6.4.3   Ecosystem-Scale Heterogeneity

Bacteria have existed on the earth for at least 3.5 billion years, adapting and evolving to new and changing environmental conditions as they have been transported by hydrologic and tectonic processes to essentially all regions of the planet (see Chapter 12). Research over the past decade has discovered the presence of bacteria in many types of geologic materials. The term "ecosystem" is used here even though most of the microorganisms (regardless of the scale being considered) may be dormant in many regions of the subsurface (see Chapter 11). Furthermore, an interacting microbial community may not exist in many subsurface environments even if cells

TABLE 6.1
Variability of Microbiological Properties Between Subsurface Strata

Setting	Depth (m)	Property	Range of Variability (Orders of Magnitude)	Correlations	Ref.
Germany	3–49	AHPC[a]	3	nd[b]	Kolbel-Boelke et al., 1988
		AODC[c]	3	nd	
Oklahoma, U.S.	1–8 m	AHPC	4	nd	Bone and Balkwill, 1988
Denmark	4–31 m	AHPC	3	ni[d]	Albrechtsen and Winding, 1992
		AODC	2	Organic content (+) and grain size (−)	
South Carolina, U.S.	8–265 m	AHPC	5	Grain size (+)	Fredrickson et al., 1989; Sinclair and Ghiorse, 1989
		Sulfate-reducers	6	Grain size (+)	Jones et al., 1989
		Σ(aerobic, anaer. MPN[e])	2[f]	Hydraulic conductivity (+)	Phelps et al., 1994b
		^{14}C-acetate into lipids	2[f]	Hydraulic conductivity (+)	
South Carolina, U.S.	365–467	AHPC	5	Hydraulic conductivity (+)	Fredrickson et al., 1991
		AHPC	2.5[g]	Hydraulic conductivity (+)	
Washington, U.S.	174–197	AODC	1[h]	Organic carbon (+)	Kieft et al., 1995
		Basal respiration	2–3[h]	Organic carbon (+)	
		^{14}C-glucose mineralization	0.5–1.5[h]	Organic carbon (+)	
		^{14}C-succinate mineraliz.	1–2[h]	Organic carbon (+)	
		Fe-reducing bacteria	2–3[h]	Organic carbon (+) and Fe(III) (+)	Brockman, unpublished; McKinley and Stevens, unpublished
Nevada, U.S.	50–450	AHPC	4	ni	Haldeman and Amy, 1993
		AODC	2	ni	
		Denitrifiers	3	ni	
Peru margin, Pacific Ocean	3–80	Σ(anaerobic MPN)	2.5	ni	Parkes et al., 1990 (Phil)
Japan Sea, Pacific Ocean	3–518	AODC	2	Availability of organic C, sulfate, and methane	Cragg et al., 1992 (ODP)
		Σ(anaerobic MPN)	3	Availability of organic C, sulfate, and methane	

[a] Aerobic heterotrophic plate count.
[b] nd, not discussed.
[c] Acridine orange direct counts.
[d] ni, none identified, i.e., correlations with geochemical and/or physical properties were tested but correlations were not identified.
[e] Most probable number.
[f] Difference in mean for 15 clay-rich samples and 22 sand-rich samples with hydraulic conductivities >200 μm/s.
[g] Difference in mean for 12 samples from one stratum and 13 samples from another stratum.
[h] Difference in mean for lacustrine stratum (n = 14 samples) as compared to paleosol (n = 7) and fluvial sand (n = 5) strata.

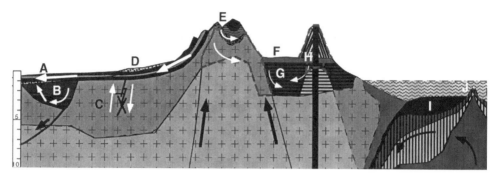

FIGURE 6.1

Hydrogeologic systems supporting microbial populations. The selection of subsurface geologic environments represented in the figure is not exhaustive, but is meant to provide an impression of the wide
variety of habitats containing viable microorganisms. Black arrows represent hydrothermal circulation of
water, white arrows represent circulation of meteoric water. A: Heterotrophic communities living in highly
permeable sediments (Savannah River, South Carolina). B: Saline-tolerant, thermophilic, heterotrophic
communities residing in deeply buried, low-permeability, sedimentary rocks (Taylorsville Triassic Basin,
Virginia). C: Mixed autotrophic/heterotrophic communities residing in igneous and metamorphic rocks
(Precambrian SRIPA granite, Sweden). D: Heterotrophic community in vadose zones with extremely low
recharge (Columbia Basin, Washington). E: Mixed heterotrophic/autotrophic communities colonizing
low-permeability sediments sterilized during deep burial, but habitable due to active uplift (Piceance
Basin, Colorado). F: Mesophilic, heterotrophic communities in low-permeability sediments (Columbia
Basin, Washington) and volcanic tuffs (Nevada Test Site). G: Hydrogen-based, autotrophic community
residing in buried fractured basalt (Columbia Basin, Washington). H: Heterotrophic community colonizing
sedimentary rock sterilized by volcanic intrusions (Cerro Negro, New Mexico). I: Heterotrophic community occupying low-permeability marine sediments (Pacific Ocean). J: Thermophilic, saline-tolerant, au-
totrophic/heterotrophic community associated with submarine hydrothermal vents (Black smokers,
Pacific Ocean). (Figure courtesy of T.C. Onstrott, Princeton University.)

are metabolically active, because they may exist as isolated individual cells or microcolonies that are not connected to one another by diffusion or pore-scale hydrologic
or gaseous flow paths.

Some of the hydrogeologic systems or ecosystems where microorganisms have
been found are shown in Figure 6.1. Several of the important factors that differentiate
these systems are surface climate (i.e., amount of recharge and quality and amount of
associated dissolved nutrients), sediment-associated organic and inorganic chemical
properties (i.e., alternative sources of microbial nutrients), and sediment physical
properties (i.e., where ground water travels and how long it takes to arrive at a subsurface location). Together, these factors may influence or control the flux, distribution, and availability of microbial nutrients.

Two complementary approaches to measure the physiological status of cells in different ecosystems are culturability and metabolic activity. Figure 6.2 presents the percent of viable cells (estimated from phospholipid fatty acid) that were cultured and
the percent of acridine orange direct-count cells that were cultured (F. Brockman,
D. Ringelberg, and D. White, unpublished data). The percentages decreased as the
relative flux of water in the ecosystems decreased. Aerobic aquifers in climates with
high precipitation had the highest percentages and vadose sediments in arid climates
possessed percentages four orders of magnitude lower. Percentages of cultured/via-
ble cells were similar to percentages of cultured/AODC cells, indicating that the
PLFA method for estimating numbers of viable cells was valid. It is possible that the
percentage of cells cultured decreased because of our lack of knowledge regarding
how microorganisms from low-flux environments can be successfully cultured. For
this reason, metabolic activity measurements are also important. Microbial activity as
measured by turnover time for substrates was two to three orders of magnitude lower
in ecosystems with low permeabilities or extremely low recharge than in the aerobic

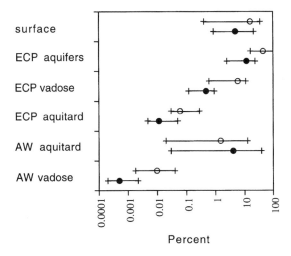

FIGURE 6.2

Percent of viable and total cells that were cultured in various subsurface ecosystems in the U.S. The number of viable cells can be estimated by extracting and measuring the cellular phospholipid fatty acids, and using an empirical conversion factor to convert mass of phospholipid to number of viable (cultured plus uncultured) cells. Cultured cells are the sum of autotrophs and aerobic and anaerobic heterotrophs grown on approximately 10 to 15 media. Open circle, median % of viable cells that were cultured. Closed circle, median % of acridine orange direct-count cells that were cultured. Error bars represent the 25th percentile and 75th percentile. ECP, eastern coastal plain; AW, arid western regions. Data for these calculations were provided by over 20 investigators in the U.S. Department of Energy Subsurface Science Program.

TABLE 6.2

Median Turnover Times of Radiolabeled Substrates (Supplied as 1 to 100 nmol/g Sediment) in Sediments From Various Subsurface Ecosystems in the U.S.

	Turnover Time in Days (Aerobic Incubation/Anaerobic Incubation)			
Substrate	ECP Aquifers	AW Aquitard	AW Saturated Consolidated Rock[a]	AW Vadose[b]
Glucose mineralization	1/3	80/120	nd	4000/nd
Succinate mineralization	nd[c]	140/nd	nd	nd
Acetate mineralization	1/7	nd	1000/200	4000/nd
Acetate incorp. into lipids	3/nd	1100/nd	nd	nd

Note: Turnover times are based on rates of mineralization or uptake during the first 24 h (except for deep vadose samples, 7 d), and only samples with measurable activity were used.

[a] Approx. 40% and 20% of samples were below detection, respectively, under aerobic and anaerobic conditions.

[b] Approx. 70% of the samples were below detection at 7 d.

[c] nd, not determined.

aquifer ecosystem (Table 6.2). These data highlight the major differences in physiological status that exist between subsurface ecosystems with different water and, by inference, nutrient fluxes.

While studies that characterize the variability of microbiological properties at the centimeter, stratum, and ecosystem scales have identified some of the environmental factors controlling microbiological properties, an understanding of how to deal with the variability for the purpose of improved characterization approaches or for developing quantitative predictive capabilities at sites is lacking. To advance towards these goals, there is a need to apply sampling designs and statistical approaches that have been used in other fields of research to the study of subsurface microbiology.

6.5 Geostatistical Analysis

Geostatistics is a specialized form of statistics that focuses on the patterns of spatially distributed data, providing tools for characterizing spatial distributions and estimating the value of variables at unsampled locations. The techniques of geostatistics developed in the mining industry about 20 to 30 years ago, but have since been applied in the petroleum industry, in hydrogeological and soils studies, for environmental characterization and risk assessment, and in agriculture, ecology, geography, and other fields. The geostatistical approach is based on the use of probabilistic models. The probabilistic model is necessary because we usually do not have access to the exhaustive spatial data distribution underlying a given data set, especially when studying the subsurface. For example, the distribution of microorganisms in the subsurface is the result of the interaction between biological, geochemical, and geological processes that are extremely complex. This complexity means that knowledge of the population density at one location may not allow us to predict the density at an unsampled location with certainty, even if the unsampled location is located near one or more data points. Geostatistics handles that uncertainty by using data from many sampling locations to build probabilistic models of the spatial continuity of the data.

6.5.1 Variogram Analysis

The variogram is the geostatistical tool most commonly used to estimate the continuity of spatially distributed data (Isaaks and Srivastava, 1989). The variogram is similar to the paired-quadrat-variance measure discussed in the statistical ecology text of Ludwig and Reynolds (1988). The experimental variogram is estimated by calculating half the average squared difference between all pairs of points separated by a given vector:

$$\gamma(h) = \frac{1}{2N(h)} \sum_{(i,j)} \left[z(x_i) - z(x_i + h) \right]^2 \tag{6.1}$$

where $\gamma(h)$ is the estimated variogram value for a vector separation of h, z is the variable of interest (e.g., the number of culturable aerobes at a sample point), x_i and $x_i + h$ are a pair of locations in the field approximately separated by the vector h, and $N(h)$ is the number of pairs approximately separated by that vector distance. The vector h separating the pairs of points may be specified by both distance and direction, and is known as the lag. The variogram is calculated as a function of h, or lag distance. For variables that exhibit spatial continuity, pairs of points that are near to one another tend to have low variogram values, with the value of the variogram increasing as the average vector distance between the pairs of points increases (Figure 6.3). The variogram may eventually level off at a value known as the sill, which is often equal to the total sample variance. The distance at which the variogram reaches the sill is known as the range. Points separated by a distance greater than the range are uncorrelated. The value of the variogram is defined to be zero at a lag separation of zero. However, when the experimental variogram points are plotted and a line fit to the first several data points, that line often does not project back to the origin. The variogram value intersected by projecting back to the origin is known as the nugget. The nugget reflects variability in the data set at distances smaller than the shortest lag interval. It

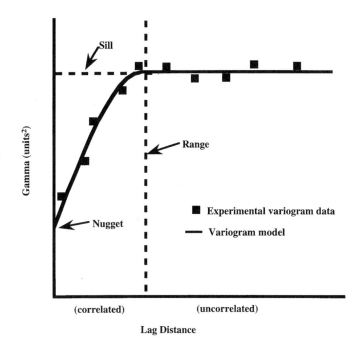

FIGURE 6.3

The variogram is a geostatistical tool that measures the average squared difference between pairs of data values separated by a given lag distance. At distances less than the *range*, the variogram (labeled Gamma on the y-axis) is a function of distance related to the degree of spatial correlation. Points separated by distances greater than the range are uncorrelated. The variogram is constant beyond the range at a value termed the sill, which is often equal to the total variance of the data set. The value of the variogram is defined to be zero for a lag distance of zero, but the nugget can be used to model short-range variability that exists at distances smaller than the sampling interval.

arises from two main sources, one of which is spatial variability that is present at very short scales below the sampling interval. The other source is variability due to measurement error or sampling error. The use of the term "nugget" is a carryover from the early days of geostatistics, when the techniques were developed and applied in the mining industry. At that time it was noticed that certain gold deposits had non-zero variogram values even for very closely spaced sampling intervals and it was discovered that these were caused by the presence or absence of gold nuggets in nearby samples. The relative size of the nugget to the sill is important, because it provides a measure of the proportion of the variance in the data set that can be explained by spatial dependence.

Figure 6.4 illustrates the importance of spatial continuity in characterizing a data set. Figure 6.4A plots the aerobic mineralization data from a vertical borehole at a site in Virginia, while Figure 6.4B is a plot of the same data set after randomly shuffling the depth coordinate. The two data sets have exactly the same univariate statistics (e.g., mean, variance, histogram), however the variograms calculated from the two data sets are extremely different. The variogram of the original data has a very well-defined variogram, fit by a variogram model with a nugget of 15% and a range of 120 cm (Figure 6.4A). The range of the variogram is related to the thickness of the zone of high microbial activity that occurs starting at a depth of about 350 cm. The variogram calculated for the randomly ordered data set does not have any discernible spatial structure (Figure 6.4B). Obviously, the ecological implications of the two different spatial distributions are extremely different. The true spatial distribution (Figure 6.4B) shows that the microorganisms are not randomly distributed, but are

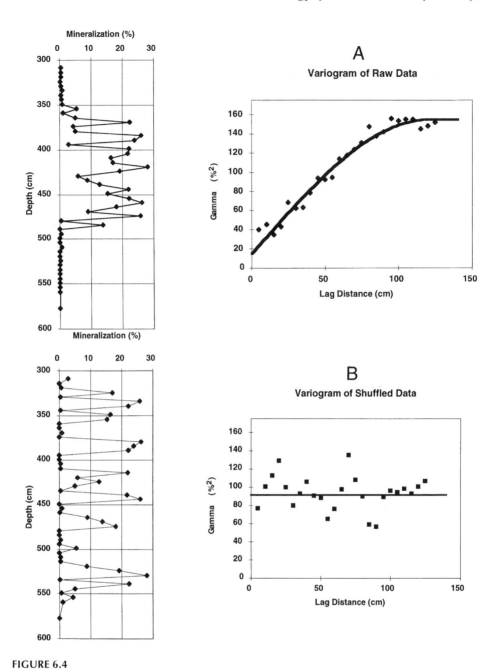

FIGURE 6.4
(A) Plot of aerobic mineralization vs. depth for samples from a vertical borehole drilled in Virginia, together with the variogram calculated from the data (F. Brockman and C. Murray, unpublished data). (B) A plot of the same aerobic mineralization data after randomly shuffling the depth coordinates. Although the two data sets have exactly the same univariate statistics (mean, variance, etc.) and could not be distinguished from one another using many of the common statistical techniques, the variogram reveals profound differences between the spatial structures of the two data sets.

concentrated in a single zone. The distribution of the microorganisms may be controlled by minor differences in the geologic properties of layers in the sediments, an effect which had been noted in subsurface studies of microbiological heterogeneity at the White Bluffs in Washington state (Murray et al., 1995).

In addition to the variogram, two geostatistical tools that can be useful in studying the spatial continuity of variables are the non-ergodic covariance and correlation functions (Deutsch and Journel, 1992; Isaaks and Srivastava, 1989). The covariance function is estimated using the calculation:

$$C(h) = \frac{1}{N(h)} \sum_{i=1}^{N(h)} z(x_i)z(x_i + h) - m_{-h}m_{+h} \tag{6.2}$$

where m_{-h} and m_{+h} refer to the mean of the head and tail values of each lag vector. Thus, the covariance function does not implicitly assume that the means of the head and tail values of the data pairs are identical, allowing some compensation for local variations in the mean. The correlation function, also known as the correlogram, is estimated as:

$$\rho(h) = \frac{C(h)}{\sigma_{-h}\sigma_{+h}} \tag{6.3}$$

which standardizes the covariance function developed above by the standard deviations of the head and tail values of each lag vector, thus compensating for variability in the standard deviation. Thus, these two tools compensate in part for fluctuations in the local mean and standard deviation which might be present in the data. Rossi et al. (1992) presents several cases where the covariance function and correlogram have been useful in ecological studies.

Indicator variograms are another geostatistical method of studying spatial continuity that have potential in the study of subsurface microbiology. Indicator variograms are a form of nonparametric geostatistics based on binary (0,1) coding of variables (Isaaks and Srivastava, 1989). Indicator data can be coded from data of several different types that cannot be directly analyzed using the geostatistical tools discussed above. These data types include presence/absence, ordinal (e.g., low, medium, and high activity), or categorical data (e.g., gram reaction). Indicator variables can also be derived from an indicator transform of an originally continuous variable. For example, aerobic mineralization potential (z) measured at several locations, designated x, could be transformed into a series of indicator variables using several cutoffs, z:

$$I(x, z) = 0 \text{ if } Z(x) \leq z$$
$$= 1 \text{ if } Z(x) > z \tag{6.4}$$

After the data are coded as an indicator variable, the spatial continuity can be analyzed using the variogram, covariance, or correlogram. The indicator variogram allows one to examine portions of the data distribution; for example, to examine the spatial continuity of samples with high aerobic mineralization potential. One major advantage of indicator variograms is that they are resistant to the presence of outliers in the data (Isaaks and Srivastava, 1989). Outliers are common in ecological data and can make it difficult to interpret the traditional variogram (Rossi et al., 1992). Rossi et al. (1992) provide several examples of the application of indicator geostatistics to ecological data.

6.5.2   Variogram Modeling

Models must be fit to experimental variograms if the user wishes to estimate or simulate the value of a variable at unsampled locations. The experimental variogram data (Figure 6.3) provide an estimate of the variogram at a discrete number of distances. A continuous model must be fit to the experimental points if we want to use kriging or a geostatistical simulation technique to provide a value at a point that is not located at a distance equal to one of those that can be calculated using the available data. In order to be used in the simulation or estimation of unsampled locations, the variogram models cannot be arbitrarily chosen. The models must meet the mathematical condition of positive definiteness (Isaaks and Srivastava, 1989). However there are a number of mathematical functions that have this property including the exponential, spherical, and Gaussian models (Isaaks and Srivastava, 1989). Most experimental variograms can be modeled using those functions or linear combinations of them, since linear combinations of those models can also be shown to be positive definite.

Variogram modeling can be used to detect the presence of anisotropy, an important property of a spatial distribution. Anisotropy indicates changes in spatial continuity of a variable with search direction, and is usually reflected by a longer variogram range in the direction of maximum continuity (geometric anisotropy), and/or by a difference in the level of the sill (zonal anisotropy [Isaaks and Srivastava, 1989]). Knowledge of the major directions of anisotropy can provide clues to the underlying processes that are controlling the spatial distribution of the microorganisms. Halvorson et al. (1994) found directional anisotropy present in the experimental variograms of total inorganic nitrogen and soil microbial biomass in a study conducted of microbial populations associated with sagebrush plants. They found a maximum spatial correlation occurring along a compass orientation of 45° and a minimum range of correlation perpendicular to it at 135°, and related the directional anisotropy to preferential distribution of leaf litter beneath the sagebrush plants by the prevailing southwesterly winds.

6.5.2.1   *Variogram Modeling of Soil Microbiology*

Variogram analysis of microbiological properties and potentially related soil chemical properties has been studied in a range of soil settings (Table 6.3). Together, the studies illustrate several important concepts relevant to considering spatial continuity in subsurface environments. In studies of tilled and bare soil, microbiological spatial continuity was not detected at the lag distances examined. This may well have been due to the very short assay period. The inability to define spatial continuity may also have been due to formation of cracks and uneven incorporation of plant residues (Rochette et al., 1991) or disruption and homogenization of the soil environment by tillage. Spatial continuity was also not detected in three of four experiments in a corn field. In more natural settings (pasture and natural grassland) with relatively high precipitation, microbial nitrogen transformations exhibited spatial continuity over distances of meters to tens of meters. The spatial continuity observed in these studies was probably related to topographic features that control the availability and cycling of microbial nutrients (Robertson et al., 1988). Two studies of sagebrush-steppe systems with varying levels of precipitation yielded substantially different results, even though the sampling designs and analytical methods were very similar. Vegetation, while sparse, is more continuous at the site with higher precipitation. Greater variability might be expected at the sagebrush site with lower precipitation because vegetation is more discontinuous and nutrients exist in "resource islands" due to the dominance of

TABLE 6.3

Spatial Continuity of Microbiological Properties in Soils

Setting	Lag Distances (m)	Spatial Continuity			Ref.
		Property (Incubation Time)	Range	Relative Nugget[a]	
Tilled field, no crop	0.15–300	Respiration (few minutes)[b]	nd[c]	100%	Rochette et al., 1991
Tilled field, no crop[d]	1–36	Denitrification (30 min)[b]	nd	100%	Folorunso and Rolston, 1984
		Respiration (30 min)[b]	nd	100%	
Corn field	0.1–1.5	Denitrification (18 h)	Linear model	~65%	Parkin et al., 1987
	0.1–5	Denitrification (18 h)	nd/1.0 m[e]	100%/32%[e]	
Pasture	0.125–20	Nitrification (8 h)	3.3 m	34%	Bramley and White, 1991a
	0.3–3	Nitrification (8 h)	0.8 m	56%	Bramley and White, 1991b
	2.5–22.5	Nitrification (8 h)	nd	100%	Bramley and White, 1991b
Grassland, 80 cm annual precipitation	1–50	Respiration (24 h)	nd	100%	Robertson et al., 1988
		Denitrification (24 h)	30 m[f]	27%	
		Nitrogen mineralization (45 d)	25 m[f]	37%	
		Nitrification (45 d)	35 m[f]	32%	
Sagebrush, 47 cm annual precipitation	0.125–7.5	Respiration (10 d)	nd	100%	Jackson and Caldwell, 1993a,b
		Nitrogen mineralization (10 d)	nd	100%	
		Nitrification potential (10 d)	nd	100%	
Desert with sagebrush, 17 cm annual precipitation	0.25–1.37	Biomass carbon (7 d)	0.7 m	10%	Smith et al., 1994;
		Carbon mineralization (21 d)	0.9 m	20%	Halvorson et al., 1994
		Nitrogen mineralization (21 d)	0.6 m	9%	

Note: The spatial continuities of associated environmental properties were also measured in most studies, but are not shown in the table.

[a] Relative nugget equals nugget divided by sill expressed as a percent.

[b] Measured with a small chamber placed over the surface of the field.

[c] nd, not detected; indicates analyses were performed but spatial continuity could not be detected at the lag distances investigated.

[d] Water and nitrate added to field immediately before the activity assays.

[e] Two experiments differing in tillage practice gave different results.

[f] Practical range estimated from Figure 3 of Robertson et al., 1988.

concentrating biological transport mechanisms over dispersive physical transport mechanisms (Garner and Steinberger, 1989). Well-defined variograms with interpretable ranges were reported at the sagebrush site with lower precipitation. The lack of detectable spatial continuity for variograms of microbiological properties from the site with higher precipitation may be due to the homogenizing effects of the more continuous ground cover at the site, or to the effects of biological mixing by burrowing rodents that are mentioned by Jackson and Caldwell (1993a,b). The importance of an appropriate sample separation distance in identifying spatial continuity was demonstrated in the studies at the pasture site and discussed in some detail by Bramley and White (1991a,b).

   With the exception of the desert soil with 17 cm precipitation, relative nuggets for microbiological properties with defined ranges were 27 to 34% indicating that a significant proportion of the variability occurred at scales less than the minimum sample separation distance (Table 6.3). This variability may result from measurement error, sample location error, and/or the existence of a different pattern of spatial heterogeneity at smaller, uncharacterized scales. The lower relative nuggets in the soil receiving 17 cm precipitation is probably due to the strong gradient between resource islands beneath and near the sagebrush plants and adjacent soil with very little vegetative cover.

6.5.2.2 *Variogram Modeling of Subsurface Microbiology*

Variogram analysis has been applied to the microbiology data from vertical and horizontal boreholes at the White Bluffs site discussed in Section 6.4.1. The variogram range of aerobic mineralization data sampled at 5-cm spacing from the vertical borehole was approximately 15 cm (C. Murray and F. Brockman, unpublished data). This range is approximately equal to that of several geochemical variables measured in the vertical boreholes. The variogram range of the microbiological data appears to be related to physical and chemical variations within the bedded sediments that occur at distances smaller than the average bed thickness of about 30 cm. The variogram of aerobic mineralization data from a horizontal borehole drilled in silty sediments at the same site indicates that the horizontal continuity is greater than the vertical continuity. This is not surprising in light of the greater spatial continuity of physical and chemical properties in the horizontal borehole, and is an effect that is often seen in bedded sedimentary sequences.

   Figure 6.4A, which was used as an example in the discussion of the variogram in Section 6.5.1, is a plot of the aerobic mineralization data and the resulting variogram from a vertical core at a site in Virginia that was sampled at 5-cm intervals. The variogram range of 120 cm is almost an order of magnitude greater than the variogram range modeled at the White Bluffs site. The greater range appears to be related to the greater thickness of sediment beds at the Virginia site.

6.5.3 Spatial Cross-Correlation

In ecological studies we are often interested in the spatial distribution of associated environmental variables and how those variables are correlated to the spatial distribution of the organism being studied. The cross-correlogram, cross-covariance, and cross-variogram can be used to examine the spatial relationships between pairs of variables.

   For example, the cross-correlogram provides an estimate of the correlation coefficient between two variables as a function of lag distance:

$$\rho_{uz}(h) = \frac{C_{uz}(h)}{\sigma_{u_{-h}} \cdot \sigma_{z_{-h}}} \tag{6.5}$$

which is simply the cross-covariance between two variables (u and z) for a given lag normalized by the product of the standard deviations of each variable for that lag. Note that for a lag of zero, the cross-correlogram is equal to the standard Pearson product-moment correlation coefficient between the two variables.

In addition to providing information on the environmental variables controlling microbial distribution, the analysis of spatial cross-correlation could be performed with another goal in mind. The analysis of microbiological variables is often a labor-intensive and costly process and this might preclude the detailed three-dimensional characterization of microbiological variables that would be useful for an *in situ* bioremediation study. If spatial dependence exists between the microbiological and environmental variables at a site, then geostatistical tools such as co-kriging or co-simulation (Deutsch and Journel, 1992) could be used to estimate or simulate the distribution of a microbiological variable. This could be done using a relatively small number of measurements of the expensive microbiological variable and a larger number of measurements of the environmental variable that is spatially correlated with the microbiology, thus lowering the overall cost of the characterization effort.

6.5.3.1 Spatial Cross-Correlation in Soils

Cross-correlegram analysis of microbiological data appears to have been conducted at only one soil study site. Biomass carbon was spatially related to the occurrence of sagebrush (approximate range, 0.5 m) and grasses (range, 0.12 to 0.37 m), carbon mineralization was related to sagebrush (range, 0.5 m) and grasses (range, 0.2 m), and nitrogen mineralization was related to grasses (range, 0.4 m) (Halvorson et al., 1994; Smith et al., 1994). Smith et al. (1994) also observed spatial cross-correlation between substrate-induced respiration and carbon mineralization potentials that extended to distances of 60 cm, which they related to resource cycling beneath sagebrush plants in the ecosystem.

6.5.3.2 Spatial Cross-Correlation in Subsurface Sediments

In a recent study of subsurface data from horizontal and vertical boreholes at the White Bluffs, spatial cross-correlation (Figure 6.5) was detected between aerobic mineralization potential and the concentration of extractable sulfate (C. Murray and F. Brockman, unpublished data). The distribution of aerobic mineralization potential is also associated with sediments containing high amounts of moisture and extractable nitrate, and low permeability. These associations with aerobic mineralization potential suggest the preferential preservation of small numbers of microorganisms in sediments with small pore throats that have retained greater moisture under the extremely low recharge conditions present at the site (about 15 μm per year). Thus, the correlation of aerobic mineralization potential with sulfate is probably related to physical and chemical conditions that favored long-term survival of dormant microorganisms, and not due to present-day interactions between microorganisms and their environment. Analysis of the spatial cross-correlation between variables can thus be used to infer the identity of ecological processes responsible for spatial distribution of microorganisms. These relationships can then be examined further, either through field sampling or the design of laboratory experiments to verify the importance of the processes.

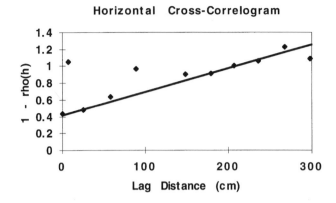

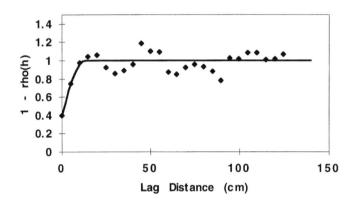

FIGURE 6.5

Cross-correlograms between aerobic mineralization and extractable sulfate data from horizontal and vertical boreholes at the White Bluffs. For convenience, the cross-correlograms have been reexpressed as standardized variograms by subtracting the correlation coefficient for each lag from 1. The horizontal cross-correlogram was fit by a linear trend model, and reflects the correlated increases in both variables with depth in the borehole. The vertical cross-correlogram was fit by a standard transitional model incorporating a sill, and has a range of 13 cm. This is less than the average thickness of the sedimentary beds in the section, which is about 30 cm.

6.5.4   Multivariate Analysis

Often there are several environmental variables that could potentially influence the distribution and activity of microorganisms in the subsurface, so evaluating pairs of variables may not be very informative. Halvorson et al. (1995) discuss a nonparametric form of multivariate geostatistics based on a multiple-variable indicator transform (MVIT), which they apply to a study of soil microbiology. They transformed their data (total N, total inorganic N, water-soluble C, soil microbial biomass, metabolic quotient, and cumulative net respiration) into indicator variables using the local median for each variable (i.e., each of the six variables was examined at each location in three experimental plots, and if the value of the variable exceeded the median of the variable for that experimental plot, then the indicator transform of that variable was set to 1, otherwise it was set to 0). They then created several MVITs by an additive model, based on the number of variables that exceeded the local median at each location. For example, if three out of the six variables at a location exceeded the median, then a MVIT designated COMB3 would be set to 1. Similar variables labeled COMB4 through COMB6 were generated from the data. The spatial continuity of the MVITs were then studied through the use of correlograms (Halvorson et al., 1995), and maps were created using indicator kriging that show the probability that samples from a given location would meet the additive criteria for that MVIT (e.g., that three of six variables would exceed the median for COMB3). In addition, they examined the spatial cross-correlation between the MVITs and vegetation type and determined that the MVITs had a higher positive cross-correlation with sagebrush than with grass species, and that the highest cross-correlations occurred in the areas downwind of the sagebrush plants. This technique appears to be quite flexible, since one can use it to examine the relationships between microbial populations or activity and several limiting resources.

In addition to the nonparametric method discussed above, there are also parametric forms of multivariate analysis. One way to analyze the variance of a multivariate data set is to perform a principal components analysis (Johnson and Wichern, 1988; Wackernagel, 1995). Principal components analysis proceeds by an eigenvalue-eigenvector decomposition of the variance-covariance or correlation matrix of the data. Each eigenvector or principal component is a linear combination of the original variables and is orthogonal to the remaining principal components (and therefore uncorrelated with them). The first principal component is the linear combination of the original variables that accounts for the largest proportion of the total variance in the data set. The second principal component is orthogonal to the first and accounts for the second largest proportion of the total variance, and so forth. With many multivariate data sets it is possible to identify a small number of principal components that account for the vast majority of the variance in the data set. Because the principal components are not correlated with one another, an important application of the technique is in multiple regression, where collinearity of multivariate data is often a problem (Chatterjee and Price, 1977). Thus the technique could be used in a multiple regression attempting to predict the response of a microbiological variable to variations in several correlated environmental properties.

Canonical correlation analysis (Johnson and Wichern, 1988; Wackernagel, 1995) is a related technique that can be applied to study of the correlation structure of multivariate data sets. Canonical correlation analysis identifies linear combinations of one group of variables that has the maximum correlation with a second group of variables. This technique could be used to study the effect of a group of environmental properties on a group composed of several microbiological variables.

The preceding discussion of multivariate analysis has not examined the spatial dependence of the correlation structure. One possibility is that a set of variables possesses a given correlation structure for a lag of 0, which can be calculated by the correlation matrix, but that there is no spatial correlation present. This spatial independence is in fact the assumption made for classical multivariate regression and implies that the variograms of each variable and all cross-variograms are best fit by a pure nugget effect. As noted earlier, the probability that a given set of subsurface environmental variables will not be spatially correlated is low, because those variables result from continuous geologic processes that operate over a volume of the subsurface whose spatial extent can be quite large.

The next simplest type of multivariate spatial structure involves a situation where the spatial covariance structure, $\mathbf{C(h)}$, is composed of a variance-covariance matrix of the variables, $\mathbf{V}$, and the spatial dependence between pairs of points is provided by a single spatial correlation structure $\rho\mathbf{(h)}$:

$$C(h) = V \cdot \rho(h) \tag{6.6}$$

This multivariate spatial correlation structure is labeled *intrinsic correlation* by Wackernagel (1995). Wackernagel (1995) provides several methods for determining if a data set is intrinsically correlated. One of the simplest is to calculate cross-variograms between the principal components of the data set. The principal components are uncorrelated at lag 0, by construction. If they remain uncorrelated for all lags (i.e., the cross-variograms can be modeled as pure nugget effects for all lags), then the data set exhibits intrinsic correlation. The presence of intrinsic correlation in a data set has important implications for spatial analysis and mapping of the data, since it indicates that only a single spatial structure needs to be identified and modeled. If that is the case then the principal components can each be mapped independently using the geostatistical estimation technique known as kriging (Wackernagel, 1995).

If the multivariate data set is not intrinsically correlated, then its spatial structure must be analyzed to determine a linear model of coregionalization (Wackernagel, 1995). The main reason for this is that the correlations between variables in the data set vary with the spatial scale. Wackernagel (1995) discusses an example in which the concentrations of zinc, lead, and copper were measured for soil samples from various locations in a forest. In that study it was determined that there were three different spatial scales present in the study area with important differences between the correlations of the metal concentrations at each scale. There are spatial extensions of the multivariate techniques described previously that can be used in such situations, including regionalized principal component analysis and regionalized canonical coefficient analysis (Wackernagel, 1995). Regionalized multivariate analysis techniques have been applied in a number of soil science and hydrogeology studies (see Wackernagel [1995] for a list), and several of those studies appear to have aspects in common with studies of the factors controlling the spatial distribution of microorganisms. These multivariate geostatistical techniques are currently being applied to studies of the relationships of subsurface microbiology to geological and geochemical variables, but published results are not yet available. One potential application of this approach is the generation of spatially cross-correlated grids of microbiological and geological variables that could be used as input to reactive flow and transport models. Those models could potentially influence research and applications in the field of *in situ* bioremediation.

6.5.5 Sampling for Spatial Characterization

Sampling plans for geostatistical analysis and spatial characterization have special design considerations. While a random sample may be sufficient for estimation of some properties, e.g., an estimate of the global average, it is unlikely that a randomly chosen group of sample locations will provide information needed for estimation of the variogram or for mapping the property across the area. Olea (1984) demonstrated that regular sampling grids provide the most efficient method of sampling for spatial variability, relative to random, stratified, and clustered sampling patterns. If directional anisotropy of the microbial distribution is expected, then a triangular grid provides the most efficient grid for determination of the anisotropy direction (McBratney et al., 1981). If the orientation of the anisotropy can be identified in advance, either by a preliminary sampling campaign, or by analysis of the processes responsible for the anisotropy, then a sampling grid can be designed that takes advantage of the

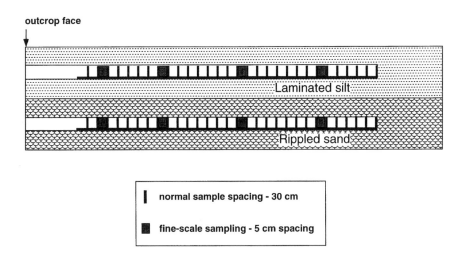

FIGURE 6.6
Nested sampling plan for the horizontal boreholes at the White Bluffs. The design includes regularly spaced samples with a sample interval of 30 cm, and four zones with closely spaced samples to examine the fine-scale spatial structure.

anisotropy, reducing the number of samples necessary, especially for mapping purposes (Flatman et al., 1988).

In order to capture both long- and short-range spatial variability, nested sampling patterns are often employed (Flatman et al., 1988). These consist of widely spaced samples interspersed with areas of closely spaced samples. Halvorson et al. (1994, 1995) used a nested approach in the design of the sampling grid for their study of the spatial distribution of microbes and resources surrounding sagebrush plants. The geostatistical studies of horizontal spatial continuity of microbiological and environmental variables performed at the White Bluffs also employ a nested sampling approach (Figure 6.6). Because of the high cost of sample analysis, the sampling at the White Bluffs has been performed along transects by sampling horizontal and vertical boreholes. McBratney et al. (1981) noted that transect sampling was well suited to sampling for variogram analysis. In the White Bluffs sampling and the sampling by Halvorson et al. (1994), the distribution of both widely and closely spaced samples is regular. Warrick and Myers (1987) present an alternative method for optimizing the number of variogram pairs entering into each lag, which produces an irregularly spaced distribution of samples.

The number of samples that need to be taken for proper variogram analysis can be a problem, especially for microbiological variables that are very expensive to analyze. A rule of thumb commonly used for design of sampling grids and transects is to ensure that at least 30 sample pairs are available for each lag in the variogram (Journel and Huijbregts, 1978), which can be attained with fewer than 25 data points arranged on a square grid in 2-D. However, other practitioners have proposed that at least 80 sample pairs should be available for each lag (Webster and Oliver, 1993). They suggest that at least 100 samples should be taken for variogram analysis in two dimensions, and perhaps as many as 150 to 250 samples, based on their studies of the representativeness of experimental variograms (Webster and Oliver, 1993). This provides one of the major incentives for performing studies to identify geological or geochemical variables that can serve as proxies for the population density or activity of microorganisms. If the proxy variables can be analyzed for lower cost, then dense sampling schemes to identify the spatial continuity would not be a problem.

6.6   Applications of Geostatistics to Subsurface Microbiology

The applications of geostatistics to subsurface microbiogy discussed in this chapter concentrate on the use of variograms and cross-correlograms to build quantitative models of the spatial distribution of the microorganisms and to attempt to identify the environmental properties controlling that distribution. There are other applications that can be pursued once the spatial models are available.

One possible application is the identification of sets of samples that can be used with classical statistical methods of hypothesis testing and analysis of variance. The classical methods make the assumption that the sample data used in the analyses are independent. Halvorson et al. (1994), in their study of soil microbiology, discuss the use of the variogram to identify the range of spatial continuity, and to then identify sets of samples located at distances greater than the range that are spatially independent. They then performed an analysis of variance and showed that the mean concentrations of total inorganic nitrogen, water-soluble carbon, and soil microbial biomass were all significantly greater near sagebrush plants than they were away from the plants.

An important use of the spatial continuity models, which has been alluded to in this chapter, is for the estimation of microbiological variables at locations where samples are not available. Kriging is the geostatistical technique that can be used for the estimation process. Kriging is a generalized form of linear regression that employs the variogram model to determine the weights to assign to the nearby data in estimating the value of a variable at an unsampled location (Deutsch and Journel, 1992; Isaaks and Srivastava, 1989; Journel, 1989). The output from the kriging process is an interpolated map of the variable of interest, usually on a regular grid. Halvorson et al. (1994, 1995) used several forms of kriging to map the distribution of microbiological and supporting environmental properties in their soils studies.

In addition to providing an estimate of the variable across the area of interest, kriging also provides an error estimate, the kriging variance, which can be used as an estimate of the uncertainty attached to an interpolated value at a particular location on the grid. If certain assumptions hold concerning the distribution of the kriging errors, then the kriging variance can also be used to construct confidence intervals for the kriged estimates at each location on the map (Journel, 1989). There are also extensions of the kriging algorithm, e.g., co-kriging and kriging with an external drift (Deutsch and Journel, 1992; Wackernagel, 1995), that can be used for estimation of a variable when related information is available and measured on more dense sample spacing than the primary variable that is being mapped.

One feature of kriging is that the resulting estimate is a smoothed interpolation of the variable of interest. The smoothing, which is similar to the smoothing seen in linear regression models fit to bivariate data, is a common feature of most local interpolation algorithms. A smoothed interpolation of a variable is fine for many applications, and it would be perfectly suitable for mapping microbial population density. However, there are applications for which smoothing is not desired, because it may smooth out some spatial features of the variable, particularly those related to the continuity of extreme values (Deutsch and Journel, 1992; Journel, 1989). If the grid of mapped values is to be used as input to a reactive flow and transport code, then the smoothing associated with the kriging algorithm may result in poor estimates of the transport times for contaminant plumes. Stochastic simulation is a geostatistical Monte Carlo technique that provides multiple realizations, or maps, of a variable and is an alternative to kriging for those situations (Deutsch and Journel, 1992; Journel,

1989). Each realization honors the conditioning data and the spatial dependence incorporated in the variogram model and can reproduce the spatial continuity of extreme values (e.g., associated with fractures or organic-rich zones). A suite of such simulations, when used as input to reactive flow and transport programs, can be used to build probabilistic models for the uncertainties associated with fluid transport and with microbial metabolism of organic solutes in aquifers.

6.7   Concluding Remarks

The application of geostatistical techniques to quantifying microbiological properties and processes could greatly improve the accuracy of ecosystem process modeling and contaminant transport models that include biodegradation. Fundamental knowledge of the distribution of microbiological properties in subsurface systems and the appropriate averaging scale is required to design improved sampling schemes. In addition, studies focused on linked measurements of microbiological, geochemical, and physical heterogeneity may assist in defining controlling factors. This could lead to the possibility of using more easily and cheaply measured properties (e.g., geophysical or remote sensing data) as proxies for microbiological properties. In the long term, knowledge of the physical and chemical properties that influence microbial abundance, activity, and diversity is important for characterizing and prioritizing sites for bioremediation. Such knowledge will determine the possibility and desirability of delivering electron donors and acceptors or other nutrients to various locations, and may make it possible to predict the overall performance of bioremediation in particular physical-chemical settings.

Acknowledgments

This work was supported by the Subsurface Science Program, U.S. Department of Energy. We thank the editors and other reviewers for their excellent suggestions.

References

Adrian, N. R., J. A. Robinson, and J. M. Suflita. 1994. Spatial variability in biodegradation rates as evidenced by methane production from an aquifer. *Appl. Environ. Microbiol.*, 60:3632.

Aitchinson, J. and J. A. C. Brown. 1957. *The Lognormal Distribution*. Cambridge University Press, London.

Albrechtsen, H.-J. and A. Winding. 1992. Microbial biomass and activity in subsurface sediments from Vejen, Denmark. *Microb. Ecol.*, 23:303.

Biggar, J. W. and D. R. Nielsen. 1976. Spatial variability of the leaching characteristics of a field soil. *Water Resour. Res.*, 12:78.

Bone, T. L. and D. L. Balkwill. 1988. Morphological and cultural comparison of microorganisms in surface soil and subsurface sediments at a pristine study site in Oklahoma. *Microb. Ecol.*, 16:49.

Bramley, R. G. V. and R. E. White. 1991a. An analysis of variability in the activity of nitrifiers in a soil under pasture. I. Spatially dependent variability and optimum sampling strategy. *Aust. J. Soil Res.*, 29:95.

Bramley, R. G. V. and R. E. White. 1991b. An analysis of variability in the activity of nitrifiers in a soil under pasture. II. Some problems in the geostatistical analysis of biological soil properties. *Aust. J. Soil Res.*, 29:109.

Brockman, F. J., T. L. Kieft, J. K. Fredrickson, B. N. Bjornstad, S.-M. L. Li, W. Spangenburg, and P. E. Long. 1992. Microbiology of vadose zone paleosols in south-central Washington State. *Microb. Ecol.*, 23:279.

Brockman, F. J., R. P. Griffiths, C. J. Murray, S. W. Li, C. S. Spadoni, and J. Braby. 1995. Spatial heterogeneity of microbial activity in subsurface sediments. *Abstr. 94th General Meeting*, American Society for Microbiology, Wasington, D.C., p. 348.

Chatterjee, S. and B. Price. 1977. *Regression Analysis by Example.* John Wiley & Sons, New York.

Cragg, B. A., S. M. Harvey, J. C. Fry, R. A. Herbert, and R. J. Parkes. 1993. Bacterial biomass and activity in the deep sediment layers of the Japan Sea, Hole 798B. Proceedings of the Ocean Drilling Program, Scientific Results 127:761.

Deutsch, C. V. and A. G. Journel. 1992. *GSLIB: Geostatistical Software Library and User's Guide.* Oxford University Press, New York.

Englund, E. and N. Heravi. 1993. Conditional simulation: Practical application for sampling design optimization. *Geostatistics Troia '92.* Soares, A., Ed., Kluwer Academic, Amsterdam, pp. 613-624.

Flatman, G. T., E. J. Englund, and A. A. Yfantis. 1988. Geostatistical approaches to the design of sampling regimes. *Principles of Environmental Sampling.* L. H. Keith, Ed., American Chemical Society, Washington, D.C., pp. 73-84.

Folorunso, O. A. and D. E. Rolston. 1984. Spatial variability of field-measured denitrification gas fluxes. *Soil Sci. Soc. Am. J.*, 48:1214.

Fredrickson, J. K., T. R. Garland, R. J. Hicks, J. M. Thomas, S.-W. Li, and K. M. McFadden. 1989. Lithotrophic and heterotrophic bacteria in deep subsurface sediments and their relation to sediment properties. *Geomicrobiol. J.*, 7:53.

Fredrickson, J. K., D. L. Balkwill, J. M. Zachara, S.-M. W. Li, F. J. Brockman, and M. A. Simmons. 1991. Physiological diversity and distributions of heterotrophic bacteria in deep cretaceous sediments of the Atlantic coastal plain. *Appl. Environ. Microbiol.*, 47:402.

Garner, W. and Y. Steinberger. 1989. A proposed mechanism for the formation of 'fertile islands' in the desert ecosystem. *J. Arid Environ.*, 16:257.

Groffman, P. M. and J. M. Tiedje. 1989. Denitrification in north temperate forest soils: spatial and temporal patterns at the landscape and seasonal scales. *Soil Biol. Biochem.*, 21:613.

Gullett, C. D. 1993. Bacterial Modulation in Diagenesis of the Black Creek Formation, South Carolina, Ph.D. dissertation. University of South Carolina.

Haldeman, D. L. and P.S. Amy. 1993. Bacterial heterogeneity in deep subsurface tunnels at Rainier Mesa, Nevada Test Site. *Microb. Ecol.*, 25:183.

Haldeman, D. L., P. S. Amy, D. Ringelberg, and D. C. White. 1993. Characterization of the microbiology within a 21 m^3 section of rock from the deep subsurface. *Microb. Ecol.*, 26:145.

Halvorson, J. J., H. Bolton, Jr., J. L. Smith, and R. E. Rossi. 1994. Measuring resource islands using geostatistics. *Great Basin Nat.*, 54:313.

Halvorson, J. J., J. L. Smith, H. Bolton, Jr., and Rossi, R. E. 1995. Evaluating shrub-associated spatial patterns of soil properties in a shrub-steppe ecosystem using multiple-variable geostatistics. *Soil Soc. Am. J.*, 59:1476.

Hirano, S. S., E. V. Nordheim, D. E. Arny, and C. D. Upper. 1982. Lognormal distribution of epiphytic bacterial populations on leaf surfaces. *Appl. Environ. Microbiol.*, 44:695.

Isaaks, E. H. and R. M. Srivastava. 1989. *An Introduction to Applied Geostatistics.* Oxford University Press, New York.

Jackson, R. B. and M. M. Caldwell. 1993a. Geostatistical patterns of soil heterogeneity around individual perennial plants. *J. Ecol.*, 81:683.

Jackson, R. B. and M. M. Caldwell. 1993b. The scale of nutrient heterogeneity around individual plants and its quantification with geostatistics. *Ecology*, 74:612.

Johnson, R. A. and D. W. Wichern. 1988. *Applied Multivariate Statistical Analysis*, 2nd ed., Prentice-Hall, Englewood Cliffs, NJ.

Jones, R. E., R. E. Beeman, and J. M. Suflita. 1989. Anaerobic metabolic processes in the deep terrestrial subsurface. *Geomicrobiol. J.*, 7:117.

Jones, A. J. and R. J. Wagenet. 1984. In situ estimation of hydraulic conductivity using simplified methods. *Water Resour. Res.*, 20:1620.

Journel, A. G. 1989. Fundamentals of geostatistics in five lessons. *AGU Short Course in Geology*, v. 8, American Geophysical Union, Washington, D.C.

Journel, A. G. and C. J. Huijbregts. 1978. *Mining Geostatistics*. Academic Press, London.

Kieft, T. L., J. K. Fredrickson, J. P. McKinley, B. N. Bjornstad, S. A. Rawson, T. J. Phelps, F. J. Brockman, and S. M. Pfiffner. 1995. Microbiological comparisons within and across contiguous lacustrine, paleosol, and fluvial subsurface sediments. *Appl. Environ. Microbiol.*, 61:749.

King, G. M. and W. J. Wiebe. 1978. Methane release from soils of a Georgia salt marsh. *Geochim. Cosmochim. Acta*, 42:343.

Kölbel-Boelke, J., E.-M. Anders, and A. Nehrkorn. 1988. Microbial communities in the saturated groundwater environment. II. Diversity of bacterial communities in a pleistocene sand aquifer and their *in vitro* activities. *Microb. Ecol.*, 16:31.

Loper, J. E., T. V. Suslow, and M. N. Schroth. 1984. Lognormal distribution of bacterial populations in the rhizosphere. *Phytopathology*, 24:1454.

Ludwig, J. A. and J. F. Reynolds. 1988. *Statistical Ecology*. John Wiley & Sons, New York.

McBratney, A. B., R. Webster, and T. M. Burgess. 1981. The design of optimal sampling schemes for local estimation and mapping of regionalized variables. *Comput. Geosci.*, 7:331.

McKinley, J. P., T. O. Stevens, J. K. Fredrickson, J. M. Zachara, F. C. Colwell, K. B. Wagnon, S. A. Rawson, and B. N. Bjornstad. The biogeochemistry of anaerobic lacrustine and paleosol sediments within an aerobic unconfined aquifer, *Geomicrobiol. J.*, in press.

Murphy, E. M., J. A. Schramke, J. K. Fredrickson, H. W. Bledsoe, A. J. Francis, D. S. Sklarew, and J. C. Linehan. 1992. The influence of microbial activity and sedimentary organic carbon on the isotope geochemistry of the Middendorf aquifer. *Water Resour. Res.*, 28:723.

Murray, C. J., F. J. Brockman, and B. N. Bjornstad. 1995. Spatial continuity of bacterial properties in subsurface sediments from south-central Washington. *GSA Abstr. Prog.*, 27:191.

Nielsen, D. R., J. W. Biggar, and K. T. Erh. 1973. Spatial variability of field-measured soil-water properties. *Hilgardia*, 42:215.

Olea, R. A. 1984. Systematic sampling of spatial functions. Kansas Series on Spatial Analysis No. 7, Kansas Geological Survey, Topeka.

Parkes, R. J., B. A. Cragg, J. C. Fry, R. A. Herbert, J. W. T. Wimpenny. 1990. Bacterial biomass and activity in deep sediment layers from the Peru margin. *Philos. Trans. R. Soc. London*, 331:139.

Parkin, T. B. 1987. Soil microsites as a source of denitrification variability. *Soil Sci. Soc. Am. J.*, 51:1194.

Parkin, T. B. 1993. Spatial variability of microbial processes in soil — a review. *J. Environ. Qual.*, 22:409.

Parkin, T. B., J. L. Starr, and J. J. Meisinger. 1987. Influence of sample size on measurement of soil denitrification. *Soil Sci. Soc. Am. J.*, 51:1492.

Parkin, R. B. and J. A. Robinson. 1989. Stochastic models of soil denitrification. *Appl. Environ. Microbiol.*, 55:72.

Phelps, T. J., E. M. Murphy, S. M. Pfiffner, and D. C. White. 1994a. Comparison between geochemical and biological estimates of subsurface microbial activities. *Microb. Ecol.*, 28:335.

Phelps, T. J., S. M. Pfiffner, K. A. Sargent, and D. C. White. 1994b. Factors influencing the abundance and metabolic capacities of microorganisms in eastern coastal plain sediments. *Microb. Ecol.*, 28:351.

Robertson, G. P., M. A. Huston, F. C. Evans, and J. M. Tiedje. 1988. Spatial variability in a successional plant community: patterns of nitrogen availability. *Ecology*, 69:1517.

Rochette, P. L., R. L. Deshardins, and E. Pattey. 1991. Spatial and temporal variability of soil respiration in agricultural fields. *Can. J. Soil Sci.*, 71:189.

Ronen, D., M. Magaritz, and I. Levy. 1987. An *in situ* multilevel sampler for preventive monitoring and study of hydrochemical profiles in aquifers. *GWMR*, Fall:69.

Ronen, D., B. Berkowitz, and M. Magaritz. 1993. Vertical heterogeneity in horizontal components of specific discharge: case study analysis. *Ground Water*, 31:33.

Rossi, R. E., D. J. Mulla, A. G. Journel, and E. H. Franz. 1992. Geostatistical tools for modeling and interpreting ecological spatial patterns and appraising risk. *Ecol. Monogr.*, 62:277.

Russell, C. E., R. Jacobson, D. L. Haldeman, and P. S. Amy. 1994. Heterogeneity of deep subsurface microorganisms and correlations to hydrogeological and geochemical parameters. *Geomicrobiol. J.*, 12:37-51.

Severson, K. J., D. L. Johnstone, C. Kent Keller, and B. D. Wood. 1991. Hydrogeologic parameters affecting vadose-zone microbial distributions. *Geomicrobiol. J.*, 9:197.

Sinclair, J. L. and W. C. Ghiorse. 1989. Distribution of aerobic bacteria, protozoa, algae, and fungi in deep subsurface sediments. *Geomicrobiol. J.*, 7:15.

Smith, J. L., J. J. Halvorson, and H. Bolton, Jr. 1994. Spatial relationships of soil microbial biomass and C and N mineralization in a semi-arid shrub-steppe ecosystem. *Soil Biol. Biochem.*, 26:151.

Smith, R. L., B. L. Howes, and J. H. Duff. 1991. Denitrification in nitrate-contaminated groundwater: occurrence in steep vertical geochemical gradients. *Geochim. Cosmochim. Acta*, 55:1815.

Stevens, T. O. and B. S. Holbert. 1995. Variability and density dependence of bacteria in terrestrial subsurface samples: implications for enumeration. *J. Microbiol. Methods*, 21:283.

Tabor, J. A., A. W. Warrick, D. E. Myers, and D. A. Pennington. 1985. Spatial variability of nitrate in irrigated cotton. II. Soil nitrate and correlated variables. *Soil Sci. Soc. Am. J.*, 49:390.

Ugland, K. I. and J. S. Gray. 1982. Lognormal distributions and the concept of community equilibrium. *Oikos* 39:171.

Vroblesky, D. A. and F. H. Chapelle. 1994. Temporal and spatial changes of terminal electron-accepting processes in a petroleum hydrocarbon-contaminated aquifer and the significance for contaminant biodegradation. *Water Resour. Res.*, 30:1561.

Wackernagel, H. 1995. *Multivariate Geostatistics*. Springer-Verlag, Berlin.

Warrick, A. W., G. J. Mullen, and D. R. Nielsen. 1977. Scaling field measured soil hydraulic properties using a similar media concept. *Water Resour. Res.*, 13:355.

Warrick, A. W. and D. E. Meyers. 1987. Optimization of sampling locations for variogram calculations. *Water Resour. Res.*, 23:496.

Webster, R. and M. A. Oliver. 1993. How large a sample is needed to estimate the regional variogram adequately? *Geostatistics Troia '92*. Soares, A., Ed., Kluwer Academic, Amsterdam, 155.

White, R. E., R. A. Haigh, and J. H. Macduff. 1987. Frequency distributions and spatially dependent vaiability of ammonium and nitrate concentrations in soil under grazed and ungrazed grassland. *Fert. Res.*, 11:193.

Section II

Microbial Ecology and Related Methods

7

Identity and Diversity of Microorganisms Cultured from Subsurface Environments

David L. Balkwill and David R. Boone

CONTENTS

KEY WORDS: *microbial diversity, subsurface microbiology, aquifer microbiology, culturable microorganisms, phylogeny, chemoheterotrophs, anaerobes,* Arthrobacter, Comamonas, Acinetobacter, Gordona, Bacillus.

7.1 Introduction

This purpose of this chapter is to summarize information on the diversity and probable identities of microorganisms that have been cultured from terrestrial subsurface environments. It is not meant to be a comprehensive review of this topic; rather, it focuses on what has been learned about the identities of microorganisms isolated in connection with the Deep Microbiology Program (DMP) within the U.S. Department of Energy (DOE) Subsurface Science Program (U.S. DOE, 1991, 1994). Over 10,000 strains of subsurface microorganisms (mostly bacteria) have been isolated since the

DMP was initiated in 1986 (Fliermans and Balkwill, 1989; USDOE, 1988). These organisms have come from geochemically and hydrologically diverse materials, including: saturated Atlantic coastal plain aquifers and confining zones in South Carolina (Ghiorse, 1989); unwelded volcanic tuff in Nevada (Amy et al., 1992; Haldeman and Amy, 1993a; Haldeman et al., 1993); unsaturated and saturated sediments and basalts in arid regions of Idaho and Washington (Brockman et al., 1992; Kieft et al., 1993); and saturated lacustrine sediments, fluvial sands and gravels, and paleosols in Washington (Fredrickson et al., 1995).

Shortly after the DMP was initiated, it was decided that microbial isolates obtained by the investigators in the program should be preserved in an organized culture collection so that they could be made available to interested researchers. As a result, DOE established the Subsurface Microbial Culture Collection (SMCC) (Balkwill, 1993) at Florida State University in 1987. Beginning in 1990, strictly anaerobic bacteria isolated in the program were preserved and distributed from a second SMCC facility located at the Oregon Graduate Institute of Science and Technology, but the perservation and distribution of aerobic and facultatively anaerobic bacteria continues at the Florida State University site.

Many of the 10,000+ strains in the SMCC are largely uncharacterized because there have not been sufficient time and resources to examine them closely. Nevertheless, several significant groups of aerobic isolates and a small number of the strict anaerobes have now been studied extensively. The results of these studies provide at least a preliminary picture of the types and variety of bacteria that can be cultured from terrestrial subsurface environments.

7.2 Aerobic Chemoheterotrophs from Atlantic Coastal Plain Sediments

7.2.1 Source of Isolates

Research in the DMP initially concentrated on the microbiology of saturated Atlantic coastal plain sediments at DOE's Savannah River Site (SRS) in Aiken, South Carolina. Samples of aquifer sands and confining-zone clays were obtained by drilling a series of boreholes, several kilometers apart, to a maximum depth of approximately 500 m below land surface (Fliermans and Balkwill, 1989; Sargent and Fliermans, 1989). The results of chemical and microbiological analyses of samples from three of these boreholes were described in detail in a special issue of the *Geomicrobiology Journal* (Ghiorse, 1989). Briefly, sizable (10^6 to 10^7 cells per gram) and diverse microbial populations were detected in the aquifer samples. Among the bacterial forms detected were aerobic and facultatively anaerobic chemoheterotrophs (Balkwill, 1989; Balkwill et al., 1989; Sinclair and Ghiorse, 1989); denitrifiers (Francis et al., 1989); methanogens and sulfate reducers (Jones et al., 1989; Madsen and Bollag, 1989); sulfur-oxidizing, nitrifying, and nitrogen-fixing bacteria (Fredrickson et al., 1989); streptomycetes and other spore formers (Balkwill, 1989); and even some unicellular cyanobacteria (Sinclair and Ghiorse, 1989). Fungi and protozoa (amoebae and flagellates, but no ciliates) were also detected in many samples (Sinclair and Ghiorse, 1989).

Despite the overall diversity of the microbial populations in the SRS aquifer samples, the most abundant culturable forms in these samples (95 to 99% of the organisms detected by plating, MPN enrichments, etc.) were the aerobic (or facultative anaerobic) chemoheterotrophic bacteria. For the most part, it was only these bacteria that were eventually isolated and preserved in the SMCC. Interestingly, these easily cultured chemoheterotrophs appeared to be representative of the numerically predominant total populations in the samples; comparison of direct and viable counts indicated that 50 to 95% of bacteria seen microscopically were recovered by aerobic plating on complex media. (This admittedly unusual phenomenon has not been seen in the case of other subsurface environments examined in the DMP.)

7.2.2 Morphological and Physiological Diversity

Although most of the isolates from the SRS were aerobic chemoheterotrophs, analysis of colony morphological traits implied that there was a considerable amount of diversity within this group of organisms. Between 11 and 62 distinct colony types (average = 24) were isolated from each aquifer sample (Balkwill, 1989). Moreover, different samples often yielded morphologically distinct types of isolates. Diversity among the aerobic chemoheterotrophs was further assessed by examination of selected physiological traits. The API NFT testing system (BioMérieux-Vitek, Inc., Hazelwood, Missouri), which tests for 9 specific enzymatic capabilities and the ability to aerobically utilize 12 different compounds as sole sources of carbon, was selected for this purpose because it was the most effective method available at the time. (A relatively simple and rapid approach was needed because more than 5000 strains were to be examined in a short period of time.) The isolates again appeared to be quite diverse in terms of the number of different ways in which they responded to the 21 tests in the API system. For example, one group of 1112 strains produced no fewer than 626 distinct response patterns (Balkwill et al., 1989). Moreover, the isolates from different depths within a single borehole or from the same geological formation in different boreholes also tended (as groups) to yield distinct API response patterns, implying that largely distinct groups of strains were isolated from different samples (Balkwill et al., 1989; Fredrickson et al., 1991).

Morphological and physiological data implied that the SRS isolates were quite diverse, but the true extent and nature of the diversity was unclear because the differences between strains could not be interpreted in the context of traditional taxonomic units (e.g., strain-level vs. species-level vs. genus-level differences). Not surprisingly, the API NFT kits, which were designed for identification of nonfermentative, Gram-negative bacteria of clinical importance, identified only a very small percentage of the subsurface isolates. Later attempts to identify the isolates with the Biolog GN system (Biolog, Inc., Hayward, California) were also largely unsuccessful (most tests resulting in no identification or matches of very low certainty). Amy et al. (1992) reported similar results with a group of aerobic isolates from volcanic tuffs in the Ranier Mesa at DOE's Nevada Test Site. Moreover, the API NFT and Biolog systems seldom assigned the same identity to the relatively small percentage of isolates that were identified in that study.

7.2.3 Phylogenetic and Taxonomic Diversity

To better assess the diversity and possible identity of the aerobic chemoheterotrophs from the SRS site, 16S ribosomal RNA gene nucleotide base sequences were determined and analyzed to learn how the isolates were related phylogenetically to one another and to previously described bacterial taxa (Reeves et al., 1995). The procedures

TABLE 7.1

Tentative Taxonomic Assignments of Aerobic
Chemoheterotrophic Bacteria from the SRS Based on
Phylogenetic Analysis of 16S Ribosomal RNA Gene Sequences

Genus or Other Taxonomic Unit	No. of Strains Assigned to Date
Alpha-*Proteobacteria:*	
Agrobacterium	1
Blastobacter	1
Sphingomonas	1
Uncertain affiliation[a]	7
Beta-*Proteobacteria:*	
Alcaligenes	6
Comamonas	25
Zoogloea	1
Uncertain affiliation[a]	2
Gamma-*Proteobacteria:*	
Acinetobacter	25
Pseudomonas	10
Uncertain affiliation[a]	1
High-G+C Gram-positive bacteria:	
Arthrobacter	25
Micrococcus	3
Terrebacter	6

[a] Isolate falls within the indicated larger taxonomic unit but cannot
be assigned to a specific genus at this time.

for sequencing and phylogenetic analysis of sequences, as well as the kinds of information that can be obtained with this approach, are detailed in Chapter 10.

To date, more than 100 isolates from the SRS have been sequenced and analyzed. These strains could be assigned to four major taxonomic groups (the high-G+C Gram-positive bacteria and the alpha-, beta-, and gamma-*Proteobacteria*) and at least 11 genera (Table 7.1). The majority of the strains fell into just three genera (*Acinetobacter, Arthrobacter*, and *Comamonas*), thereby indicating that there was less diversity among the isolates than might be inferred from the API physiological test data (Balkwill et al., 1989; Fredrickson et al., 1991). In this regard, Reeves et al. (1995) reported that groups of SRS strains having slightly different API test response patterns sometimes turned out be closely related phylogenetically (clearly members of a single genus and, possibly, members of a single species) when compared by analysis of 16S rRNA gene sequences. It should be realized, however, that only a very small proportion (approximately 2%) of the SRS isolates have been sequenced to date. A wider range of genera may be detected as additional strains are characterized, even though most of the strains are likely to fall into a limited number of taxa.

Most of the aerobic chemoheterotrophs from the SRS examined thus far appear to belong to a limited number of genera, but within those groups of isolates there may be an appreciable amount of diversity at species level. For example, Figure 7.1 shows a phylogenetic tree that was generated by analyzing 16S rRNA gene sequences for 10 SRS *Arthrobacter*-like isolates and various comparison strains (including most previously described species of *Arthrobacter*). The subsurface isolates form at least four distinct clusters (labeled A, B, C, and D in Figure 7.1), each of which probably represents a distinct species. Three of the clusters may represent new species of *Arthrobacter*, as they are well separated from all of the established species included in the

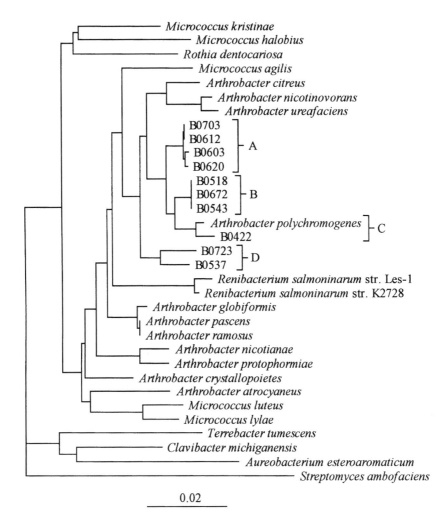

FIGURE 7.1
Phylogenetic tree generated by analysis of 16S rRNA gene sequences for 10 strains of *Arthrobacter*-like subsurface isolates from the SRS and selected previously described species (for comparison). Sequences for previously described species were obtained from the Ribosomal Data Project (Maidak et al., 1994) and GenBank databases. The phylogenetic analysis was performed using a distance matrix method (see Chapter 10). The scale bar represents 2 substitutions per 100 bases.

analysis for comparison. Similar results have been obtained for a group of *Sphingomonas* SRS isolates that degrade aromatic compounds (Fredrickson et al., 1995). In this case, analyses of 16S rRNA gene sequences, cell lipid compositions, and DNA-DNA reassociation values indicated that five isolates could be classified as three new species of *Sphingomonas* (Balkwill et al., 1996).

Phylogenetic analysis of the *Arthrobacter* isolates (Figure 7.1) indicated that strains B518, B543, and B672 (cluster B) had virtually identical 16S rRNA gene sequences. However, there were minor distinctions between the strains in clusters A and D that might represent strain-level differences. Strain-level differences are better detected by methods like restriction endonuclease analysis (REA) (Reeves et al., 1995; Chapter 10) or BOX-PCR fingerprinting (Louws et al., 1994; Martin et al., 1992). Preliminary results from REA (Reeves, 1996) and BOX-PCR (Balkwill and Crocker, 1996) analyses of large numbers of SRS *Arthrobacter* strains indicate that only a small proportion of the isolates are likely to be identical at the strain level. However, much additional research will be required to fully assess strain-level diversity among aerobic chemoheterotrophs from the SRS.

7.3   Aerobic Chemoheterotrophs from Volcanic Tuff

7.3.1   Source of Isolates

D. L. Haldeman and P. S. Amy at the University of Nevada-Las Vegas, in collaboration with D. B. Ringelberg and D. C. White at the University of Tennessee, recently performed an elegant and detailed study of the culturable aerobic chemoheterotrophs living in samples of volcanic tuff (consolidated volcanic fly ash) from the Rainier Mesa tunnel system at DOE's Nevada Test Site (Haldeman et al., 1993). The tuff samples were taken from selected locations within a 21-m^3 section of rock that was located within a perched water zone approximately 400 m beneath the caprock of the mesa. The main purpose of this study was to define the distribution of the subsurface microbiota on a small, three-dimensional scale and to determine how related groups of culturable aerobic heterotrophs were distributed throughout the rock section. However, the study also provided a considerable amount of information on the diversity and possible identities of the culturable subsurface microorganisms at this study site.

7.3.2   Taxonomic Diversity Based on Analysis of Fatty Acids

A total of 210 bacterial strains were isolated from 18 rock samples (one example of each morphologically distinct colony type from each sample; see Haldeman and Amy, 1993a, 1993b) in the aforementioned Rainier Mesa tunnel system study. Each of these isolates was analyzed for fatty acid methyl esters (FAME) using the Microbial Identification System (MIDI; see Chapter 8). There were 28 strains nonculturable under the growth conditions required for a cluster analysis, but the remaining isolates formed 29 clusters, corresponding approximately to groups of organisms related at the genus level; 11 of these clusters could not be related to any of the established genera included in the MIDI database. The remaining clusters (Table 7.2) were related to at least 11 genera in 4 major taxonomic groups (high- and low-G+C Gram-negative bacteria and alpha- and gamma-*Proteobacteria*). There were some interesting similarities between these results and those obtained thus far for SRS isolates (above), in that 3 genera accounted for at least 80 of the 182 Rainier Mesa isolates that could be analyzed and 2 of these genera (namely *Acinetobacter* and *Arthrobacter*) also accounted for relatively large proportions of the SRS isolates. However, there were also some notable differences between the two sites. The most abundant genus among the Rainier Mesa isolates *(Gordona)* has yet to be detected among the SRS isolates, and one of the most abundant groups of SRS strains *(Comamonas)* was not detected among the Rainier Mesa isolates.

Haldeman et al. (1993) also used the MIDI cluster analysis to separate the Rainier Mesa isolates into clusters corresponding approximately to groups of strains related at the species level. As was the case at the SRS (above), this analysis provided evidence that there may be significant diversity among groups of isolates related to a single genus at the species level. For example, the isolates that were related to the genus

TABLE 7.2

Genus-Level Taxonomic Assignments of Aerobic
Chemoheterotrophic Bacteria from Rainier Mesa (Nevada
Test Site, U.S.) Samples Based on MIDI Analysis[a]

Genus or Other Taxonomic Unit	No. of Strains Assigned
Alpha-*Proteobacteria:*	
Methylobacterium	2
Gamma-*Proteobacteria:*	
Acinetobacter	14
Xanthomonas	2
High-G+C Gram-positive bacteria:	
Arthrobacter	24
Corynebacterium	4
Gordona	36
Micrococcus	8
Nocardoides	4
Low-G+C Gram-positive bacteria:	
Bacillus	2
Staphylococcus	1
Other categories[b]	
Acidovorax	1
Mixed Gram-positives (*Corynebacterium, Aureobacterium, Clavibacter, Curobacterium*)	6
Mixed Gram-positives (*Arthrobacter, Micrococcus, Staphylococcus, Aureobacterium*)	29
Mixed Gram-negatives (*Acidovorax, Hydrogenophaga, Pseudomonas*)	16

[a] Data summarized from Haldeman et al. (1993).

[b] Clusters of isolates containing more than one genus or assigned to
    genera that have not yet been placed in one of the larger taxonomic
    categories according to 16S rRNA gene sequence analysis.

Arthrobacter could be related to at least three species in that genus, and the isolates
related to *Acinetobacter* could be related to at least two species.

7.4   Aerobic Chemoheterotrophs from "GEMHEX" Samples

7.4.1   Source of Isolates

Another group of reasonably well-characterized DMP isolates was obtained from the
"GEMHEX" (**GE**ological **M**icrobiological **H**ydrological **EX**periment) interval in a
borehole drilled at DOE's Hanford site (near Richland, Washington) in 1992 (Fredrickson et al., 1995). Samples were taken from a series of saturated-zone sediments
(including paleosols, lacustrine sediments, and fluvial sands and gravels) ranging in
depth from 173 to 218 m below land surface. The sample interval was selected to permit the comparison of microbial populations in physically and chemically distinct
types of subsurface materials. In contrast to what was seen with the aquifer sand samples from the SRS, only a very small percentage of the microorganisms detected in the
"GEMHEX" samples by direct microscopic counts could be cultured. Aerobic (or facultatively anaerobic) chemoheterotrophic bacteria were, once again, the numerically

predominant cultured forms. However, the total number of isolates from all samples was comparatively low, so it has been possible to characterize almost all of them by phylogenetic analysis of their 16S rRNA gene sequences. The results of this analysis, then, can provide a reasonably complete picture of the probable identities and diversity of the aerobic chemoheterotrophs cultured from the "GEMHEX" samples.

7.4.2   Phylogenetic and Taxonomic Diversity

Tentative taxonomic assignments of the GEMHEX strains at the genus level (based on analysis of 16S rRNA gene sequences) are summarized in Table 7.3. The isolates could be placed in 6 major taxonomic groups and at least 25 different genera. Many of the Gram-positive strains could be assigned to the genus *Arthrobacter*, as was the case with isolates from the SRS and the Rainier Mesa samples (see above). However, the next most frequently represented Gram-positive genera were *Bacillus* and *Staphylococcus*. These genera have not been detected (so far) among the SRS isolates, and they accounted for only a very small proportion (one strain each) of the Rainier Mesa isolates.

The Gram-negative bacteria isolated from the SRS consisted mostly of *Comamonas* and *Acinetobacter* species (Table 7.1), and the largest group of Gram-negative strains from the Rainier Mesa samples also consisted of *Acinetobacter* species (Table 7.2). In contrast, no genera were overly predominant among the Gram-negative strains isolated from the "GEMHEX" samples, and only one of those strains was assigned to the genus *Acinetobacter*. Both the Gram-positive and the Gram-negative bacteria from the "GEMHEX" samples were considerably more diverse at the genus level than were those from the SRS (compare Tables 7.1 and 7.3), but some of this difference could arise from the fact that a much greater percentage of the "GEMHEX" isolates have been characterized.

Research is currently in progress to assess the diversity of the aerobic chemoheterotrophs from the "GEMHEX" samples at the species and strain levels. Preliminary results from detailed analyses of 16S rRNA gene sequences indicate that the relatively large group of *Arthrobacter*-like isolates may include as many as seven different species (Balkwill and Crocker, 1996). As was the case with the SRS isolates, only a small percentage of these strains appear to be closely related to previously described *Arthrobacter* species, and most of them are likely to differ at the strain level.

7.5   Strict Anaerobes from the Deep Subsurface

7.5.1   General Information and Isolation Strategy

As noted earlier, DOE established a special unit of the SMCC in 1990, as the DMP began to examine subsurface environments in which anaerobes were likely to be the predominant microbial forms. To date, approximately 80 strains of strict anaerobes have been isolated from subsurface materials and preserved in the SMCC/W. An intensive effort is currently underway to characterize these isolates by analysis of physiological, phylogenetic, and other traits.

The microbial breakdown of organic compounds in anoxic environments usually depends on collaborations between two or more species, in which H_2 or formate serves as an extracellular intermediate in the process. The fate of the H_2 or formate is

TABLE 7.3

Tentative Taxonomic Assignments of Aerobic
Chemoheterotrophic Bacteria from the "GEMHEX"
Interval at the Hanford Site Based on Phylogenetic
Analysis of 16S Ribosomal RNA Gene Sequences

Genus or Other Taxonomic Unit	No. of Strains Assigned to Date
Alpha-*Proteobacteria:*	
Blastobacter	1
Caulobacter	5
Erythromicrobium	2
Rhodobacter	1
Sphingomonas	5
Uncertain affiliation[a]	4
Beta-*Proteobacteria:*	
Comamonas	7
Leptothrix	1
Rhodocyclus	1
Telluria	3
Variovorax	6
Zoogloea	1
Uncertain affiliation[a]	2
Gamma-*Proteobacteria:*	
Acinetobacter	1
Pseudomonas	4
Flexibacter-Cytophaga-Bacteroides Phylum:	
Uncertain affiliation[a]	1
High-G+C Gram-positive bacteria:	
Arthrobacter	40
Clavibacter	2
Frankia	1
Gordona	2
Micrococcus	6
Mycobacterium	2
Nocardioides	1
Rhodococcus	2
Rothia	3
Streptomyces	4
Uncertain affiliation[a]	9
Low-G+C Gram-positive bacteria:	
Bacillus	21
Staphylococcus	15
Uncertain affiliation[a]	7

[a] Isolate falls within the indicated larger taxonomic unit but
cannot be assigned to a specific genus at this time.

oxidation (i.e., to H^+ or bicarbonate, respectively), which is linked to the reduction of a terminal electron acceptor in the environment. The most energetically favorable electron acceptor available is generally the one that is used in this situation, commonly nitrate, nitrite, manganese (IV), iron (III), sulfate, sulfur, or CO_2 (in that order). The strategy for investigating the anaerobic microflora of deep subsurface environments, then, has been to enumerate populations of terminal bacteria in the anaerobic food web that use H_2 or formate (as well as other fermentation acids such as lactate) and the various available electron acceptors. Habitat-simulating media containing the appropriate electron acceptor were developed to perform most-probable-number (MPN) analyses of anaerobic microbial populations of various physiological groups, including bacteria that reduce nitrate, iron, manganese, sulfate, sulfur, and CO_2. The highest dilutions from the MPN analyses in which growth appeared served as enrichment cultures that, in turn, could be utilized as the inocula for isolation of the numerically predominant anaerobes from various subsurface materials. A special effort was directed toward the isolation and characterization of bacteria from enrichment cultures that were likely to contain novel organisms, such as iron-reducing bacteria.

7.5.2   Examples of Characterized Subsurface Anaerobes

The first group of subsurface anaerobes to be characterized thoroughly was a group of six thermophilic, iron-reducing strains that are the sole known members of a new species of bacteria, *Bacillus infernus* (Boone et al., 1995). This bacterium not only represents a new species, but as a strict anaerobe assigned to a genus defined as aerobic, has led to an emendation of the definition of the genus *Bacillus*. *Bacillus infernus* appears to be the major iron-reducing bacterium in deep terrestrial environments like those sampled in the Taylorsville Triassic Basin in Virginia, at depths from 2.65 to 2.77 km below land surface (Balkwill et al., 1994). The range of reduced substrates that this bacterium can utilize appears to be limited, so *Bacillus infernus* appears to exist as part of a catabolic consortium including other bacteria that degrade more complex organic molecules and supply *B. infernus* with H_2 or the simple reduced organic compounds it needs for growth.

A second group of anaerobic bacteria from the Taylorsville Triassic Basin also appears to belong in a previously undescribed species. These are thermophilic, sulfate-reducing bacteria with similarities to members of the genus *Desulfotomaculum*. However, the deep subsurface isolates differ phylogenetically from known species, their closest relative, *Desulfotomaculum nigrificans*, having a 16S rRNA gene sequence similarity of only 94.6% (certainly too dissimilar to classify the subsurface isolates as *D. nigrificans*). The deep subsurface strains also differ from established species of *Desulfotomaculum* in some of their physiological characteristics, including the ability to use methanol as a reduced catabolic substrate and the ability to grow over a wide pH range (from pH 6.0 to 7.8).

Other metal-ion reducing subsurface bacteria are still being characterized. These include a thermophilic manganese-reducing isolate from the Taylorsville Triassic Basin site that grows very slowly (not yet well characterized) and a group of mesophilic iron-reducing bacteria. Analyses of 16S rRNA gene sequences indicate that the iron-reducers are closely related (phylogenetically) to the iron-reducing species *Shewanella alga* and *S. putrefaciens*. It is not yet clear whether the subsurface strains belong to one of these previously described species or represent a new species of *Shewanella* (Boone and Balkwill, unpublished data).

7.6   Conclusions

Research on the identity and diversity of culturable subsurface microorganisms has shown that a substantial variety of bacteria (representing at least six major taxonomic categories) can be isolated from deep subsurface environments. Among the most frequently isolated Gram-positive genera are *Bacillus*, *Gordona*, *Staphylococcus*, *Micrococcus*, and *Arthrobacter* (which might occur in nearly all subsurface environments that are not fully anoxic). Frequently isolated Gram-negative genera include *Comamonas*,

Acinetobacter, and *Pseudomonas.* Nevertheless, there is still much to be learned about culturable microorganisms from the deep subsurface. Many of the strains in the SMCC remain uncharacterized, so very little is known about the actual identities of the bacteria isolated from the some of the sites examined in the DMP. In addition, information on species- and strain-level diversity (even among relatively well-characterized groups of isolates) is just beginning to appear. It should also be realized that, with the exception of the aquifer samples from the SRS, the great majority of the microorganisms in the subsurface materials examined to date have never been cultured. Very little is known about the possible identities or diversity of these noncultured populations.

Although much remains to be learned about the diversity and identity of microorganisms cultured from terrestrial subsurface environments, an important trend has been seen in the results of the research completed thus far. Most of the subsurface bacteria that have been thoroughly characterized can be assigned to previously described genera, but the majority of them appear to be novel species within those genera. As efforts to characterize subsurface microbial isolates continue, then, they can be expected to provide new insight into the total diversity of life on the Earth.

Acknowledgments

We thank Robert H. Reeves, Jane Y. Reeves, Gwendolyn R. Drake, and Fiona H. Crocker for providing some of the 16S rRNA gene sequence data and analytical findings that were used to construct Figure 7.1 and Tables 7.1 and 7.3. The research described herein was supported by the U.S. Department of Energy Subsurface Science Program (Office of Energy Research, Office of Health and Environmental Research).

References

Amy, P. S., Haldeman, D. L., Ringelberg, D., Hall, D. H., and Russell, C. 1992. Comparison of identification systems for classification of bacteria isolated from water and endolithic habitats within the deep subsurface. *Appl. Environ. Microbiol.,* 58:3367-3373.

Balkwill, D. L. 1989. Numbers, diversity, and morphological characteristics of aerobic, chemoheterotrophic bacteria in deep subsurface sediments from a site in South Carolina. *Geomicrobiol. J.,* 7:33-52.

Balkwill, D. L. 1993. DOE makes subsurface cultures available. *ASM News,* 59:504-506.

Balkwill, D. L., and Crocker, F. H. 1996. Unpublished data.

Balkwill, D. L., Boone, D. R., Colwell, F. S., Griffin, T., Kieft, T. L., Lehman, R. M., McKinley, J. P., Nierzwicki-Bauer, S., Onstott, T. C., Tseng, H. V., Gao, G., Phelps, T. J., Ringelberg, D., Russell, B., Stevens, T. O., White, D. C., and Wobber, F. J. 1994. DOE seeks origin of deep subsurface bacteria. *Eos,* 75:385, 395-396.

Balkwill, D. L., Drake, G. R., Reeves, R. H., Fredrickson, J. K., White, D. C., Ringelberg, D. B., Chandler, D. P., Romine, M. F., Kennedy, D., and Spadoni, C. M. 1996. Taxonomic study of aromatic-degrading bacteria from deep saturated subsurface sediments and description of *Sphingomonas aromaticivorans* sp. nov., *Sphingomonas subterrae* sp. nov., and *Sphingomonas stygialis* sp. nov. (submitted).

Balkwill, D. L., Fredrickson, J. K., and Thomas, J. M. 1989. Vertical and horizontal variations in the physiological diversity of the aerobic chemoheterotrophic bacteria microflora in deep Southeast Coast Plain subsurface sediments. *Appl. Environ. Microbiol.,* 55:1058-1065.

Boone, D. R., Liu, Y., Zhao, Z.-J., Balkwill, D. L., Drake, G. R., Stevens, T. O., and Aldrich, H. C. 1995. *Bacillus infernus* sp. nov., an Fe(III)- and Mn(IV)-reducing anaerobe from the deep terrestrial subsurface. *Int. J. Syst. Bacteriol.,* 45:441-448.

Brockman, F. J., Kieft, T. L., Fredrickson, J. K., Bjornstad, B. N., Li, S. W., Spangenberg, W., and Long, P. E. 1992. Microbiology of vadose zone paleosols in south-central Washington state. *Microb. Ecol.,* 23:279-301.

Fliermans, C. B. and Balkwill, D. L. 1989. Microbial life in deep terrestrial subsurface. *Bio-Science,* 39:370-377.

Francis, A. J., Slater, J. M., and Dodge, C. J. 1989. Denitrification in deep subsurface sediments. *Geomicrobiol. J.,* 7:103-116.

Fredrickson, J. K., Balkwill, D. L., Drake, G. R., Romine, M. F., Ringelberg, D. B., and White, D. C. 1995. Aromatic-degrading *Sphingomonas* isolates from the deep subsurface. *Appl. Environ. Microbiol.,* 61:1917-1922.

Fredrickson, J. K., Balkwill, D. L., Zachara, J. M., Li, S. W., Brockman, F. J., and Simmons, M. A. 1991. Physiological diversity and distributions of heterotrophic bacteria in deep cretaceous sediments of the Atlantic coastal plain. *Appl. Environ. Microbiol.,* 57:402-411.

Fredrickson, J. K., Garland, T. R., Hicks, R. J., Thomas, J. M., Li, S. W., and McFadden, K. M. 1989. Lithotrophic and heterotrophic bacteria is deep subsurface sediments and their relation to sediment properties. *Geomicrobiol. J.,* 7:53-66.

Fredrickson, J. K., McKinley, J. P., Nierzwicki-Bauer, S. A., White, D. C., Ringelberg, D. B., Rawson, S. A., Li, S. W., Brockman, F. J., and Bjornstad, B. N. 1995. Microbial community structure and biogeochemistry of Miocene subsurface sediment: implications for long-term microbial survival. *Molec. Ecol.,* 4:619-626.

Ghiorse, W. C. (ed.) 1989. Special issue on deep subsurface microbiology. *Geomicrobiol. J.,* 7:1-135.

Haldeman, D. L. and Amy, P. S. 1993a. Bacterial heterogeneity in deep subsurface tunnels at Rainier Mesa, Nevada Test Site. *Microb. Ecol.,* 25:183-194.

Haldeman, D. L. and Amy, P. S. 1993b. Diversity within a colony morphotype: implications for ecological research. *Appl. Environ. Microbiol.,* 59:933-935.

Haldeman, D. L., Amy, P. S., Ringelberg, D., and White, D. C. 1993. Characterization of the microbiology within a 21 m³ section of rock from the deep subsurface. *Microb. Ecol.,* 26:145-159.

Jones, R. E., Beeman, R. E., and Suflita, J. M. 1989. Anaerobic metabolic processes in the deep terrestrial subsurface. *Geomicrobiol. J.,* 7:117-130.

Kieft, T. L., Amy, P. S., Brockman, F. J., Fredrickson, J. K., Bjornstad, B. N., and Rosacker, L. L. 1993. Microbial abundance and activities in relation to water potential in the vadose zones of arid and semiarid sites. *Microb. Ecol.,* 26:59-78.

Louws, F. J., Fulbright, D. W., Stephens, C. T., and deBruijn, F. J. 1994. Specific genomic fingerprints of phytopathogenic *Xanthomonas* and *Pseudomonas* pathovars and strains generated with repetitive sequences and PCR. *Appl. Environ. Microbiol.,* 60:2286-2295.

Madsen, E. L. and Bollag, J. M. 1989. Aerobic and anaerobic microbial activity in deep subsurface sediments from the Savannah River Plant. *Geomicrobiol. J.,* 7:93-101.

Maidak, B. L., Larsen, N., McCaughey, M. J., Overbeek, R., Olsen, G. J., Fogel, K., Blandy, J., and Woese, C. R. 1994. The ribosomal database project. *Nucleic Acids Res.,* 22:3485-3487.

Martin, B., Humbert, O., Camara, M., Guenzi, E., Walker, J., Mitchell, T., Andrew, P., Prud-homme, M., Alloing, G., Hakenbeck, R., Morrison, D. A., Boulnois, G. J., and Claverys, J. 1992. A highly conserved repeated DNA element located in the chromosome of *Streptococcus pneumoniae. Nucleic Acids Res.,* 20:3479-3483.

Reeves, R. H. 1996. Personal communication.

Reeves, R. H., Reeves, J. Y., and Balkwill, D. L. 1995. Strategies for phylogenetic characterization of subsurface bacteria. *J. Microbiol. Methods,* 21:235-251.

Sargent, K. A., and Fliermans, C. B. 1989. Geology and hydrology of the deep subsurface microbiology sampling sites at the Savannah River Plant, South Carolina. *Geomicrobiol. J.,* 7:3-13.

Sinclair, J. L., and Ghiorse, W. C. 1989. Distribution of aerobic bacteria, protozoa, algae, and fungi in deep subsurface sediments. *Geomicrobiol. J.,* 7:15-31.

U.S. DOE. 1988. Microbiology of Subsurface Environments: Program Plan and Summary. DOE/ER-0293. Office of Health and Environmental Research, Office of Energy Research, U.S. Department of Energy, Washington, D.C.

U.S. DOE. 1991. Subsurface Science Program: Program Overview and Research Abstracts, FY 1989 — FY 1990. DOE/ER-0432. Office of Health and Environmental Research, Office of Energy Research, U.S. Department of Energy, Washington, D.C.

U.S. DOE. 1994. Subsurface Science Program: Program Overview. DOE/ER-0640. Office of Health and Environmental Research, Office of Energy Research, U.S. Department of Energy, Washington, D.C.

8

Utility of the Signature Lipid Biomarker Analysis in Determining the In Situ Viable Biomass, Community Structure, and Nutritional/Physiologic Status of Deep Subsurface Microbiota

D. C. White and D. B. Ringelberg

CONTENTS

KEY WORDS: *signature lipid biomarkers, viable biomass, community composition, physiological status,* in situ *analysis, nonculturable microbes, detection of viable biomass, phospholipid ester-linked fatty acids, poly beta-hydroxy alkanoate, stress biomarkers, unbalanced growth, toxicity response.*

8.1   Introduction

The classical microbiological approach which was so successful in public health for the isolation and culture of pathogenic species is clearly less than satisfactory for subsurface sediment samples. It has been repeatedly documented in the literature that viable or direct counts of bacteria from various environmental samples may represent only 0.1 to 10% of the extant community (White et al., 1993; Tunlid and White, 1991; White, 1983, 1986, 1988). Moreover, microbes may still be metabolically active and potentially infectious even though they are not culturable. Classical microbial tests are time-consuming and provide neither an indication of the nutritional status nor evidence of toxicity within the extant community.

The signature lipid biomarker (SLB) analysis does not depend on growth or morphology for identification. Instead, microbial biomass and community structure are determined in terms of universally distributed lipid biomarkers which are characteristic of all cells. Lipids, recoverable by extraction in organic solvents, are an essential component of the membranes of all viable cells. SLB have been shown to be readily extractable from most environments, including the deep subsurface. The extraction process provides for both purification and concentration of "signature" lipid biomarkers from the cell membranes and walls of microorganisms (Guckert et al., 1985).

8.1.1   Viable Biomass

A determination of the total ester-linked phospholipid fatty acid (PLFA) content provides a quantitative measure of the viable or potentially viable biomass. A viable organism will have an intact membrane containing PLFA. Upon cell death or a cell lysis, cellular enzymes hydrolyze phospholipids releasing the polar head groups. The hydrolysis can occur within minutes to hours of cell lysis (White et al., 1979). The lipid moiety remaining which is called a diglyceride, contains the same signature fatty acids as the phospholipids (Figure 8.1). An estimation of the total nonviable and the total viable biomass can be made by measuring diglyceride fatty acids and phospholipid fatty acids, respectively. A careful study of subsurface sediment showed that the viable biomass, as determined by PLFA, was equivalent (with a smaller standard deviation) to estimations based on intercellular ATP, cell wall muramic acid, and carefully performed acridine orange direct counts (Balkwill et al., 1988). Since phospholipids are found in reasonably constant amounts in all cellular membranes in a variety of bacterial cells, have a high natural turnover rate, and are rapidly degraded in nonviable cells, the measurement of PLFA provides an accurate estimation of the viable or potentially viable microbial biomass.

8.1.2   Community Structure

The analysis of SLB by capillary gas chromatography/mass spectrometry (GC/MS) provides sufficient information for the identification of specific subsets of the microbial community. Specific groups of microbes often contain characteristic lipids, in particular, fatty acids (White et al., 1993; Tunlid and White, 1991; White, 1983, 1986, 1988). For example, PLFA prominent in the hydrogenase-containing *Desulfovibrio* sulfate-reducing bacteria are distinctly different from those found in the *Desulfobacter* sulfate-reducing bacteria (Edlund et al., 1985; Dowling et al., 1986). The SLB analysis is

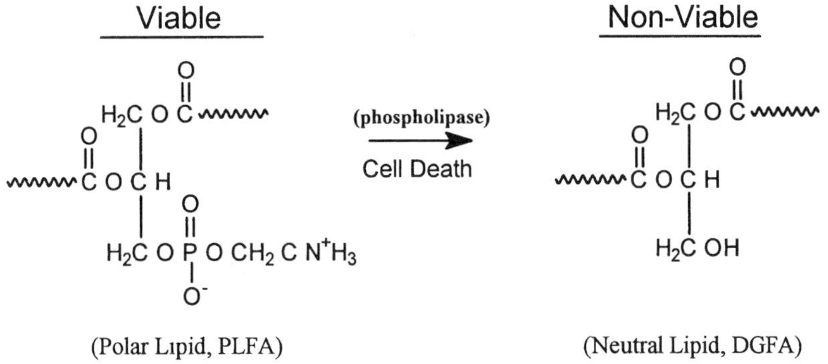

FIGURE 8.1
Illustration showing the conversion of a phospholipid (PL) to a diglyceride (DG) as a result of cell death.

unable to identify every species of microorganism in an environmental sample as many species contain overlapping PLFA patterns. However, many species or physiologically similar groups of microorganisms are readily detected in environmental samples. Species such as *Francisella tualrensis*, *Vibrio cholera*, *Syntrophomonas wolfei*, *Flavobacterium* spp., members of the genera *Vitreoscilla*, *Flexibacter*, *Filibacter*, *Thiobacilli*, *Frankia*, and groups of organisms such as the Archae, actinomycetes, acetogens, and type I and type II methanotrophs have all been detected in complex environmental matrices by analyses of the respective PLFA patterns (Nichols et al., 1985, 1986, 1987; Guckert et al., 1986; Kerger et al., 1986, 1987; Ringelberg et al., 1988; Phelps et al., 1991; Coleman et al., 1993; Tunlid et al., 1989).

Distinctive patterns of fatty acids (largely from phospholipids and lipopolysaccharides) released from cultured bacteria are currently used for pure culture identifications or to group them by fatty acid relatedness. Descriptive patterns of prominent ester-linked fatty acids recovered from isolated microbes grown on standardized media have been developed for over 650 species (Microbial Identification System, MIDI, Newark, DE) (Welch, 1991). Patterns of PLFA can also be used in identifying single species. For example, a comparison of PLFA patterns of 18 species of methane-oxidizing bacteria and 21 species of Gram-negative sulfate-reducing bacteria by hierarchical cluster analyses paralleled the relationships obtained from the analysis of 16S rRNA sequence homologies (Guckert et al., 1991; Kohring et al., 1994). Since identifications are based on the isolation and culture of microorganisms, unculturable microbes go undetected.

The analysis of other lipids such as sterols for microeukaryotes (nematodes, algae, protozoa) (White et al., 1980), glycolipids for Gram-positive bacteria and phototrophs, and hydroxy fatty acids from the Lipid A portion of a lipopolysaccharide (LPS-OHFA) for Gram-negative bacteria (Parker et al., 1982) can provide a more detailed analysis of the microbial community structure (Figure 8.2). Species of the genera *Planctomyces*, *Pirella*, *Geobacter*, *Desulfomonile*, and *Legionella* have all been readily detected and differentiated by analysis of their LPS-OHFA (Kerter et al., 1988; Lovely et al., 1992; Ringelberg et al., 1993; Walker et al., 1993). In a specific application, LPS-OHFA was utilized to differentiate between contamination of sediments by an enteric bacterium from that by a pseudomonad (Parker et al., 1982).

The SLB analysis provides a pattern of PLFA which characterizes the entire microbial community. By quantifying differences between PLFA patterns obtained from environmental samples, inferences can be made regarding shifts in community composition. Multivariant statistical analyses, such as hierarchical cluster and principal components analysis, have proven to be valuable tools in identifying these differences.

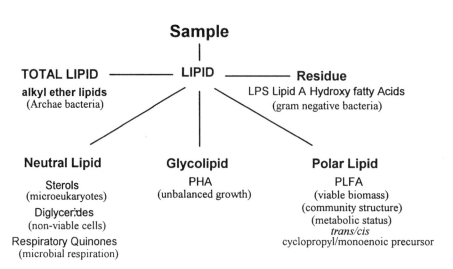

FIGURE 8.2
Schematic representation of the signature lipid biomarker analysis.

Variance among a specific PLFA or group of PLFA within the sample set can then be interpreted based on what is currently known regarding fatty acid biosynthesis and bacterial membrane PLFA compositions.

8.1.3   Nutritional/Physiological Status

In bacteria, specific patterns of PLFA can indicate physiological stress. Exposures to toxic environments can lead to minicell formation and a relative increase in *trans* monoenoic PLFA, when compared to the *cis* homologues. it has also been shown that for increasing concentrations of phenol toxicants, the bacteria *Pseudomonas putida* forms increasing proportions of *trans* PLFA (Heipieper et al., 1992). The ratio of *trans/cis* PLFA in toxic sediments fluctuates around 0.2, and in rapidly growing "healthy" bacteria it is around .01. Similarly, ratios of cyclopropane fatty acids to their monoenoic homologues (as well as *trans/cis* ratios) were shown to increase in response to a change in environmental conditions, i.e., starvation (Guckert et al., 1986; Kieft et al., 1994) (see Chapter 11).

In addition to PLFA, other lipid components provide insight into a microbial community's nutritional or physiological status. The accumulation of poly β-hydroxyalkanoic acids (PHA) in bacteria (Nickels et al., 1979) and triglycerides in microeukaryotes (Gehron and White, 1982), when compared to PLFA concentrations, are two such compounds. Bacteria exposed to adequate carbon and terminal electron acceptors form PHA when they are unable to divide. Under ambient environmental conditions, a cessation of cell division is generally due to the absence (or lack of) an essential nutrient or nutrients (phosphate, nitrate, trace metal, etc.).

The analysis of the respiratory quinone structure provides an estimation of the degree of aerobic microbial activity (Figure 8.3) (Hedrick and White, 1986). Environments with high potential terminal electronic acceptors (oxygen, nitrate) induce the formation of benzoquinones in bacteria whereas bacteria respiring on organic substrates form naphthoquinones. Fermentative organisms, such as many *Lactobacillus*

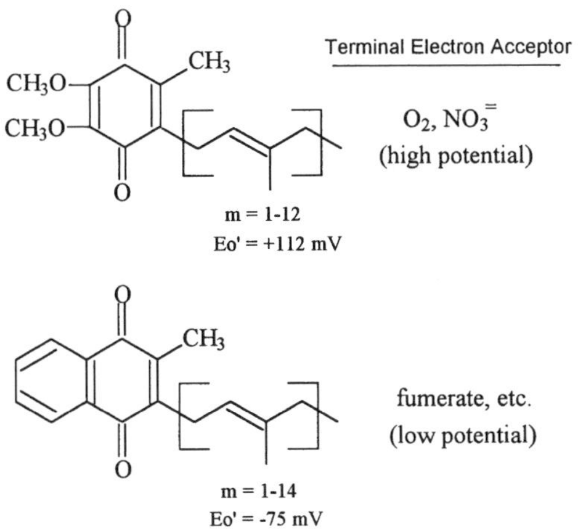

FIGURE 8.3
Illustration of bacterial respiratory quinone structure where high potential terminal electron acceptor refers to benzoquinones and Gram-negative bacteria, and low potential to menaquinones and both Gram-negative and Gram-positive bacteria.

sp., may actually form no respiratory quinones. By determining the quinone content within a given sample, proportions of aerobic, anaerobic respiratory, and/or anaerobic fermentative bacteria can be estimated. When compared to viable biomass estimates and community structure determinations the analysis of respiratory quinones, polyhydroxy alkanoates, and the ratios of specific PLFA (*trans/cis* and cyclopropyl to monoenoic precursor) provide valuable insights into the nutritional and physiological status of the extant microbiota.

8.2   Materials and Methods

8.2.1   Lipid Analyses

Frozen or lyophilized cells from subsurface isolates or sediments and/or filter retentates from ground waters can be quantitatively extracted for PLFA and other lipid components. Prior to extraction, sediments are homogenized under sterile conditions. Samples are saturated with a single-phase organic solvent system comprised of chloroform, methanol, and aqueous 50 mM PO_4 (pH 7.4) buffer (1:2:0.8, v:v:v) (White et al., 1979). After agitation and an extraction period of approximately 3 h in duration, equal volumes of chloroform and nanopure water are added to the extractant, resulting in a two-phase system with a final composition of 1:1:0.9, chloroform:methanol:water, v:v:v. In the case of the sediments, the samples are centrifuged (~650 × g for 30 min) and the supernatant decanted into a separatory funnel prior to the addition of chloroform and water. The chloroform is used in washing the sediment, followed by recompaction, prior to combination with the original extractant and water in the separatory funnel. A complete separation of phases usually occurs within a 12-h period. The lower organic (lipid-containing) phase is then collected and evaporated at 37°C either under a gentle stream of N_2 or by rotary evaporation. the upper phase and residue (lipopolysaccharide) are also collected for analysis of the LPS-OHFA as described by Parker et al. (1982). The organic (lipid-containing) phase is then fractionated on a silicic acid column into neutral lipids, glycolipids, and polar lipids (Tunlid et al., 1989). The polar lipid fraction, recovered in methanol, is subjected to a transesterification for the recovery of PLFA (Tunlid et al., 1989). PLFA are then further separated and quantified by GC/MS with selected ion recording using either positive ion chemical (Tunlid et al., 1989) or electron impact ionizations (Kieft et al., 1994). The neutral lipid fraction may be processed for the collection of diglyceride fatty acids (Kieft et al., 1994), triglycerides (Gehron and White, 1982), and sterols (Nichols et al., 1983). The glycolipid fraction may be processed for the collection of PHA (Findlay and White, 1983) and other complex carbohydrates such as those found in cell walls of certain Gram-positive bacteria.

8.2.1.1 Fatty Acid Nomenclature

Fatty acids are designated by the total number of carbon atoms followed by the total number of double bonds beginning with the position of the double bond closest to the methyl end (ω) of the molecule. The configuration of the double bond is designated by either a c for *cis* or t for *trans*. For example, 16:1ω7c is a PLFA with a total of 16 carbons, 1 double bond located 7 carbons from the methyl end of the molecule in the *cis* configuration. Branched fatty acids are designated as i for *iso*, a for *anteiso*, if the methyl branch is one or two carbons from the ω end of the molecule (i.e., i15:0 or a15:0), or by a number indicating the position of the methyl group from the acid end of the molecule (i.e., 10me16:0). Methyl branching at undetermined positions is indicated by a "br". Cyclopropyl fatty acids are designated by the prefix "cy". For hydroxy fatty acids, the position of the hydroxyl group is numbered from the acid end of the molecule followed by the prefix "OH" (i.e., 3-OH14:0).

8.2.1.2 Statistical Analyses

Typically, results of a PLFA analysis are expressed as a molar percentage and subjected to an analysis of variance to address differences between population means. The same PLFA profiles can also be treated as multivariant data. Applications such as hierarchical cluster and principal components analyses often provide information regarding community shifts which were not identified in an ANOVA. It is generally accepted that the multivariant statistics be performed on ratios and that the values be arcsin transformed prior to analysis. A hierarchical cluster analysis will provide a dendrogram which illustrates linkages between samples based on similarities in PLFA profiles. The significance of these similarities (usually expressed as Euclidean distances) is dependent on the size and nature of the sample set being analyzed. A principal component analysis reduces a number of correlated variables into only a few mutually independent variables (i.e., principal components). Results of the analysis are generally plotted in two-dimensional space and indicate not only how samples vary in PLFA composition, but also to what extent each PLFA accounts for this variance. Comparisons of the PLFA which describe specific subsets of samples can then be used in assessing shifts in microbial community composition (Frostegaard et al., 1993).

8.2.1.3 Procedural Considerations

In order to detect PLFA extracted from microbial cells recovered from the deep subsurface, it is necessary that background levels of PLFA (those introduced in sample workup) be kept at a consistently low level. Since the sample size is often restricted when working in the deep subsurface, optimizing the signal-to-noise ratio becomes necessary. In order to do so, it is usually necessary to obtain the highest-purity solvents and chemicals available. In addition, all glassware should be cleaned and muffled in a furnace at ~450°C for a minimum of 4 h prior to use. Care must be taken at all times to minimize the introduction of lipids during sample workup. Some common sources of contamination are the skin, plastics, condensation, vacuum grease, and pump oil vapors. By establishing laboratory practices which check solvent quality on a routine basis and incorporate procedural blanks with each set of samples processed, sources of contamination can be more quickly identified and eliminated, minimizing damage to the research results.

8.3   Results and Discussion

8.3.1   Establishment of Uncontaminated Deep Subsurface Samples

The recovery of sediments from the deep subsurface entails the use of much more complex drilling procedures than those employed in collecting samples from shallow aquifers (see Chapter 3). In many instances drilling muds and make-up waters are required. The utilization of drilling muds under high pressure imposes a great challenge in assuring the quality of samples recovered from the deep subsurface. In 1986 the Department of Energy Science Program began an interdisciplinary program (Wobber) to develop the technology necessary for the recovery of deep subsurface samples suitable for microbiological analysis. The technology that resulted carefully defined the levels of possible contamination through the innovative use of tracers and the control of both drilling and sample handling procedures (Phelps et al., 1989; Russell et al., 1992; Colwell et al., 1992).

SLB proved to be particularly useful in establishing the validity of microbial communities recovered from deep subsurface sediments (Lehman et al., 1996). Table 8.1 highlights differences in PLFA abundance among samples of drilling muds, make-up waters, corehole cuttings, and sample cores recovered from a site on the eastern coast of the U.S. Samples of drilling mud contained between 3 to 15 times more viable biomass than that recovered form the sample core. The sample core itself resulted after three consecutive parings of outer materials. Analyses of PLFA content of each of these parings showed biomass to progressively increase from the outside in (Lehman et al., 1996). Analysis of the PLFA profiles showed a concurrent increase in the proportions of terminally branched saturated PLFA, with the highest proportion occurring in the sample core (Lehman et al., 1996). Since there were substantial differences in both viable biomass and community composition between the outer paring and the inner core material, it was difficult to conceive of how there could have been significant microbiological contamination from the drilling muds. A comparison of PLFA mole percent profiles representative of drilling muds, make-up waters, and cuttings and cores by hierarchical cluster analysis showed that distinct differences in PLFA composition existed between the different sample types (Figure 8.4). There was a remarkable parallel between the result of this cluster analysis and that generated through the application of a community-level physiological profile (CLPP) (Lehman et al., 1996). This profile is based on the results of a sole-carbon-source utilization test using 95 different sole-carbon sources (and nutrients). The parallelism between the two assays reinforces the assumption that there was essentially no transference of microbial contamination from the drilling muds into the sample cores. Used together or separately, the two assays, best described as inherent traces, provided an additional and valuable measure of sample microbiological integrity.

TABLE 8.1

Comparison of Viable Biomass Estimates (PLFA) Between Sediment
Samples Recovered From the Drilling of a Corehole

Depth[a]	Drilling Mud	Make-Up Water	Cutting	Sample Core
1725	147[b]	2		
1890	121			
2027	1	15	66	
2217	186	24		
2251	234			
2329			38	
2408	18		8	
2502		12		
2669	33		19	
2743	65			
2793	297			
2798	136			16
2831	98			
3005	188	7		
avg ± sd[c]	124 ± 88	12 ± 8	33 ± 25	16

[a] Depth expressed in meters.
[b] Values expressed as picomole PLFA per gram of sample except for make-up waters which are expressed as picomole PLFA per milliliter of water filtered.
[c] Values expressed as the average ± a standard deviation.

8.3.1.1 Use of SLB to Validate Sample Integrity

In studies of the origins of deep subsurface microbes it is important to be as sure as possible that organisms isolated from deep subsurface cores are truly representative of the extant microbiota. Two studies which utilized SLB analysis have graphically illustrated the pitfalls of sample storage (see Chapter 5). Subsurface sediments recovered from a recently exposed landslide scarp paleosol in the arid western U.S. were incubated in jars for 32 weeks as either intact cores or desegregated material (by mortar and pestle and passed through a 2-mm sieve). The samples were incubated at 15°C under atmospheres composed of 0.5% O_2, 0.03% CO_2, and the balance as N_2. A marked difference existed in the responses of the intact or homogenized sediments over the 32-week incubation period (Table 8.2). Whole samples showed 4 times the total viable microbial biomass (compared to the homogenized cores) by week 15 of the incubation period. Throughout the time course, there was no observable increase in total viable biomass in the homogenized cores.

The increase in biomass observed in the whole cores (0.5 to 2.0 pmol PLFA per gram from week 3 to 32, respectively) was also associated with a shift in microbial community composition. Differences in the percentages of monounsaturated (mono) and terminally branched saturated (terbrsat) PLFA between week 3 and 32 were greater in magnitude in the whole than in the homogenized samples (Table 8.2). In the whole cores, both mono and terbrsat PLFA increased in percentage between weeks 3 and 5 of the incubation, although the magnitude of the increase was greater with the terbrsat PLFA. The monounsaturated PLFA, which are typically attributed to Gram-negative bacteria, reached a maximum around week 9 of the incubation then dropped slightly by week 32, ending at a value twice that observed at the beginning of the incubation. The terminally branched saturated PLFA, typically attributed to Gram-positive bacteria, did not reach a maximum until week 16 of the incubation, also followed by a

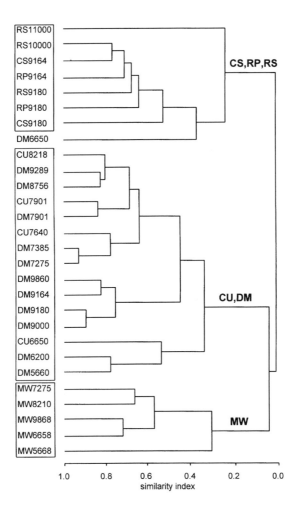

FIGURE 8.4
A dendrogram illustrating the results of a hierarchical cluster analysis relating sample core (including core parings), drilling mud plus corehole cutting, and make-up water PLFA mole % (arcsin transformed) profiles. The similarity index (SI) is based on an Euclidean distance (ED) where SI = 1-ED/ED(max). Sample type abbreviations are as follows: RP = inner sample paring, CS = outer sample paring, RS = sample core, CU = corehole cutting, DM = drilling mud, and MW = make-up water. Each abbreviation is followed by the depth in meters from which the sample was recovered.

slight decrease in percentage at week 32. The decreases in mono and terbrsat percentages at week 32 of the incubation correlated with a decrease in the viable biomass.

The homogenized cores did not exhibit similar responses in community composition to the whole cores. Monounsaturated PLFA were measured at similar percentages for 16 weeks of the incubation period followed by an order of magnitude drop at week 32. Terminally branched saturated PLFA increased in percentage to a maximum at week 9 of the time course, followed by a twofold decrease at week 32. Again, the decreases in percentages at week 32 correlated with a drop in the viable biomass.

The ability to measure changes in PLFA patterns and abundance over time provided insight into the effects of disturbance on microbial community composition. Data obtained from viable counts indicated that, initially, few organisms were culturable, whereas at the end of the incubation period culturable numbers increased by two to three orders of magnitude in both the whole and homogenized cores (data

TABLE 8.2

Comparison of PHA/PLFA Ratios in Whole
and Homogenized Subsurface Sediments
Over a 32-Week Incubation Period

Time[a]	Whole[b]	Homogenized[c]
3	0.05 ± 0.09	0.02 ± 0.03
5	0	0
9	0.01 ± 0.00	0
16	10.7 ± 13.5	5.45 ± 9.45
32	66.7 ± 112	175 ± 275

Note: Values expressed as an average of the ratio
of PHA/PLFA ± a standard deviation.
[a] Time expressed in weeks.
[b] Intact sediment cores.
[c] Sediment cores homogenized with mortar and
pestle and sieved (2 mm).

TABLE 8.3

Viable Microbial Biomass and Community Structural Shifts With Time in Whole (Intact)
and Homogenized Sediment Cores Incubated at 15°C

Incubation[a]	Biomass[b]		Monosaturates		Terminally Branched Saturates	
	Whole[c]	Homogenized[d]	Whole	Homogenized	Whole	Homogenized
3	0.5 ± 0.4	1.1 ± 1.4	2.9 ± 1.4	2.5 ± 1.8	1.9 ± 2.4	2.1 ± 2.0
5	0.6 ± 0.1	0.6 ± 0.6	5.8 ± 1.4	1.8 ± 1.9	13.6 ± 11.2	3.3 ± 4.0
9	0.6 ± 0.3	0.7 ± 0.6	7.4 ± 1.4	2.2 ± 3.1	25.6 ± 7.2	17.2 ± 20.6
16	4.3 ± 3.7	1.5 ± 0.6	6.6 ± 0.8	2.3 ± 2.1	43.3 ± 8.0	5.2 ± 5.5
32	2.0 ± 1.8	0.3 ± 0.2	4.8 ± 1.7	0.9 ± 1.5	35.9 ± 19.4	9.6 ± 12.2

Note: Values expressed as average ± a standard deviation (n = 3).
[a] Incubation time in weeks at 15°C.
[b] Biomass expressed as picomole PLFA per gram.
[c] Whole indicates an intact core.
[d] "Homogenized" indicates a core after grinding by mortar and pestle and sieving (2 mm).

provided by F.J. Brockman, Battele/PNL). The ratio of PHA/PLFA indicated that
with the increase in biomass and changes in community composition, a portion of the
extant microbiota was experiencing unbalanced growth. This imbalance was likely
due to nutrient limitations, a characteristic of most subsurface environs (Table 8.3).
This study demonstrated that changes in the microbiota can occur in intact sediment
cores over time and that the point in time a sample is cultured may dictate the results
obtained.

Samples of deep subsurface volcanic rock showed large changes in the numbers of
viable organisms that were recoverable following 1 week of storage at 4°C and perturbation (Haldeman et al., 1994). Morphologically distinct colonies recovered before
and after the incubation were analyzed for fatty acid content and API rapid NFT
strips. The analyses showed that while recoverability increased, diversity generally
decreased, and that those cultures appearing postincubation showed a greater ability
to utilize the carbon sources. The changes in community composition and the fact that
some isolates were only recovered postincubation suggested that bacterial outgrowth
had occurred. Again, the time at which the cores were sampled resulted in a different
description of the extant microbiota.

8.3.2 Near Surface and Deep Subsurface Microbial Communities

8.3.2.1 *Near Surface Microbial Communities*

Initial applications of SLB methods to subsurface aquifer materials were with near surface fine clays recovered from both vadose and saturated zones from sites in TX, OK, and FL (Smith et al., 1986; White et al., 1983). Results showed that the subsurface microbiota contained intact cytoplasmic membranes composed of PLFA, thus indicating the presence of viable cells. Results also indicated that the abundance of viable subsurface microorganisms were sparse when compared to surface soils. In terms of community composition, the subsurface microbiota was characterized by a greater proportion of Gram-positive and Gram-negative anaerobic bacteria, as compared to surface soils or to estuarine sediments. This was indicated by the detection of an increased abundance of terminally branched saturated PLFAs. In addition, there was an absence of SLB for microeukaryotic organisms (i.e., algae, protozoa, fungi, and micrometazoans) in the subsurface cores. This was indicated by the absence or low abundance of polyunsaturated PLFA. With respect to microbial nutritional status, the subsurface microbiota was characterized by a higher propensity for unbalanced growth, as indicated by greater ratios of PHA/PLFA when compared to the surface soils. Surface soils and subsurface sediments, even those collected from near the surface, contained distinctive characteristics which were identified with the SLB analyses. The characteristics identified in the studies in TX, OK, and FL have also been identified at other sites. Different characteristics between surface and subsurface microbiology are, however, likely to occur over large areas.

8.3.2.1.1 *Effects of Aquifer Contamination on Microbial Communities*

Samples acquired from near surface aquifers contaminated with biodegradable compounds like phenols and creosols showed distinct and reproducible differences in microbial community composition and physiological status between uncontaminated and contaminated sediments (Smith et al., 1986). The differences were attributed to specific responses by the extant microbiota to the presence of the biodegradable contaminant. The contaminated sediments were characterized by an increase in the total viable biomass, a shift in the microbial community composition toward Gram-negative heterotrophs, and a decrease in microbial-related environmental stress. These assumptions were based on the detection of greater total PLFA abundance, an increased percentage of monounsaturated and normal saturated PLFA, and a decrease in the ratios of cyclopropane PLFA to the monounsaturated analogues.

In two integrated DOE bioremediation demonstration programs involving either petroleum (Walker and Walker, 1994) or trichloroethylene (TCE) (Pfiffner et al., 1995) contamination, SLB analyses showed that an increase in microbial abundance occurred as a result of treatments involving nutrient additions. Microbial community shifts were also detected in each study which were related to both the presence and nature of the contaminant. For example, SLB specific for type II methanotrophs were observed to increase after methane gas additions in ground waters collected from the TCE-contaminated site. Similar results were recorded with exposures to propane and air (Ringelberg et al., 1988). Results such as these suggest that the SLB analyses can be used in predicting the success of *in situ* bioremediation when correlations can be made with successful feasibility studies. In addition, SLB analyses provide evidence for bioremediation effectiveness through the monitoring of changes in microbial community composition and abundance (see Chapter 14).

TABLE 8.4

A Comparison of Microbial Biomass in Surface, Near Subsurface, and Deep Subsurface Sediments Recovered in the U.S.

Site	Surface[a]	Near Subsurface[b]	Deep Subsurface[c]
ID	140		2–7
OK	4180	180	
SC	2400		4–12
TX		600	
WA	145		1–16

Note: Values expressed as picomoles PLFA per gram.
[a] Collected from a depth of 0–10 cm.
[b] Collected from a depth of 4 m.
[c] Collected from a depth of >5 m.

8.3.2.2 Deep Subsurface Microbial Communities

Deep subsurface sediments recovered from the two sites in the arid western U.S. (ID and WA) (Colwell et al., 1992) and one in the southeastern coastal plain (SC) (White et al., 1991) showed very low numbers of microorganisms — considerably lower than those observed in near subsurface samples (Table 8.4). A multivariant statistical analysis of the PLFA profiles recovered from these same sediments showed three distinct groups, indicating differences in microbial community composition (Figure 8.5). As defined, group 1 contained the surface samples from all three sites. The PLFA profiles from these samples were enriched in PLFA biomarkers characteristic of soil Actinomycetes (i.e., 10me18:0 or tuberculosteric acid). Also included in this group were the higher biomass (>8 pmol PLFA per gram) vadose zone samples from both the ID and WA sites. Loadings from the analysis indicated that 16:1ω7c and 18:1ω7c were significant in defining group 1. These two PLFA are end products of bacterial anaerobic desaturase fatty acid biosynthesis which is a pathway typically used by Gram-negative bacteria, including the aerobic Gram-negative bacteria which were prevalent in the surface soils. As defined, group 2 was comprised of the majority of deep subsurface vadose zone sediments from the two western sites. These samples were characterized by a very low microbial biomass. The samples collected from around the water table (indicated by the *) showed evidence of toxicity. The *trans/cis* ratios of these samples were greater than 0.1 proposed threshold (Guckert et al., 1986). As defined, group 3 was largely comprised of the SC site samples. Results of the principal components analysis showed two PLFA, 14:0 and 18:2ω6, to be significant in the definition of this group. The dienoic, 18:2ω6, is a prominent fatty acid in the cell membranes of fungal species, whereas the normal saturate, 14:0, has been shown to be prominent in a number of species of microalgae. Although neither fatty acid alone indicates the absolute presence of either of these functional groups, taken together they do suggest a microeukaryotic presence in these sediments and that the presence was greater in the coastal plain than in the arid west.

In summary, a range of sediments recovered from the deep subsurface showed viable microbial biomass levels on the order of one magnitude less than those observed in near surface sediments. The diversity of PLFA detected decreased as well, due in part to the low biomasses. However, differences in microbial community structure between deep subsurface communities could be identified and related to differences in fatty acid biosynthesis and nutritional/physiological status.

8.3.2.2.1 Relationships of Microbial Community Structure and Geochemical Gradients

Another example for the utility of the SLB analysis is in discerning the *in situ* microbial ecology along geochemical gradients. One example is the research performed in

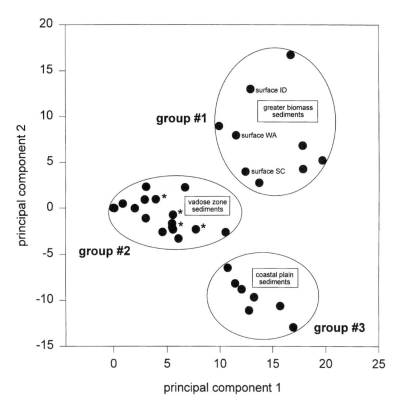

FIGURE 8.5

A 2-D representation of a principal components analysis of PLFA mole % profiles obtained from deep subsurface sediments recovered from sites in SC, ID, and WA. Group #1 contains sediments exhibiting a viable biomass >8 pmol g^{-1} including the surface soils. Group #2 contains almost all of the vadose zone samples from the two western sites as well as those sediments exhibiting a high (>0.1) *trans/cis* ratio (indicated by an *). Group #3 contains the subsurface coastal plain sediments.

support of the Geological Microbiological Hydrogeological Experiment (GEMHEX) conducted at the Hanford site near Richland, WA (Fredrickson et al., 1995). Deep subsurface sediment samples were recovered over a depth interval of 174 to 195 m that included three distinct geological zones: a lacustrine, a paleosol, and a fluvial sand. Viable microbial biomass estimates peaked in the lacustrine sediments, reaching a maximum of 45 pmol PLFA per gram after starting at background levels of <1 pmol PLFA per gram. Values were observed to gradually decrease through the paleosol, reaching background levels by the end of this zone and throughout the fluvial sands. The peak in viable microbial biomass was characterized by an increase in the proportions of both terminally branched saturated and monounsaturated PLFA. A more detailed analysis of the terminally branched saturated PLFA indicated that the ratio of *iso/anteiso* heptadecanoic acid (17:0) was greater than twice that observed in the adjacent sediments. Ratios of i17:0/a17:0, at the levels observed in the lacustrine sediments, were found to be characteristic of certain *Desulfovibrio* species of sulfate-reducing bacteria. Corroborative evidence for the presence of sulfate-reducing bacterial populations was found through the positive hybridization of recovered DNA with a 16S rRNA probe specific for sulfate-reducing bacteria in the lacustrine sediments. The lacustrine sediments exhibited a very low hydraulic conductivity (10^{-7} to >10^{-9} cm s^{-1}), a total organic content of ~1%, a higher carbonate content, and lower pore water sulfate than the surrounding sediments, all of which suggest *in situ* microbial metabolism. The finding of siderite, which apparently can be formed by sulfate-reducing bacteria under conditions which are sulfate limiting (Coleman et al., 1993), added further credence to the likely presence and activity of these organisms in the 6- to 8-million-year-old impermeable sediments.

8.3.3 Characterization of Deep Subsurface Isolates by SLB Analysis

An extensive study of microbial isolates recovered from volcanic tuff in tunnels at Rainier Mesa, Nevada Test Site was carried out by Amy and Haldeman at the University of Nevada, Las Vegas (Haldeman et al., 1993). The characterization included the use of membrane lipids as a phenotypic description of the isolates. A comparison of the fatty acid profiles obtained for each isolate was then interpreted in terms of microbial diversity. The comparison was made through the application of a hierarchical cluster analysis. Assuming that an Euclidean distance of <25 indicated a genus, 29 genera were identified within a 21-m^3 section of the volcanic tuft. A comparison of the fatty acid profiles with those contained in a bacterial library database (Microbial Identification, Inc., Newark, DE) indicated the genera *Arthrobacter, Gordona,* and *Acinetobacter* to be abundant. Two genera were also indicated, representing 16 of the isolates recovered, which did not show any similarity to the ~650 isolate fatty acid profiles contained in the database. These two proposed genera represent a potential for the discovery of unique bacteria which would have possibly gone undetected under standard microbiological analyses. The potential, in reality, is greater since direct bacterial counts were several orders of magnitude higher than viable counts (see Chapter 11). Finally, there appeared to be no clear pattern to the distribution of the bacteria within the rock cube. Most of the isolates were, however, recovered from three specific sites within the rock cube, highlighting the fact that spatial distribution in the deep subsurface is a concern when attempting to recover viable organisms (see Chapter 6).

8.4 Summary

By application of the signature lipid biomarker technique to sediments or rocks recovered from the deep subsurface, insight into the abundance, nature, and distribution of subsurface microorganisms can be obtained. Microbial abundance can be quantified through the recovery of membrane lipid moieties (PLFA) which are reflective of only the viable community. Since the chemical measure is independent of the bias inherent in classical culturing techniques, both the culturable and nonculturable microbiota are measured. The *in situ* measurement provides a description of the nature of the microorganisms present. This description, based on the relative patterns of PLFA recovered, identifies the composition of the extant microbial communities. The distribution of microorganisms in the deep subsurface is reflected in both the abundance and patterns (viable biomass and community composition, respectively) of PLFA recovered. By comparing patterns of PLFA recovered from different geologic horizons with respect to storage conditions or in association with pollutants, the impact of the environmental influence can be defined in terms of the microbial ecology.

Acknowledgments

We would like to thank the Department of Energy, Office of Health and Safety Research, Subsurface Science Program, administered by Dr. F. Wobber, for funding the research described in this chapter. We would also like to acknowledge the fine analytical work of Ms. S. Bean, Ms. S. Sutton, Ms. H. Pinkart, and Ms. J. Stair, all past and present employees of the Center for Environmental Biotechnology at the University of Tennessee, Knoxville in producing the majority of the results discussed.

References

Balkwill, D. L., Leach, F. R., Wilson, J. T., McNabb, J. F., and White, D. C. 1988. Equivalence of microbial biomass measures based on membrane lipid and cell wall components, adenosine triphosphate, and direct counts in subsurface sediments. *Microb. Ecol.,* 16: 73.

Coleman, M. L., Hedrick, D. B., Lovely, D. R., White, D. C., and Pye, K. 1993. Reduction of Fe(III) in sediments by sulfate-reducing bacteria. *Nature,* 361: 436.

Colwell, F. S., Stromberg, G. J., Phelps, T. J., Birnbaum, S. a., Mckinley, J., Rawson, S. A., Veverka, C., Goodwin, S., Long, P. E., Russell, B. F., Garland, T., Thompson, D., Skinner, P., and Glover, S. 1992. Innovative techniques for collection of saturated and unsaturated subsurface basalts and sediments for microbiological characterization. *J. Microbiol. Methods,* 15: 279.

Dowling, N. J. E., Widdel, F., and White, D. C. 1986. Phospholipid ester-linked fatty acid biomarkers of acetate-oxidizing sulfate reducers and other sulfide forming bacteria. *J. Gen. Microbiol.,* 132: 1815.

Edlund, A., Nichols, P. D., Roffey, R., and White, D. C. 1985. Extractable and lipopolysaccharide fatty acid and hydroxy acid profiles from *Desulfovibrio* species, *J. Lipid Res.,* 26: 982.

Findlay, R. H. and White, D. C. 1983. Polymeric beta-hydroxy alkanotes from environmental samples and *Bacillus megaterium, Appl. Environ. Microbiol.,* 45: 71.

Fredrickson, J. K., Mckinley, J. P., Nierzwicki-Bauer, S. A., White, D. C., Ringelberg, D. B., Rawson, S. A., Shu-mei, Li, and Brockman, F. J. 1995. Microbial community structure and biogeochemistry of Miocene subsurface sediments. *Mol. Ecol.,* 4: 619.

Frostegaard, A., Baath, E., and Tunlid, A. 1993. Shifts in the structure of soil microbial communities is limed forests as revealed by phospholipid fatty acid analysis, *Soil Biol. Biochem.,* 25: 723.

Gehron, M. J and White, D. C. 1982. Quantitative determination of the nutritional status of detrital microbiota and the grazing fauna by triglyceride glycerol analysis. *J. Exp. Mar. Biol.,* 64: 145.

Guckert, J. B., Antworth, C. P., Nichols, P. D., and White, D. C. 1985. Phospholipid, ester-linked fatty acid profiles as reproducible assays for changes in prokaryotic community structure of estuarine sediments. *FEMS Microbiol. Ecol.,* 31: 147.

Guckert, J. B., Hood, M. A., and White, D. C. 1986. Phospholipid, ester-linked fatty acid profile changes in nutrient deprivation of *Vibrio cholera*. Increase in the *trans/cis* ratio and proportions of cyclopropyl fatty acids. *Appl. Environ. Microbiol.,* 52: 794.

Guckert, J. B., Ringelberg, D. B., White, D. C., Henson, R. S., and Bratina, B. J. 1991. Membrane fatty acids as phenotypic markers in the polyphasic taxonomy of methyltrophs within the proteobacteria, *J. Gen. Microbiol.,* 137: 2631.

Haldeman, D. L., Amy, P. S., Ringelberg, D. B., and White, D. C. 1994. Changes in bacteria recoverable from surface volcanic rock samples during storage at 4°C. *Appl. Environ. Microbiol.,* 60: 2697.

Haldeman, D. L., Amy, P. S., Ringelberg, D. B., and White, D. C. 1993. Characterization of the microbiology within a 21 m^3 section of rock from the deep subsurface. *Microb. Ecol.*, 26: 145.

Hedrick, D. B. and White, D. C. 1986. Microbial respiratory quinones in the environment. A sensitive liquid chromatographic method. *J. Microbiol. Methods*, 5: 243.

Heipieper, H. J., Diffenbach, R., and Keweloh, H. 1992. Conversion of *cis* unsaturated fatty acids to *trans*, a possible mechanism for the protection of phenol degrading *Pseudomonas putida* P8 from substrate toxicity. *Appl. Environ. Microbiol.*, 58: 1847.

Kerger, B. D., Mancuso, C. A., Nichols, P. D., White, D. C., Langworthy, T., Sittig, M., Schlessner, H., and Hirsch, P. 1988. The budding bacteria, *Pirellula* and *Planetomyces*, with a typical 16S-rRNA and absence of peptidoglycan, show eubacterial phospholipids and unusually high proportions of long-chain beta-hydroxy fatty acids in the lipopolysaccharide Lipid A. *Arch. Microbiol.*, 149: 255.

Kerger, P. D., Nichols, P. D., Antworth, C. P., Sand, W., Bock, E., Cox, J. C., Langworthy, T. A., and White, D. C. 1986. Signature fatty acids in the polar lipids of acid producing *Thiobacilli*: methoxy, cyclopropyl, alpha-hydroxy-cyclopropyl, and branched and normal monenoic fatty acids. *FEMS Microbiol. Ecol.*, 38: 67.

Kerger, B. D., Nichols, P. D., Sand, W., Bock, E., and White, D. C. 1987. Association of acid producing *Thiobacilli* with degradation of concrete: analysis by "signature" fatty acids from the polar lipids and lipopolysaccharide, *J. Ind. Microbiol.*, 2: 63.

Kieft, T. L., Ringelberg, D. B., and White, D. C. 1994. Changes in ester-linked phospholipid fatty acid profiles of subsurface bacteria during starvation and desiccation in a porous medium. *Appl. Environ. Microbiol.*, 60: 3292.

Kohring, L. L., Ringelberg, D. B., Devereux, D., Stahl, M., Mittleman, M. W., and White, D. C. 1994. Comparison of phylogenetic relationships based on phospholipid fatty acid profiles and ribosomal RNA sequence similarities among dissimilatory sulfate-reducing bacteria. *FEMS Microbiol. Lett.*, 119: 303.

Lehman, R. M., Colwell, F. S., Ringelberg, D. B., and White, D. C. 1996. Drilling mud microbial communities as fortuitous tracers for sample collection in deep terrestrial habitats. *J. Microbiol. Methods*, in press.

Lovely, D. R., Giovannoni, S. J., White, D. C., Champine, J. E., Phillipis, E. J. P., Gorby, Y. A., and Goodwin, S. 1992. *Geobacter metallireducans* gen. nov. sp. nov., a microorganism capable of coupling the complete oxidation of organic compounds to the reduction of iron and other metals. *Arch. Microbiol.*, 159: 363.

Nickels, J. S., King, J. D., and White, D. C. 1979. Poly-beta-hydroxybutyrate accumulation as a measure of unbalanced growth of the estuarine detrital microbiota, *Appl. Environ. Microbiol.*, 17: 459.

Nichols, P. D., Mancuso, C. A., and White, D. C. 1987. Measurement of methanotroph and methanogen signature phospholipids for use in assessment of biomass and community structure in model systems. *Org. Geochem.*, 11: 451.

Nichols, P. D., Mayberry, W. R., Antworth, C. P., and White, D. C. 1985. Determination of monounsaturated double bond position and geometry in the cellular fatty acids of the pathogenic bacteria *Francisella tularensis. J. Clin. Microbiol.*, 21: 738.

Nichols, P. D., Smith, G. A., Antworth, C. P., Hanson, R. S., and White, D. C. 1985. Phospholipid and lipopolysaccharide normal and hydroxy fatty acids as potential signatures for the methane-oxidizing bacteria. *FEMS Microbiol. Ecol.*, 31: 327.

Nichols, P. D., Stulp, B. K., Jones, J. G., and White, D. C. 1986. Comparison of fatty acid content and DNA homology of the filamentous gliding bacteria *Vitreoscilla, Flexibacter,* and *Filibacter. Arch. Microbiol.*, 146: 1.

Nichols, P. D., Volkman, J. K., and Johns, R. B. 1983. Sterols and fatty acids of the marine alga, FCRG51. *Phytochemistry*, 22: 1447.

Parker, J. H., Smith, G. A., Fredrickson, H. L., Vestal, J. R., and White, D. C. 1982. Sensitive assay, based on hydroxy-fatty acids from lipopolysaccharide Lipid A for Gram-negative bacteria in sediments. *Appl. Environ. Microbiol.*, 44: 1170.

Pfiffner, S. M., Ringelberg, D. B., Hedrick, D. B., Phelps, T. J., and Palumbo, A. V. 1995. Subsurface microbial communities and degradative capacities during trichloroethylene bioremediation, in Proc. 3rd Int. Symp. *in situ* Bioreclamation, San Diego, CA.

Phelps, T. J., Fliermans, C. B., Pfiffner, S. M., and White, D. C. 1989. Recovery of deep subsurface sediments for microbiological studies. *J. Microbiol. Methods*, 9: 267.

Phelps, T. J., Schram, R. M., Ringelberg, D. B., Dowling, N. J. E., and White, D. C. 1991. Anaerobic microbial activities including hydrogen mediated acetogenesis within natural gas transmission lines. *Biofouling*, 3: 265.

Ringelberg, D. B., Davis, J. D., Smith, G. A., Pfiffner, S. M., Nichols, P. D., Nickels, J. B., Hensen, J. M., Wilson, J. T., Yates, M., Kampbell, D. H., Reed, H. W., Stocksdale, T. T., and White, D. C. 1988. Validation of signature polar lipid fatty acid biomarkers for alkane-utilizing bacteria in soils and subsurface aquifer materials, *FEMS Microbiol. Ecol.*, 62: 39.

Ringelberg, D. B., Townsend, T., Dewaard, K. A., Suflita, J. M., and White, D. C. 1993. Detection of the anaerobic dechlorinator *Desulfomonile tiedjei* in soil by its signature lipopolysaccharide branched-long-chain hydroxy fatty acids, *FEMS Microbiol. Ecol.*, 14: 9.

Russell, B. F., Phelps, T. J., Griffin, W. T., and Sargent, K. A. 1992. Procedures for sampling deep subsurface microbial communities in unconsolidated sediments, *Ground Water Monitoring Remediation*, (Winter), p. 96.

Smith, G. A., Nickels, J. A., Kerger, B. D., Davis, J. D., Collins, S. P., Wilson, J. T., McNabb, J. F., and White, D. C. 1986. Quantitative characterization of microbial biomass and community structure in subsurface material. A prokaryotic consortium responsive to organic contamination. *Can. J. Microbiol.*, 32: 104.

Tunlid, A., Hoitink, H. A. J., Low, C., and White, D. C. 1989. Characterization of bacteria that suppress *Rhizoctania* damping-off in bark compost media by analysis of fatty acid biomarkers. *Appl. Environ. Microbiol.*, 55: 1368.

Tunlid, A., Ringelberg, D. B., Phelps, T. J., Low, C., and White, D. C. 1989. Measurement of phospholipid fatty acids at picomolar concentrations in biofilms and deep subsurface sediments using gas chromatography and chemical ionization mass spectroscopy. *J. Microbiol. Method*, 10: 139.

Tunlid, A. and White, D. C. 1991. Biochemical analysis of biomass, community structure, nutritional status, and metabolic activity of the microbial communities in soil, in *Soil Biochemistry*, Bollag, J. M. and Stotzky, G., Eds., Marcel Dekker, New York, 229.

Walker, J. R. and Walker, A. B. 1994. The Kwajalein bioremediation demonstration: Final technical report, National Technical Information Services, ORNL/TM-12871. U.S. Department of Commerce, Washington, D.C., 64.

Walker, J. T., Sonesson, A., Keevil, C. W., and White, D. C. 1993. Detection of *Legionella pneumophila* in biofilms containing a complex microbial consortium by gas chromatography-mass spectrometric analysis of genus-specific hydroxy fatty acids. *FEMS Microbiol. Lett.*, 113: 139.

Welch, D. F. 1991. Applications of cellular fatty acid analysis, *Clin. Microbiol. Rev.*, 4: 422.

White, D. C. 1983. Analysis of microorganisms in terms of quantity and activity in natural environments, in *Microbes in Their Natural Environments*, Slater, J. H., Whittenbury, R., and Wimpenny, J. W. T., Eds., Society for General Microbiology, London, 34: 37.

White, D. C. 1986. Environmental effects testing with quantitative microbial analysis. Chemical signatures correlated with *in situ* biofilm analysis by FT/IR, *Toxicity Assessment*, 1: 315.

White, D. C. 1988. Validation of quantitative analysis for microbial biomass, community structure, and metabolic activity. *Adv. Limnol.*, 31: 1.

White, D. C., Bobbie, R. J., Nickels, J. S., Fazio, S. D., and Davis, W. M. 1980. Nonselective biochemical methods for the determination of fungal mass and community structure in estuarine detrital microflora. *Bot. Mar.*, 23: 239.

White, D. C., Davis, W. M., Nickels, J. S., King, J. D., and Bobbie, R. J. 1979. Determination of the sedimentary microbial biomass by extractable lipid phosphate. *Oecologia*, 40: 51.

White, D. C., Ringelberg, D. B., Guckert, J. B., and Phelps, T. J. 1991. Biochemical markers for *in situ* microbial community structure, in *Proc. First Int. Symp. Microbiology of the Deep Subsurface*, Fliermans, C. B. and Hazen, T. C., Eds., WSRC Information Services, Aiken, SC, p. 4-45.

White, D. C., Ringelberg, D. B., Hedrick, D. B., and Nivens, D. E. 1993. Rapid identification of microbes from clinical and environmental matrices by mass spectrometry, in *Identification of Microorganisms by Mass Spectrometry*, Fenselau, C., Ed., Symp. Ser. 541, American Chemical Society, Washington, D.C., p. 8.

White, D. C., Smith, G. A., Gehron, M. J., Parker, J. H., Findlay, R. H., Martz, R. F., and Fredrickson, H. L. 1983. The ground water aquifer microbiota: biomass, community structure, and nutritional status. *Dev. Ind. Microbiol.*, 24: 201.

Wobber, F. J., *Microbiology of Subsurface Environments*, DOE/ER-0293, U.S. Department of Energy, Washington, D.C.

9

Life in the Slow Lane: Activities of Microorganisms in the Subsurface

T. L. Kieft and T. J. Phelps

KEY WORDS: *direct viable counts, INT (iodonitrotetrazolium chloride), rRNA probes, LIVE/DEAD BacLight, nalidixic acid, microautoradiography, phospholipid fatty acids (PLFAs), diglyceride fatty acids (DGFAs), microcosms, radiotracers, enzyme assays, enrichment cultures, in situ activities, geochemical modeling aerobic heterotrophs, anaerobes, fermentation, denitrification, iron reduction, sulfate reduction, methanogenesis, chemoautotrophy, iron-oxidizing bacteria, hydrogen-oxidizing bacteria, trichloroethylene (TCE), nutrient limitation, nitrous oxide, triethyl phosphate (TEP), bioremediation, Savannah River Site, coastal plain sediments, paleosols, vadose zone, shales, sandstones, aquifers, aquitards.*

9.1   Introduction

Reports of microorganisms in deep subsurface environments have raised questions regarding the degree to which these microorganisms are active *in situ* and, by extension, the extent to which they influence the geochemistry of their surroundings. Unfortunately, our abilities to answer these questions are hampered by methodological limitations. Multiple techniques are available to assess the abundance or total biomass of microorganisms in subsurface samples, but it is more problematic to determine the proportion of cells that are viable or metabolically active. Moreover, there exists a variety of tests based on different criteria for assessing cell viability and activity (Rusack and Colwell, 1987). Even if one knows the percent of cells capable of metabolic activity, one would ideally like to know the *in situ* rate of that metabolic activity. To understand the role of microorganisms as mediators of geochemical change in subsurface environments, one should assess the *in situ* metabolic activity for each of the major microbial functional groups in that environment. This chapter describes the available methods for assessing rates of metabolic activity in the subsurface and current understanding of the *in situ* activities of important functional groups of subsurface microbes.

Methods used for determining the activities of microorganisms in subsurface environments include: (1) quantifying the number of viable cells using direct viable count procedures (Brockman et al., 1992; Chapelle et al., 1987, 1988; Kieft et al., 1993, 1995; Marxsen, 1988; Rusterholz and Mallory, 1994), (2) measuring rates of a particular microbial activity (e.g., respiration) in laboratory microcosm experiments (Brockman et al., 1992; Hicks and Fredrickson, 1989; Kieft et al., 1993, 1995; Phelps et al., 1989b), and (3) using geochemical assessments to estimate and bound *in situ* rates of microbial activity (Chapelle and Lovley, 1990; Murphy et al., 1992; Phelps et al., 1994a; Wood et al., 1993). Combinations of these methods may provide an understanding of carbon and energy flow from which activities may be constrained (i.e., minimum or maximum values may be estimated) in particular environments. These methods also allow comparisons of relative activities between subsurface environments.

Measuring rates of microbial activities in microcosms is by far the most commonly employed method for assessing microbial activities, or more accurately, potential activities of subsurface microbial communities. Advantages of microcosm experiments include: (1) the assays are relatively easy to perform; (2) they can be highly sensitive, especially when radiotracers are used; (3) they can be performed on samples from multiple sites and multiple strata, thereby enabling comparisons; (4) they can give indications of the relative importance of the activities of various functional groups of microbes (e.g., aerobic heterotrophs, denitrifiers, iron reducers, sulfate reducers, methanogens, etc.); and (5) they can be used to determine factors that limit subsurface microbial activity. Disadvantages of the laboratory microcosm approach stem from the fact that the collection of subsurface samples invariably results in disturbance to those samples, consequently altering the rates of microbial metabolism (Findlay et al., 1985; Phelps et al., 1994b). This alteration of rates can be a decrease (e.g., when strict anaerobes are exposed to even small amounts of oxygen) or, more commonly, an increase due to a physical rearrangement of the sample in which microbes come into contact with previously unavailable nutrients. Even in the case of samples being processed as intact cores (Smith and Harvey, 1991), changes in hydrostatic pressure, moisture gradients, partial pressure of oxygen, etc. increase diffusion of nutrients within the sample. Thus, laboratory microcosm estimates of microbial activity are often orders of magnitude higher than *in situ* rates (Chapelle and Lovley,

1990; Phelps et al., 1994a) and should therefore be considered to represent potential microbiological metabolic activities.

   Geochemical assessments may currently be the most appropriate means of quantifying *in situ* rates of microbial metabolism. Briefly, the approach employs information on groundwater flow (advection, diffusion, and dispersion) combined with quantitative data on the chemistry of the solid and liquid phases (and gaseous phase in unsaturated environments) to determine rates of biologically mediated chemical change along a flow path in advection-dominated systems, e.g., groundwater aquifers (Chapelle and Lovley, 1990; Chapelle et al., 1987; Lovley and Chapelle, 1995; Murphy et al., 1992; Phelps et al., 1994a), or along a diffusion path in diffusion-dominated systems, e.g., confining layers (aquitards) (Chapelle and McMahon, 1991; McMahon and Chapelle, 1991) and unsaturated (vadose) zones (Wood et al., 1993). The obvious advantage of the modeling approach is that it yields estimates of *in situ* rates, i.e., rates of biologically mediated chemical change in the subsurface that are consistent with geochemical parameters, rather than inflated potential activities. However, the method is not without problems; estimates are only as good as the assumptions and data on which they are based. For this reason, the rates estimated through geochemical modeling may be accurate to an order of magnitude at best. Another disadvantage of the modeling approach is that the calculated rates of metabolic activity are averaged over time and distance, whereas actual rates of activity are temporally and spatially heterogeneous (Adrian et al., 1994; Beeman and Suflita, 1990; Murphy et al., 1992; Stevens and Holbert, 1995). This chapter addresses the uses of the two approaches, microcosms and geochemical assessments, with references to particular subsurface habitats. Microcosm data are presented for comparisons among subsurface environments and as a means of determining the dominant microbial functional groups and biogeochemical processes present. Estimates of microbial activities derived from geochemical modeling are presented for comparisons among subsurface environments, comparisons between subsurface environments and common surface environments, and as support of the emerging theory that individual microorganisms survive in some subsurface environments for millennia while carrying out cellular metabolism at glacially slow rates.

9.2   Quantifying Microbial Activities: Approaches

Direct measures of microbial activities in subsurface environments have been elusive. Estimates of subsurface microbial activities have been approached from two directions: those based on laboratory experiments using recovered materials and estimates based on field observations. Subsurface microbial potentials estimated from recovered samples typically rely on biomass determinations, functional group enrichments, microcosm studies, or radioisotope time-course experiments. Field observations, typically coupled with geochemical models, can be used to bound subsurface microbial activities. For example, the observation of oxygenated groundwaters in confined aquifers several kilometers from recharge constrains aerobic microbial oxidation of organics to $<50~\mu M~kg^{-1}~y^{-1}$, since faster metabolic rates would remove available oxygen in the ground waters within a few kilometers of recharge or a few years of residence time in a confined aquifer. The observation of millimolar concentrations of volatile fatty acids or sulfide may be suggestive of anaerobic metabolism greater than nanomoles per kilogram per year, since slower rates would take millions of years to accumulate these products to millimolar concentrations. While not indicative of current *in situ* activities at a defined space and time, field observations coupled with geochemical models can assess or constrain activities of subsurface microorganisms on the basis of area such as aquifer or flow-path studies. Estimates of subsurface microbial activities based on retrieved materials typically involve laboratory experiments with recovered ground waters or sediments. Disturbance artifacts, redistribution of nutrients during sampling and processing, and stimulation of microbial activities during sample storage are variables which likely may result in overestimating *in situ* activities from sediment or groundwater analyses.

9.2.1   Direct Viable Counting

As in most natural environments, the majority of microorganisms in subsurface environments are nonculturable (Balkwill and Ghiorse, 1985; Colwell, 1989; Ghiorse and Wilson, 1988; Haldeman et al., 1993; Hirsch and Rades-Rohkohl, 1988; Kolbel-Boelke et al., 1988; Madsen and Ghiorse, 1993; Sinclair and Ghiorse, 1989); however, many of these nonculturable cells may be viable (Brockman et al., unpublished data; Kieft, 1997) (see Chapter 11). Enumerating live or active cells in a subsurface environment and the proportion of the total that is alive or active provides a starting point for determining the potential for *in situ* microbial activity. The available microscopic methods for determining the total number of viable cells in an environment include (1) incubation with redox dyes (e.g., tetrazolium salts) that accept electrons from microbial electron carriers, (2) use of fluorescent rRNA probes that detect active cells that have a sufficient number of ribosomes to generate a detectable fluorescence signal, (3) simultaneous use of two fluorescent dyes, one that permeates intact membrane and one that does not (LIVE/DEAD *Bac*Light, Molecular Probes, Inc., Eugene, OR), (4) microautoradiography, and (5) incubation with an organic nutrient mixture plus nalidixic acid to stimulate growth but inhibit cell division. A severe limitation of the direct viable count method is the high minimum level of detection (log 4.5 to 6.0 cells per gram) due to the need for detecting cells microscopically.

INT [2-(*p*-iodophenyl)-3-(*p*-nitrophenyl)-5-phenyl tetrazolium chloride] (Zimmerman et al., 1978) and CTC (5-cyano-2,3-ditolyl tetrazolium chloride) (Rodriguez et al., 1992) are redox dyes that are colorless in their oxidized states and can be reduced by microbial electron carriers in metabolically active cells to colored formazan compounds. When combined with microscopy, these dyes can be used to estimate the total number of actively respiring cells. In the case of CTC, the formazan product is a fluorochrome. Since the method quantifies actively metabolizing bacteria, it yields an underestimate of the total number of viable cells. However, with fresh subsurface samples, it can estimate the number of cells that are metabolically active *in situ*. INT has been used for quantifying active cells in ground water (Marxsen, 1988; Rusterholtz and Mallory, 1994), shallow aquifer sediments (Beloin et al., 1988), and unsaturated buried soils (Brockman et al., 1992).

Fluorescently labeled rRNA probes have been used in combination with epifluorescence microscopy to quantify metabolically active cells in subsurface samples (Haldeman et al., 1995a). Only cells that are metabolically active have a sufficient number of ribosomes (approximately 1000 ribosomes per cell) for a detectable fluorescent signal. Universal probes are suited for quantifying all the active microbes in an environment while specific probes can quantify specific groups or populations. The method has been used for quantifying metabolically active microorganisms in saturated lacustrine sediments in eastern Washington State and Cretaceous shales and sandstones in New Mexico (Nierzwicki-Bauer, unpublished data).

TABLE 9.1

Estimations of Percent Viability in Subsurface Microorganisms Determined by Various
Direct Microscopic Count Techniques

Sample Type	Method	Percent Viable	Ref.
Karstic ground water, Mammoth Cave, Kentucky	INT	53–58	Rusterholz and Mallory, 1994
Shallow aquifer sediments, Oklahoma	INT	12–40	Beloin et al., 1988
Shallow aquifer sediments, Germany	INT	0.66–7.4	Marxsen, 1988
Vadose zone paleosols south-central Washington State	INT	0.43–6.3	Brockman et al., 1992
Deep Cretaceous shales and sandstones, central New Mexico	rRNA probes	2.9–33	Nierzwicki-Bauer, unpublished data
Deep Cretaceous shales and sandstones, central New Mexico	LIVE/DEAD *Bac*Light	49–93	Kieft, unpublished data
Deep granitic ground waters, southeastern Sweden	Microautoradiography		Pedersen and Ekendahl, 1990
	Glucose	0–97.8	
	Leucine	9.4–17.6	
	Formate	0–24.5	
	Acetate	0–56.7	
	Thymidine	0	
Deep Atlantic coastal plain aquifer sediments, Maryland	Nalidixic acid	0.3–61	Chapelle et al., 1987
Deep Atlantic coastal plain aquifer sediments, South Carolina	Nalidixic acid	28–100	Chapelle et al., 1988

The LIVE/DEAD *Bac*Light system from Molecular Probes, Inc. tests for cell viability based on membrane integrity. Viable cell membranes allow penetration of one fluorescent dye, but exclude a second one. This is a less stringent criterion for viability than either the tetrazolium or rRNA probe methods and thus higher proportions of viable cells may be measured by this method. We have observed higher proportions of viable cells in Cretaceous shale and sandstone sediments in New Mexico with the LIVE/DEAD *Bac*Light system than were found using rRNA probes (Table 9.1).

Pedersen and Ekendahl (1990) employed microautoradiography to determine the percentages of microbes in deep groundwaters that were able to assimilate a variety of radiolabeled organic substrates. Chapelle et al. (1987, 1988) utilized the nalidixic acid technique of Kogure et al. (1979) to quantify viable cells in aquifer core samples. An advantage of autoradiographic and nalidixic-acid methods is that they can test the ability of cells to respond to individual substrates.

9.2.2 Potential Activities Indicated by Biomass Assays and Culture Techniques

Direct biomass determinations by microscopy or phospholipid analyses are typically used to estimate biomass without requiring recovery on laboratory media. Direct measures of biomass are also advantageous because they enable samples to be inhibited or fixed at the time of collection, thereby avoiding growth of microorganisms during sample processing and storage. As direct biomass measures based on molecular probes become more widely available, one may be capable of distinguishing and quantifying dozens of microbial populations in gram-sized samples.

Classical methods for estimating biomass include enumeration of colony forming units on solid media or most probable number dilution series. In general, higher values are observed in liquid and more dilute media (Balkwill, 1989; Beloin et al., 1988;

Phelps et al., 1989b). Unfortunately, these procedures require the growth of microorganisms on artificial laboratory media, a process that yields a biased underestimate of the *in situ* populations and their densities (Hirsch and Rades-Rohkohl, 1988). A biomass assay not requiring the growth of microorganisms is phospholipid fatty acid (PLFA) analysis (White et al., 1979a, 1979b, 1983). Since phosphate is generally cleaved from phospholipid molecules upon cell death, PLFA provides an estimate of living cells. When phosphate groups are cleaved from lipids, diglyceride fatty acids (DGFAs) are produced. Therefore, DGFAs may provide a measure of dead cells, and the ratio of PLFAs to DGFAs can provide an estimate of the proportion of living and dead cells (Kieft et al., 1994; Haldeman et al., 1995b). When biomass measures indicate sparse microbial densities, then activities <0.1 mM kg^{-1} y^{-1} would be expected. Conversely, when multiple measures of direct biomass indicate populations $>10^6$ cells per gram, then activities >1.0 nM kg^{-1} y^{-1} may be deduced in order to maintain the existent microbial community.

Enrichments of microbial trophic groups, colony forming units, and most probable number enumerations provide indicators of relative significance of microbial populations (Fredrickson et al., 1991, 1989; Kolbel-Boelke et al., 1988; Sinclair and Ghiorse, 1989). While it is accepted that enumerations and enrichments requiring growth on laboratory media usually underestimate *in situ* microbial numbers by one to three orders of magnitude (Balkwill and Ghiorse, 1985; Beloin et al., 1988; Colwell, 1989; Ghiorse and Balkwill, 1983; Ghiorse and Wilson, 1988; Haldeman et al., 1993; Hirsch and Rades-Rohkohl, 1988; Kolbel-Boelke et al., 1988; Madsen and Ghiorse, 1993; Sinclair and Ghiorse, 1989), enumerations can be useful as indicators of the significance of particular microbial processes. While the inability to recover specific populations in culture does not preclude their presence, the observance of populations representing high percentages of the total enumerated microorganisms is evidence of relative significance. For example, the absence of anaerobes on enrichment media does not mean that anaerobes are unimportant, whereas the detection of populations of sulfate reducers, methanogens, and fermentative microorganisms at $\geq 10^5$ cells per gram combined with cultured cells $\geq 10^6$ CFU per gram provide strong evidence that obligate anaerobes are metabolically and ecologically important *in situ*.

9.2.3 Laboratory Microcosms

The use of laboratory microcosms to measure potential rates of microbial activity is a mainstay of microbial ecology. The general scheme is to place a sample of material collected from a natural environment into a closed container (with or without additional substrate amendment), incubate under controlled conditions, and measure disappearance of a reactant and/or appearance of a product. Typically, sterile or poisoned controls are used to distinguish influences of biological from abiotic processes. Types of microcosm/activity assays used with subsurface material include disappearance of a xenobiotic compound (Wilson et al., 1983), production of CO_2 (Chapelle et al., 1988; Kieft et al., 1995; Kieft and Rosacker, 1991; Lind and Eiland, 1989; McMahon et al., 1990) and consumption of O_2 (Aelion and Bradley, 1991; McMahon et al., 1990). The activities of functional groups accomplishing various forms of anaerobic metabolism have also been measured. Denitrification activity has been quantified in subsurface sediment aquifers using the acetylene-blocking technique and analyzing N_2O production (Bradley et al., 1992; Francis et al., 1989; Morris et al., 1988; Smith and Duff, 1988) and by quantifying nitrate disappearance (Johnson and Wood, 1992). Jones et al. (1989) measured disappearance of lactate, formate, acetate, phenol, and benzoate in subsurface sediments under anaerobic conditions and also quantified concomitant rates of acetogenesis and methanogenesis. Methanogenesis was quantified in subsurface coastal plain sediment samples by Madsen and Bollag (1989).

The use of radiotracers is particularly useful in microcosm/activity measures due to the high degree of specificity and sensitivity. By labeling a particular reactant, e.g., an organic compound, with ^{14}C, one can measure the rate of transformation of that substrate as well as the rate of end product formation. This has been particularly useful for determining the potential rate of mineralization of organic pollutants to CO_2. A very sensitive assay can be achieved by adding a labeled substrate (e.g., readily degraded or assimilated compounds such as ^{14}C-glucose or ^{14}C-acetate, or more difficult to degrade compounds such as ^{14}C-phenol or ^{14}C-aniline) to microcosms and then monitoring rates of $^{14}CO_2$ evolution (Albrechtsen and Winding, 1992; Brockman et al., 1992; Ekendahl and Pedersen, 1994; Kazumi and Capone, 1994; Kieft et al., 1993, 1995; Konopka and Turco, 1991; Hicks and Fredrickson, 1989; Madsen and Bollag, 1989; McMahon and Chapelle, 1991; Phelps, 1991; Phelps et al., 1989b; Swindoll et al., 1988). In addition to measuring $^{14}CO_2$ release, McMahon and Chapelle (1991) measured production of labeled fermentation end products (formate and acetate) from added ^{14}C-glucose in aquifer and aquitard samples. It is also possible to quantify the extent and rate of radiolabeled substrate incorporation into microbial biomass, as well as that which is mineralized, thereby permitting closure of mass balance (Aelion and Bradley, 1991; Dobbins and Pfaender, 1988; Kazumi and Capone, 1994; Swindoll et al., 1988). One can also examine the uptake of radiolabeled compounds into cells, e.g., uptake of ^3H-acetate into membrane lipids (Phelps et al., 1989a, 1994b, 1989b, 1988), uptake of $^{32}PO_4^{3-}$ into lipids (Konopka and Turco, 1991), or uptake of ^3H-thymidine into DNA (Albrechtsen and Winding, 1992; Kazumi and Capone, 1994; Phelps et al., 1989b, 1988; Thorn and Ventullo, 1988). Autotrophic CO_2 fixation can be measured by the uptake of $H^{14}CO_3^-$ (Ekendahl and Pedersen, 1994). Addition of $^{35}SO_4^{2-}$ to microcosms and subsequent measurement of rates of $H^{35}S^-$ production have been used to quantify dissimilatory sulfate reduction (Ekendahl and Pedersen, 1994; Madsen and Bollag, 1989; Phelps et al., 1994a; Shanker et al., 1991). Although the rates measured by these methods should be considered to be potential rates and not actual *in situ* rates, they are useful for comparing relative activities among subsurface samples.

Enzyme activity (e.g., dehydrogenase) assays have been adapted from soil microbiology for use in subsurface microbiology (Barbaro et al., 1994; Sarbu et al., 1994). Robert Griffiths at Oregon State University has modified sensitive phosphatase and β-glucosidase activity assays, originally developed for oligotrophic marine environments, for use with subsurface sediments and rocks. These assays are based on the use of methylumbelliferone (4-MUF)-labeled substrates. The 4-MUF is covalently bound to the substrate; specific enzyme activity releases fluorescent 4-MUF, which can then be measured with a fluorimeter. Griffiths has recently applied the assays to Cretaceous sandstones and shales in New Mexico and Colorado (unpublished data). Again, while useful for making comparisons within and among sites, *in vitro* enzyme activity assays may yield results that are orders of magnitude greater than *in situ* values.

9.2.4 *In situ* Microcosms

Attempts have been made to devise *in situ* tests of biological activities in limited volumes of the subsurface. For example, a slug test has been used to estimate rates of activity. Trudell et al. (1986) injected a volume of groundwater into a shallow aquifer and later recovered water from the same well to estimate rates of denitrification. For this approach to work, advection of groundwater away from the injection/pumping well must be minimal. Gillham et al. (1990) devised an *in situ* microcosm. This device consists of a length of drill pipe that isolates a section of aquifer from its surroundings. Well screens allow movement of groundwater into and out of the microcosm. They used their *in situ* microcosm to estimate rates of denitrification and of benzene biodegradation in a Canadian aquifer. Beeman and Suflita (1990) used a similar approach; they inserted plastic pipes as gas sampling devices into a shallow anoxic aquifer at sites adjacent to a landfill to estimate *in situ* rates of methanogenesis. Another *in situ* approach is to expose minerals or other surfaces to microbial colonization and weathering in wells (Hiebert and Bennett, 1992; Hirsch and Rades-Rohkohl, 1990). These *in situ* approaches allow incubation in a physical-chemical regime that is somewhat representative of actual subsurface conditions. However, they all require the use of wells, the drilling of which severely disturbs the physical and chemical subsurface milieu. Therefore, they may not yield accurate estimates of *in situ* microbial activity.

9.2.5 Geochemical Assessments

Geochemical assessments estimating rates of *in situ* microbial activities are increasingly utilized to visualize and model bioremediation and to assess the impacts of groundwater treatments and perturbations. Typically, groundwaters and sediments are obtained from numerous formations and depths along a groundwater flow path or, in and adjacent to the habitat of interest. Geochemical assessments differ from microcosm experiments in that laboratory time course experiments are not used to extrapolate *in situ* activities. Rather than time course measurements, spatial distances are sampled and relevant chemistries are determined. For example, after an aquifer becomes confined, it no longer receives oxygenated waters vertically from the surface; thus, surface water reaches distal points in the aquifer solely via the flow path from the recharge area. Importantly, water in confined aquifers does not communicate with shallower or deeper aquifers; consequently groundwater movement can be modeled as plug flow from the recharge area. Accordingly, changes in groundwater chemistry are related to distance and time from recharge.

A variety of techniques are used to characterize sediment and groundwater chemistry. It is important to estimate groundwater age, flow regimes, types and age of sediments, and organic loading, as well as to determine if waters communicate between formations or if aquifers are confined and distal from surface recharge. Once the flow unit and path are identified, analysis of known energy sources, electron donors, electron acceptors, end products, and their sinks can be used to identify predominant metabolic pathways used by resident microorganisms. Since oxygen will not be replenished after it is utilized, assessments of biological activity often rely on decreases in dissolved oxygen concentrations with depth and distance from recharge. Oxygen utilization, coupled with the accumulation of carbonates along an aquifer flow path, is commonly used to infer microbial metabolism over time and distance (Chapelle and McMahon, 1991; Murphy et al., 1992; Wood et al., 1993). Further downgradient, alternative electron acceptors are used. Groundwater ages, and to a lesser extent the flow path, can be evaluated by looking at changes in the distribution of ionic species and stable isotopes of hydrogen, carbon dioxide, and chlorine. Extrapolating fluxes for various pools of reactants and products which are then estimated over time and space can be useful in visualizing water, energy, and electron flow in subsurface environments. By comparing spatial and temporal changes over a flow path one can speculate and constrain the potential for biological modification of ground waters and *in situ* microbial activities.

Buswell and Larson in 1937 noted explosive concentrations of methane in anaerobic Illinois ground waters and speculated that biologically mediated methanogenesis accumulates significant quantities of methane over geologic time. More recently, investigators have modeled CO_2 production and biological activity in aquifers. Iron reduction, a form of bacterial respiration not requiring oxygen, which would be predicted to occur thermodynamically after the depletion of oxygen and nitrate, was observed by Lovley et al. (1990) in oxygen-depleted Atlantic coastal plain sediments. Further analyses revealed that iron reduction may be a predominant electron-accepting reaction for bacterial metabolism at intermediate distances from recharge in confined Atlantic coastal plain aquifers. Sulfate reduction was observed further downgradient in these aquifers (Lovley et al., 1994). In confining clay zones, bacterial fermentation was deemed the predominant metabolic process (McMahon and Chapelle, 1991). Furthermore, McMahon and Chapelle (1991) modeled bacterial fermentation of lignite to short-chained acids and suggested that diffusion of volatile fatty acid fermentation end products may provide a substantial flux of electron donors to confined Atlantic coastal plain aquifers. In other coastal plain sediments in the same geological sequence as that used by Chapelle and Lovley (1990), Murphy et al. (1992) and Phelps et al. (1994a) modeled oxygen consumption as well as CO_2 production and accumulation.

Sophisticated models enable consideration of changes in redox chemistry and equilibria of inorganic species, thereby assessing relative importance of abiotic solute equilibria changes from biologically mediated reactions. Coupled with information on groundwater ages from isotopic analyses and detailed groundwater and sediment chemistry, these numeric codes can model *in situ* microbial activities such that they are consistent with groundwater inputs, deposited materials, available substrates, intermediates, and end products considered in the context of appropriate geologic time scales.

9.3   Activities of Subsurface Microbial Communities

9.3.1   Community Viability

Estimates of the total number of viable cells vary with the criteria used to define viability as well as with site conditions. Culture-based methods (plate counts, multiple-tube most probable number methods, etc.) generally grossly underestimate the numbers of viable cells (Roszak and Colwell, 1987). Estimates of percent viability based on microscopic methods for various subsurface environments (Table 9.1) vary from zero to 100%. The highest values were found for ground water, as opposed to sediment or rock samples, suggesting that unattached bacteria were more generally alive and active at those sites than attached bacteria. Available evidence also suggests that attached and free-living subsurface bacteria represent different populations (Amy et al., 1992; Hazen et al., 1991). However, Pedersen and Ekendahl (1990) have suggested that bacteria free in the ground water may be detached inactive cells derived from active attached populations. Their use of various ^{14}C-labeled substrates and microautoradiography on groundwater samples yielded nearly the widest possible range of values (0 to 97.8%). This range was found for a single substrate, glucose, applied to ground water from different depths in different boreholes. Another substrate, leucine, resulted in 17.6 to 81.2% viability for the same samples. Chapelle et al.

(1988) measured 28 to 100% microbial viability in samples from coastal plain aquifer samples in South Carolina.

Consistently low values for percent viability (0.43 to 6.3%) were observed in unsaturated buried soil (paleosol) sediments in a semiarid site in Washington State. Microorganisms in the unsaturated zones of arid and semiarid regions exist in extremely low nutrient flux environments (Kieft et al., 1993), and thus it is not surprising to find a low percentage of viable cells. One would expect the percent viabilities to be even lower in unsaturated sediments not having the residual organic matter that is present in some buried soil environments.

When viability was measured using two different methods with the same samples, the results reflected the particular criteria used for viability. Kieft's estimates for percent viability of bacteria in Cretaceous shale and sandstone samples, measured using the LIVE/DEAD *Bac*Light method, were consistently higher than Nierzwicki-Bauer's data, derived using rRNA probes for the same samples (Table 9.1). This is not surprising; one can envision relatively inactive cells in a low-nutrient environment not having the requisite number of ribosomes to be detected with rRNA probes, and yet maintaining membrane integrity and thus appearing viable when viewed with LIVE/DEAD *Bac*Light probes.

It is rare that one can test for viability and activity of subsurface microbes immediately following sample collection. Any period of handling and storage tends to increase microbial activities (Brockman et al., 1992; Fredrickson et al., 1995; Haldeman et al., 1994; Hirsch and Rades-Rohkohl, 1988). Moreover, the INT, CTC, microautoradiographic, and nalidixic acid assays require an incubation period that may further revive inactive cells. For these reasons, the published values for percent viability or percent activity of subsurface microbes may be inflations of actual *in situ* values (see Chapter 5).

Most bacteria in uncontaminated subsurface environments are relatively inactive, moribund, or dead as a result of severe nutrient deprivation; this idea is supported by multiple lines of evidence. The low adenylate energy charge data (mean = 0.44, SD = 0.19) for microbes in Atlantic coastal plain sediments (Kieft and Rosacker, 1991) and the membrane phospholipid fatty acid stress signatures, e.g., high ratios of *trans*- to *cis*-monoenoic fatty acids and saturated to unsaturated fatty acids (White et al., 1991) indicate that subsurface microbes are generally physiologically challenged and relatively inactive. The generally small size of subsurface bacteria *in situ* further suggests a nutritionally challenged state (Kieft et al., 1997; Amy et al., 1992) (see Chapter 11). Finally, the geochemical modeling data discussed below also indicate that subsurface microbial communities are relatively inactive.

9.3.2 Activities of Various Functional Groups

9.3.2.1 *Aerobic Heterotrophic Activities*

Aerobic heterotrophs have been the most-studied group of subsurface microbes, in part because they are relatively easy to study, but also because shallow uncontaminated subsurface environments tend to be aerobic with traces of organic compounds serving as energy sources. Rates of CO_2 production in laboratory-incubated microcosms of subsurface samples span at least three orders of magnitude (Table 9.2). These are rates of CO_2 production from samples that have not been amended with organic carbon. As applied to soils, this measure has been called the basal respiration rate (Insam and Domsch, 1988); the measured rates should be considered to be potential activities and may not reflect *in situ* rates. In some cases, these basal respiration rates are correlated with the organic carbon content of the samples (Kieft et al., 1995)

TABLE 9.2

Rates of CO_2 Production in Microcosms of Subsurface Materials Without Carbon or Nutrient Amendment

Sample Type	Rate of CO_2 Production ($\mu g\ CO_2\ g^{-1}d^{-1}$)	Ref.
Coastal plain aquifer sediments, South Carolina	~2.6	Chapelle et al., 1988
Coastal plain aquifer sediments, South Carolina	0.00156–0.43[a]	McMahon et al., 1990
	0–0.24[b]	
	0–0.60[c]	
Coastal plain aquifer sediments, South Carolina	0.088–1.3[b]	Morris et al., 1988
Coastal plain aquifer and aquitard sediments, South Carolina	0–25.9	Kieft and Rosacker, 1991
Sandy unsaturated sediment Denmark	~1.46	Lind and Eiland., 1989
Saturated sediments, south-central Washington State		
Lacustrine	<0.024–15.9	
Paleosol	<0.024–0.840	
Fluvial	0.0024–0.816	Kieft et al., 1995

[a] Incubated aerobically.
[b] Incubated anaerobically with nitrate added.
[c] Incubated anaerobically with sulfate added.

and in other cases they are not (McMahon et al., 1990). In the case of the lacustrine sediments collected in south-central Washington State (Kieft et al., 1995), residual lake sediment organic carbon exists in these clay-rich, low-permeability strata and is inaccessible to microbes. Disturbance caused by sample collection and handling may result in microbes coming into contact with this residual organic carbon, thus stimulating microbial respiration and reproduction. The benefits of measuring these potential rates of respiration are that they demonstrate the presence of viable heterotrophic microorganisms, they demonstrate the biological lability of the organic carbon, and they enable comparisons among different sites and different subsurface strata.

As stated above, however, the rates measured in microcosms cannot be considered to be estimates of actual *in situ* rates of heterotrophic activity. For example, Kieft et al. (1995) measured rates of CO_2 production from buried lacustrine sediments in microcosms that were as high as 0.664 $\mu g\ g^{-1}\ h^{-1}$. Given the low permeability and slow groundwater flow of these sediments, aerobic heterotrophs functioning at this rate would rapidly deplete available dissolved oxygen. Moreover, given the corresponding total organic C value of 14,700 mg kg^{-1}, available organic carbon would be depleted in less than a decade. Since the age of the organic C in these sediments is at least 6 million years, it is evident that the *in situ* rates of heterotrophic metabolism are many orders of magnitude more sluggish than those measured in laboratory microcosms.

Detection of microbial mineralization of easily degraded ^{14}C-labeled organic substrates, e.g., glucose or acetate, in microcosms of subsurface materials has demonstrated the presence of viable heterotrophic communities. In some cases glucose mineralization occurred in samples from which culturable heterotrophs were not recovered (Kieft et al., 1993). Mineralization of a variety of ^{14}C-labeled organic compounds, including relatively recalcitrant xenobiotic compounds, e.g., chlorobenzene, *p*-chlorophenol, and trichlorobenzene, by microbes in uncontaminated aquifer material has demonstrated the potential for subsurface heterotrophs to degrade a diverse array of compounds (Chapelle et al., 1988; Hicks and Fredrickson, 1989; Swindoll et al., 1988). Microbes from different environments show varied patterns of allocation of C incorporation into biomass vs. mineralization to CO_2. Swindoll et al. (1988)

found that microbes from a shallow aquifer mineralized a lower percentage of the C metabolized (40 to 75%) compared to surface water microbiota (up to 90%). Given this variability in C allocation among microbial communities, mineralization of ^{14}C alone is not an adequate measure of overall microbial metabolism. This shortcoming of the ^{14}C mineralization approach complicates the inherent limitations of microcosm methods.

9.3.2.2 Anaerobic Heterotrophic Activities

As in soils, the porous media of subsurface environments often contain a mixture of aerobic and anaerobic microsites; and thus both aerobes and anaerobes can be isolated from many subsurface environments, and the potential for various aerobic and anaerobic forms of metabolism can be measured in microcosm experiments. In general, groundwater recharge contains some amount of dissolved oxygen upon entering the water table. Thereafter, microbial activity creates a biochemical oxygen demand that slowly reduces the oxygen concentration, thereby increasing the proportion of anaerobic microsites. A predictable spatial pattern occurs along a flow path in which the dominant forms of microbial respiration (i.e., terminal-electron-accepting processes — TEAPs) follow a spatial succession after oxygen depletion, from nitrate-utilizing to CO_2-utilizing microorganisms: (1) nitrate reduction (denitrification) and manganese reduction, (2) iron (III)-reduction, (3) sulfate reduction, and (4) methanogenesis (Lovley and Chapelle, 1995; Lovley et al., 1994). This spatial pattern is also observed in other habitats, e.g., aquatic sediments and biofilms. However, in pristine aquifers, the spatial succession from systems dominated by aerobic heterotrophs to low-redox methanogen-dominated systems may span many kilometers rather than a few millimeters or centimeters (Lovley and Chapelle, 1995; Lovley et al., 1994; Murphy et al., 1992). The great distances required along a confined large-grained aquifer flow path for sequential depletion of electron acceptors may be explained by the general dearth of easily degraded organic electron donors or limiting concentrations of mineral nutrients such as bioavailable phosphate. The broad geographic scale required for electron acceptor depletion also gives strong evidence for the extremely lethargic rates of *in situ* subsurface microbial metabolism. Where concentrations of organic matter occur in the subsurface, e.g., plumes of spilled organic pollutants (Bjerg et al., 1995; Lovley et al., 1994) or pockets of buried organic carbon such as lignite or wood chips (Murphy et al., 1992), TEAPs also occur in a predictable spatial pattern, with methanogens dominating in the most strongly reduced zone immediately surrounding the organic carbon source and aerobic organisms occurring in the oxic zone furthest from the source of the organic reductant. Under these electron-donor-rich conditions, the pattern of TEAPs may cover a few millimeters to a few meters. Fermentative microorganisms occur throughout anoxic zones and are critical for initial degradation of complex organic carbon sources. Their fermentation products (organic acids and H_2) serve as electron donors for the iron-reducing, sulfate-reducing, and methanogenic bacteria.

9.3.2.2.1 Fermentation

Simple organic acids, acetate and formate, have been detected in the pore waters of low-permeability aquitards in Atlantic coastal plain sediments of South Carolina (McMahon and Chapelle, 1991). The same aquitard sediments were shown to generate H_2 upon addition of a fermentable substrate. Jones et al. (1989) detected acetate accumulation in microcosms derived from South Carolina coastal plain sediments, and Murphy et al. (1992) detected acetic acid, formic acid, and oxalic acid in the vicinity of particles of lignite within the Middendorf aquifer in South Carolina. These data suggest that fermentable organic carbon exists in these sediments and that fermentation is occurring. McMahon and Chapelle (1991) further suggested that these fermentation products diffuse from aquitards to sandy aquifers, where they are utilized by respiratory bacteria. Though not specifically studied in most subsurface investigations, fermentation is undoubtedly ubiquitous in subsurface environments containing appreciable microbial activity.

9.3.2.2.2 *Denitrification*

Nitrate is the most energetically favorable alternative to oxygen as an electron acceptor; denitrifying bacteria can use either oxygen or nitrate. Two different studies of denitrification, each using the acetylene blocking technique whereby N_2O is measured as the product of denitrification, and each using core material from deep southeastern coastal plain sediments in South Carolina, detected a potential for denitrification activity at all depths tested (Francis et al., 1989; Morris et al., 1988). In both studies, denitrification rates in the subsurface were considerably slower than those at the surface. In both studies, denitrification rates were stimulated by nitrate amendment; in the Francis et al. study (1989), addition of a carbon source (succinate) had little stimulatory effect on denitrification. These data suggest that denitrification rates are nitrate limited rather than electron-donor limited. This is especially true in the case of groundwater environments contaminated with organic pollutants, e.g., hydrocarbon mixtures, where nitrate may be depleted by denitrifying bacteria (Cozzarelli et al., 1995). However, limitation of denitrification by nitrate availability is not universal in the subsurface. Given the extremely low concentrations of organic carbon in some subsurface environments, denitrification may instead be electron-donor limited. Lind and Eiland (1989) found that N_2O production was stimulated by organic carbon amendment in samples of shallow unsaturated sediments. An aquifer that had been subjected to artificial recharge with treated wastewater effluent had high levels of nitrate and detectable denitrification activity that was limited by available organic C (Smith and Duff, 1988). Similarly, Obenhuber and Lowrance (1991) measured increased rates of denitrification in a nitrate-rich shallow aquifer when organic C was added. Bradley et al. (1992) measured denitrification in microcosms containing shallow anaerobic aquifer sediments from beneath a Florida golf course and found that the N_2O rates, which ranged from below detection to 1.58 nmol g^{-1} h^{-1}, were positively correlated with the organic content. In the aquifer sediments that were most contaminated with nitrate, the potential rates of denitrification rose as high as 60 nmol g^{-1} h^{-1} N_2O when they were amended with 1 mM glucose. Thus, microbial denitrification activity has the potential to remove nitrate from groundwater. However, aquifers containing high levels of nitrate combined with very low levels of organic carbon may lack the potential for denitrification to remove nitrate as a toxin. Similar results were found by Russell et al. (1994) in volcanic tuff of low organic carbon content.

The measured rates of denitrification in laboratory microcosms are undoubtedly high compared to *in situ* values. Even the *in situ* injection method of Trudell et al. (1986) yielded unrealistically high rates of denitrification in a shallow aquifer: 0.0078 to 0.013 g m^{-3} h^{-1} NO_3^--N. This single-well injection-withdrawal method requires that advection through the aquifer be extremely slow (confirmed with the use of Br^- as a conservative tracer). If *in situ* rates of denitrification were actually as high as those measured in this experiment, then nitrate and/or organic C would be very rapidly depleted from the aquifer. The intact-core denitrification measurements made by Smith and Harvey (1991) were 9 to 15 times lower than those measured in sediment slurry incubations, but still probably overestimated *in situ* rates.

9.3.2.2.3   *Iron Reduction*

Lovley et al. (1990) provided convincing evidence of organic carbon oxidation microbially coupled to iron (III) reduction in deep Atlantic coastal plain sediments in South Carolina. Iron-reducing bacteria were isolated from a portion of sediment samples at three sites. All of the sediment samples from which iron reducers were isolated were able to reduce ferric iron in microcosm incubations, while none of the other samples showed this capability. Acetate served as the electron donor in these microcosms. Finally, all of the sediments showing iron-reducing potential contained particles with ferric oxide coatings. Murphy et al. (1992) also detected the potential for iron reduction coupled to organic matter oxidation in an Atlantic coastal plain aquifer. Although rates of *in situ* iron-reducing activity have not been published, the evidence is strong that it occurs in anoxic subsurface sites having the appropriate combination of organic C and ferric iron. The potential for hydrocarbon degradation coupled to iron reduction has also been demonstrated in microcosm experiments using subsurface sediments (Cozzarelli et al., 1995).

9.3.2.2.4   *Sulfate Reduction*

Sulfate-reducing bacteria are strict anaerobes that are found in sulfate-containing, low-redox zones of deep aquifers (Chapelle and McMahon, 1991; Dockins et al., 1980; Johnson and Wood, 1992, 1993; Jones et al., 1989; Olson et al., 1981; Pedersen and Ekendahl, 1990; Stevens et al., 1993), the distal portions of aquifer flow paths (Lovley et al., 1994; Murphy et al., 1992), and in subsurface areas with high concentrations of organic matter, e.g., surrounding organic contaminants (Beeman and Suflita, 1990; Bjerg et al., 1995; Cozzarelli et al., 1995; Lovley et al., 1994) and surrounding naturally occurring organic C deposits such as lignite (Murphy et al., 1992). Energy sources for these organisms are fermentation end products (H_2, organic acids) of anaerobic organisms that partially degrade the organic C. Products are H_2S and sulfide minerals, e.g., pyrite; presence of sulfide in pore waters and/or pyrite within sediments can provide evidence of past or present biological sulfate reduction (Chapelle and McMahon, 1991; Pedersen and Ekendahl, 1990).

  Shanker et al. (1991) tested for sulfate reduction in microcosms containing Atlantic coastal plain sediments. They detected measurable rates in sediments from 3 out of 20 depths tested when sulfate was added; sulfate rates ranged from 0.2 to 1.8 nmol g^{-1} dry wt. day^{-1}; 5 of 20 samples showed measurable sulfate reduction when organic C (lactate, indole, or pyridine) was added, with sulfate rates as high as 5 nmol g^{-1} dry wt. day^{-1}. Thus, sulfate reduction appears to be C-limited in some sediments. In other cases, particularly subsurface environments contaminated with organic pollutants, e.g., hydrocarbons, sulfate may be the factor that limits sulfate-reducing activity (Cozzarelli et al., 1995). As with other microbial activities, sulfate reduction is undoubtedly stimulated in microcosms; actual *in situ* rates may be orders of magnitude slower, since rates of consumption of organic C and/or sulfate in subsurface environments would otherwise outstrip rates of replenishment. Modeling of geochemical processes along an aquifer flow path can provide a rough estimate of rates of sulfate reduction. For example, Plummer et al. (1990) estimated rates of organic matter oxidation in the Madison aquifer of Montana, Wyoming, and South Dakota to be 0.12 μmol l^{-1} pore water per year. It was assumed that sulfate was the dominant electron acceptor and that the majority of organic matter oxidation was coupled to sulfate reduction. Phelps et al. (1994a) dramatically demonstrated the discrepancy between microcosm and *in situ* rates of sulfate reduction. Using microcosms filled with Atlantic coastal plain sediments, measured rates of sulfate reduction ranged from <0.1 to 34 μmol kg^{-1} day^{-1} SO_4^{2-} while *in situ* rates estimated from water chemistry analyses along aquifer flow paths were <0.001 µmol kg^{-1} day^{-1} SO$_4^{2-}$. Thus, in this instance, radiotracer methods in microcosms overestimated sulfate reduction rates by as much as four orders of magnitude.

9.3.2.2.5 Methanogenesis

Biological methane generation occurs in the subsurface under the very lowest oxidation-reduction conditions where alternative electron acceptors, e.g., O$_2$, NO$_3^-$, Fe^{3+}, and SO$_4^{2-}$ have typically been depleted. Under these anaerobic conditions, the biological flow of carbon leads ultimately to methane and CO$_2$. As such, methanogenesis is the dominant process in very deep aquifers (Stevens et al., 1993), in the most distal portions of aquifer flow paths (Murphy et al., 1992), and in the zones immediately surrounding pockets of organic matter (Bjerg et al., 1995; Lovley et al., 1994; Murphy et al., 1992). As with the sulfate reducers, H$_2$ and low molecular weight organic acids serve as the electron donors, while CO$_2$ is the electron acceptor. Indications of biological methanogenesis in the subsurface include: detection of methane as >99% of the alkane gases (i.e., the sum of C2-C5 gases represents less than 1% of the gaseous hydrocarbons) (Rice, 1992), δ^{13}C that are generally lighter than –55‰ and δD values are usually in the range of –150 to –400‰ (Rice, 1992), isolation of methanogenic bacteria (Beeman and Suflita, 1990; Jones et al., 1989; Olson et al., 1981; Stevens et al., 1993), presence of methane in groundwater (Stevens et al., 1993), suitably low oxidation-reduction potential (<–100 mV), and pore water H$_2$ concentrations in a range (7 to 10 nM) typically found where methanogenesis occurs (Lovley et al., 1994).

Very few estimates have been made of *in situ* rates of methanogenesis. Beeman and Suflita (1990) and Adrian et al. (1994) followed seasonal patterns of methane generation in a shallow contaminated aquifer using *in situ* gas sampling devices and calculated rates on the basis of area. Rates varied from below detection to peak rates as high as 560 to 120,000 µmol m$^{2-1}$ δ^{-1} CH$_4$. Surprisingly, *in situ* rates of methane accumulation were similar to those measured in lab microcosms. This suggests that installation of the *in situ* gas sampling device caused some physical disturbance and thus accelerated methanogenesis. Even so, the rates of methanogenesis measured in this contaminated aquifer were one to two orders of magnitude lower than those measured in aquatic sediments. One might assume that uncontaminated deep anoxic aquifers generate methane at rates that are even slower.

9.3.2.3 Chemoautotrophy

Various chemoautotrophic (chemolithotrophic) bacteria have been cultured from deep subsurface environments, but they have not been shown to represent a dominant form of subsurface metabolism. As with other types of organisms, their isolation and culturing from a site does not indicate an appreciable level of activity. Sites where one would be likely to find chemoautotrophs include interfaces between organic-carbon-rich anoxic zones (e.g., organic-rich aquitards) and aerobic zones (e.g., sandy aquifers). The anoxic zones could be sources of reduced inorganics, e.g., H$_2$, HS$^-$, Fe^{3+}, and NH$_3$) that could be oxidized by chemoautotrophs in the presence of oxygen or nitrate. Fredrickson et al. (1989) cultured nitrifiers, sulfur oxidizers, and hydrogen oxidizers from deep coastal plain sediments in South Carolina. However, to date the activities of these chemolithotrophs have not been extensively evaluated in the deep subsurface; and in the majority of subsurface environments studied thus far, chemoautotrophic activities appear to be relatively insignificant.

Neutrophilic iron-oxidizing bacteria (e.g., *Gallionella* spp.) can occur in isolated microenvironments surrounding the well-screen of producing water wells. These organisms proliferate when a well is drilled into an anaerobic Fe(II)-rich aquifer, creating a redox gradient conducive to the growth of iron-oxidizing bacteria (Chapelle, 1993). The accumulated biomass of these chemoautotrophic bacteria may cause problems by clogging well screens. Again, rates of this activity are unavailable, but engineers at municipal drinking water plants can attest to the speed with which biomass can accumulate.

Chemoautotrophic hydrogen-utilizing bacteria, particularly methanogens, may play a unique role in some deep basaltic and granitic aquifers. It was recently observed while drilling in a basaltic subsurface region in south-central Washington State that considerable amounts of H_2 could be generated by mechanical action on basalt (Bjornstad et al., 1994). This suggests that H_2 may also be naturally generated geochemically at slow rates in basaltic subsurface environments. This has led to the intriguing hypothesis that hydrogen-utilizing bacteria may constitute a significant proportion of the subsurface microbial communities in these environments and that primary production by these bacteria may be the basis for a vast subsurface ecosystem (somewhat analogous to the deep-sea hydrothermal vent communities that are based on sulfur oxidizers as primary producers) (Stevens and McKinley, 1995). Rates of microbially mediated H_2 utilization coupled to CO_2 fixation and methanogenesis in the subsurface are currently unavailable, but may prove to be significant.

9.4 Physical and Chemical Factors Limiting Microbial Activities

Microbial activities in subsurface environments are clearly far less than those in surface environments or industrial processes. It is reasonable to assume that physical and chemical constraints limit *in situ* activities (Kieft et al., 1993; Madsen and Ghiorse, 1993; Providenti et al., 1993). Effects of particle size and permeability were shown to constrain biomass and activities of microorganisms in Atlantic coastal plain sediments of the southeastern U.S. to $<0.2\ \mu M\ y^{-1}$ (Phelps et al., 1994b). Particle size of silts and clays limits the flux of water and the flow of aqueous nutrients. However, deltaic and lacustrine sediments of well-sorted fine particle sizes may also contain high amounts of organic carbon. In arid sites of the western U.S. with low primary production and low water flow, total organic carbon (TOC) may be limiting (Kieft et al., 1993, 1995; Severson et al., 1991), and silts containing high TOC harbor significant microbial populations and relatively low activities ($\leq 1\ nM\ y^{-1}$). Phosphorus is a common mineral in many sediments, yet available phosphorus is generally $<0.05\ mg\ l^{-1}$ and appears to limit microbial activities in many subsurface environments. Other nutrients such as nitrogen, sulfate, trace minerals, and readily digestible organics may also limit microbial activities.

Investigations of limiting nutrients in subsurface environments produced unexpected results. Results from Atlantic coastal plain sediments are shown in Table 9.3, but similar trends have been noted from arid sites in the western U.S. (Brockman, personal communication). Mixing of samples appears to redistribute existing nutrients, making them available to resident microorganisms. A similar observation was made in a field demonstration for trichloroethylene (TCE) bioremediation at Westinghouse's Savannah River Site (WSRS), where vacuum extraction or air sparging caused the stimulation of subsurface microbial populations and activities (Hazen et al., 1994; Pfiffner et al., 1995). It was hypothesized that air sparging at 200 standard cubic feet per minute (scfm) or vacuum extraction redistributed nutrients and or toxicants, resulting in increased TCE degrader populations and increased microbial activities even though the total microbial MPN did not increase (Pfiffner et al., 1995).

TABLE 9.3

Effects of Various Treatments on Microbial Activities Observed in Microcosm Experiments Using Atlantic Coastal Plain Sediments From South Carolina. The Number of Replicate Microcosms Showing Stimulation in Response to a Particular Treatment is Shown as a Fraction of the Total Number of Replicates, Followed By a Relative Measure of the Degree of Stimulation (Ranging From "+" for Mild Stimulation to "++++" for Strong Stimulation of Microbial Activities)

| | Sediment Type | | | |
Treatment	Aquifer Sands		Confining Clays	
Phosphate	4/4	+	1/2	++
Nitrate	2/4	++	0/2	
Sulfate	3/4	++	0/2	
Triethylphosphate	2/2	+++	nd	
Nitrous oxide	2/2	+++	nd	
Ammonia	1/2	+	nd	
Glucose	1/4	++	1/2	+
Trace minerals	2/4	++	1/2	+
Water alone	4/4	++	2/2	++++
Mixing alone without water or nutrient amendment	4/4	nd	1/2	nd

Note: nd = not determined.
Data from Phelps et al., 1994.

The addition of excess water in laboratory experiments also contributed to redistributing existing nutrients which enabled detection of increased activities within minutes to hours (Phelps et al., 1994b) (Table 9.3). The addition of excess water to confining silts and clays, which incidentally contained high organic carbon concentrations, resulted in activities being stimulated more than an order of magnitude within hours of slurry mixing. Phosphate additions as low as 10 μM orthophosphate caused stimulation in most sediments where baseline activities were detectable, substantiating the hypothesis that bioavailable phosphorus was a limiting nutrient. Addition of triethylphosphate (TEP), which also added some carbon, caused significant stimulation and appeared to supply some bioavailable phosphorus. Surprisingly, each addition caused stimulation in at least some samples. A trace mineral solution, glucose, sulfate, nitrate, ammonia, or nitrous oxide caused stimulation in one third or more of samples examined. The observation that all additions resulted in stimulation of microbial activities within hours in some samples suggests that subsurface microorganisms may encounter multiple nutrient limitations. Accordingly, microorganisms residing in subsurface environments may endure multiple nutrient deprivation and they may attempt to store scavenged nutrients until such a time that they have sufficient resources to grow or divide. This hypothesis is supported by the observation by Balkwill and Ghiorse (1985) that bacterial cells from subsurface environments appear highly granulated.

It was noted that TEP and nitrous oxide caused a greater level of stimulation than other forms of nitrogen and phosphorus examined in limited studies. Furthermore, they had an additional advantage of being readily dispersed as a gas. Investigators at the WSRS integrated demonstration of TCE bioremediation (Hazen et al., 1994; Pfiffner et al., 1995) decided to inject gaseous N_2O and TEP along with the addition of methane with the intent of stimulating subsurface microbial activities, including the biodegradation for TCE. Within six weeks, TCE degrader populations increased and they were evidenced in greater numbers upon enrichment. Modeling, toxicant concentrations in groundwater, and chloride chemistry substantiated the observation that N_2O and TEP stimulated subsurface microbial populations resulting in increased toxicant degradation. The results indicated that <0.1% of added TEP was likely incorporated into biomass and that the stimulated microbial activities were still less than a millimole per year, with less than 1% of the total *in situ* microbial activity directed towards the cometabolic degradation of TCE and related chlorinated toxicants. Importantly, the laboratory and field experiments developed fundamental conceptual tools for assessing, predicting, and visualizing subsurface microbial activities that were capable of being stimulated and directed toward toxicant bioremediation (Hazen et al., 1994). Furthermore, the observations from this applied bioremediation project were consistent with our conceptual understanding of subsurface microbial ecology.

9.5   *In situ* Activities: Comparisons among Different Environments

Despite severe limitations of direct biomass measures and culture-based estimates, biomass determinations are very useful and efficient tools for constraining microbial biomass and activity estimates. An important consideration is establishing multiple lines of evidence which are indicative of various levels of mass and activity. The apparent absence of PLFA, CFU, and radioisotope mineralization, coupled with unsuccessful attempts to enrich multiple trophic groups, represent lines of evidence suggesting Spartan life and activities in an environment. Similarly, evidence of multiple bacterial populations >10^5 cells per gram is evidence of abundant biomass and diversity. Abundant microbial diversity coupled with electron donors, electron acceptors, and apparent end products such as sulfide or methane are multiple lines of evidence indicative of bacterial activity. Biomass in the absence of electron donors, electron acceptors, or noticeable end products may be indicative of preservation of biomass from previous activities. The observance of end products such as methane or sulfide (Dockins et al., 1980; Jones et al., 1989; Rice, 1992) may be considered indicative of current or recent biological activity, particularly if isotopic compositions support their biological origin. Methane accumulations of nonbiological origin are common in many basins and often represent the previous migration of thermally derived natural gas rather than recent biological activity. Consequently, the concept of using biomass determinations to constrain the extent of biological activity in a subsurface environment should rely on multiple lines of evidence supporting the interpretations, including a hydrodynamic basin-wide perspective (Rice, 1992).

A common theme in subsurface investigations where interpretations are drawn from recovered samples is that laboratory-derived activities do not relate to *in situ* rates. Studies of Atlantic coastal plain aquifers confirm this view; data from saturated zones of arid sites enable further comparisons spanning several orders of magnitude. Unfortunately, when the observed rates are extrapolated to three-dimensional subsurface environments, the activities are orders of magnitude faster than what available electron donors and acceptors could support. Moreover, these laboratory-derived rates of heterotrophic activity would quickly remove total particulate carbon. Phelps et al. (1994a) recently suggested that many of these laboratory-based values overestimate *in situ* rates by factors of 10^3 to 10^6 over what geochemical models and knowledge of groundwater flows would substantiate. Many investigators have reported rapid utilization of radiolabeled substrates corresponding to turnover times of hours to weeks in laboratory microcosm experiments. For example, Phelps et al. (1994a) and Chapelle and Lovley (1990) reported turnover times for glucose in aquifers to be days to weeks. Hicks and Fredrickson (1989) and Kieft et al. (1993, 1995) reported high percentages of glucose and acetate mineralization within weeks. Unfortunately, *in situ* aerobic turnover rates of years or organic oxidation rates of mil- limoles per year would likely require all of the available oxygen and render the sub- surface anaerobic, indicating that estimating *in situ* rates from laboratory-derived experiments may be inappropriate.

Studies showing the widest disparity between laboratory and *in situ* measurements may well be samples from semiarid sites with large vadose zones extending tens to hundreds of meters beneath the land surface. In these semiarid vadose zones of the western U.S., radioisotope time course experiments typically showed glucose miner- alization, acetate incorporation, and acetate mineralization transformations compat- ible with 7- to 60-day turnover times (Brockman et al., 1992; Kieft et al., 1993, 1995; Phelps, unpublished data). Based on the available pools of substrates and specific activities of the radioisotopes employed in the experiments, extrapolated *in situ* met- abolic rates ≥ 1.3 mM y^{-1} can be calculated. While such rates may seem plausible in shallow rapid flowing aquifers, they appear unreasonable in vadose zones more than 30 m thick where vertical recharge is less than 1 mm y^{-1} and TOC is <1%. According to geochemical models, the *in situ* metabolic rate for CO_2 evolution may be ≤ 1 nM y^{-1}, or several orders of magnitude slower than laboratory-based interpretations.

To put subsurface microbial activities in perspective, it may be useful to compare estimated *in situ* activities, based on the production of terminal end products, with microbial activities from other ecosystems (Figure 9.1). The rate of carbon dioxide production via microbial respiration can be used as a tool for assessing and compar- ing biological activities from various environments. Figure 9.1 shows estimates of CO_2 production rates by microorganisms in surface and subsurface environments extending to rock formations >2 km below the earth's surface. The values in the figure are based on geochemical models and/or chemical measurements extrapolated to moles of CO_2 evolved per liter volume each year. To extend these comparisons to bio- logical reactors one could consider anaerobic sludge digestors (Conrad et al., 1985; Phelps, 1991), which exhibit 10 to 100 times greater activity than surface soils, and industrial fermentors (Webb et al., 1995) with rates that are 1 to 5 orders of magnitude greater than those observed in surface soils (Chapatwala et al., 1996; Phelps et al., 1994c; Siegrist et al., 1994). Surface and sediment environments are habitats where considerable growth of microorganisms can occur. In subsurface environments pro- ducing orders of magnitude less CO_2, growth may be extremely slow and average doubling times for the microbial community may exceed decades or centuries (Phelps et al., 1994a), with survival or preservation being more important than growth.

Deep-sea waters (Williams and Carlucci, 1976) and sediments (Jahnke et al., 1982; Jorgensen, 1982) would be expected to produce less CO_2 than lake sediments (Phelps, 1991; Phelps and Zeikus, 1984, 1985) or surface soils. Subsurface aquifers produce CO_2 at less than 0.01% of the rates of surface soils. Even though subsurface aquifers contain diverse microbial communities, their metabolic rates are significantly slower than surface soils or shallow sediments. Subsurface clays, deep arid sediments, or deep rock formations — environments through which little water passes — produce low quantities of CO_2 and offer little opportunity for microbial growth. In isolated low-flux environments spatially and geologically removed from surface phenomena, the long-term preservation of microbial life far exceeds the capacity and opportunity for growth. Despite the harshness of deeply isolated environments, diverse microbial communities are observed (Boone et al., 1995; Zhang et al., 1996), and techniques that are capable of assessing microbial metabolism at rates exceeding 10^5 M y^{-1} can also be used to constrain microbial activities in deep subsurface environments to ≤ 1 pM y^{-1}.

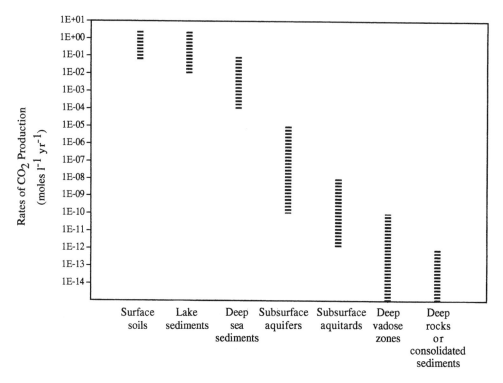

FIGURE 9.1

Ranges of rates of *in situ* CO$_2$ production for various surface and subsurface environments, as estimated by ground water chemical analyses and geochemical modeling. Data are based on references: Chapelle and Lovely, 1990; Murphy et al., 1992; Phelps et al., 1994; Lovely and Chapelle, 1995; McMahon and Chapelle, 1991; Phelps, 1991; Chapatwala et al., 1996; Phelps et al., 1994; Siegrist et al., 1994; Jahnke et al., 1982; Jorgensen, 1982; Phelps and Zeikus, 1984; E. Murphy, unpublished data; and T. J. Phelps, unpublished data.

In situ microbial activities in confined Atlantic coastal plain aquifer sediments tens to hundreds of meters beneath land surface receiving groundwater hundreds to thousands of years after recharge are estimated to be 0.1 to 10 µM y^{-1} (Chapelle and Lovley, 1990; Lovley and Chapelle et al., 1995; McMahon and Chapelle, 1991; Murphy et al., 1992; Phelps et al., 1994a). Semiarid vadose zones greater than 50 m below land surface which receive little vertical recharge and have very little organic C are likely to have microbial activities <0.1 nM y^{-1}. Although not specifically determined, microbial activities in the sedimentary rocks of the Taylorsville Triassic Basin were estimated to be less than picomoles per year. These slow rates are consistent with available electron acceptors and donors as well as little accumulation of end products such as methane, CO$_2$, HS$^-$, or reduced iron despite groundwater age of millions of years and groundwater flows of millimeters per year. Although these estimated activity rates match available groundwater and sediment chemistries they suggest extremely slow metabolic activities on a cellular basis, which could correspond to slow growth rates for the average resident microorganism.

The advent of bioremediation has heightened the significance, sophistication, and application of assessing biologically mediated metabolism in subsurface environments (see Chapter 14). In contrast to regional groundwater flow path models, bioremediation efforts typically require detailed analyses in close spatial orientation of cubic hectares or less (see Chapter 6). Consequently, multiple sample points are typically examined in and adjacent to the treatment site, with site-specific heterogeneities being more important than regional variables. The use of stable isotopes or chemical tracers, diffusion of single or multiple inert gases, and fine-scale physical and chemical analyses are often employed to characterize the site. Contaminants and their breakdown products, along with exogenously added nutrients, are factored into available electron donor and electron acceptor fluxes. Less than ideal dispersion of nutrients, alleviation of nutrient limitations, nonrandom positioning of contaminants, and related hydrophobicity and toxicity all compound the difficulties of characterization, visualization, and modeling of bioremediation. However, geochemically based visualizations and modeling are becoming important tools for planning, evaluating, assessing, and communicating impacts of treatment regimens during bioremediation.

Contaminated subsurface environments may be considered extreme for reasons that are different from pristine subsurface environments. Organic contamination, e.g., by hydrocarbons, can convert an electron donor-limited system into an extremely anaerobic electron acceptor-limited system. Contaminants such as solvents or heavy metals may also be directly toxic to the microorganisms. Despite these stresses, microbial activities occur in the vicinity of contaminant plumes and can contribute to environmental remediation (Adrian et al., 1994; Beeman and Suflita, 1990; Bjerg et al., 1995; Cozzarelli et al., 1995; Lovley et al., 1994; Phelps et al., 1988). The rates of microbial activities in these contaminated environments may be higher than those of uncontaminated environments (Phelps et al., 1989b, 1988); however, they are generally limited by environmental conditions and thus remain comparatively slow. Addition of electron acceptors, inorganic nutrients, cometabolites, etc. may stimulate beneficial biological processes, thereby accelerating detoxification efforts in subsurface environments (Hazen et al., 1994; Palumbo et al., 1995; Pfiffner et al., 1995). Quantifying *in situ* rates of biological activity in contaminated subsurface environments has rarely been attempted, but may prove useful for further refinement of bioremediation techniques.

9.6 Summary

Microbiological activities in uncontaminated subsurface habitats proceed at rates that are orders of magnitude slower than those encountered in nearly every other portion of the biosphere. As such, these environments may be considered to be among the most extreme on the planet from the standpoint of energy and nutrient flux. Given the sequestration of microbes for millennia within many of these subsurface habitats, there exists extremely strong selective pressure for long-term starvation survival. In some cases, particularly in deep unsaturated zones and fine-grained low-permeability sediments, individual microorganisms may survive through geologic time by endogenous metabolism alone. In other cases, e.g., sandy aquifers, microbial metabolism appears to be fueled by a trickle of exogenous nutrient sources; however, the rates of activity are still so slow as to be measurable only by quantifying microbially catalyzed geochemical changes occurring over tens of kilometers. Stuck in the "slow lane" as they are, subsurface microorganisms still exert profound effects on geochemical processes occurring over geological time scales (thousands to millions of years) and over long distances (tens of kilometers) within regional groundwater aquifers.

References

Adrian, N. F., J. A. Robinson, and J. M. Suflita. 1994. Spatial variability in biodegradation rates as evidenced by methane production from an aquifer. *Appl. Environ. Microbiol.*, 60:3632.

Aelion, C. M. and P. M. Bradley. 1991. Aerobic degradation potential of subsurface marine microorganisms from a jet-fuel-contaminated aquifer. *Appl. Environ. Microbiol.*, 57:57.

Albrechtsen, H.-J. and A. Winding. 1992. Microbial biomass and activity in subsurface sediments from Vejen, Denmark. *Microb. Ecol.*, 23:303.

Amy, P. S., D. L. Haldeman, D. Ringelberg, D. H. Hall, and D. Russell. 1992. Comparison of identification systems for classification of bacteria isolated from water and endolithic habitats within the deep subsurface. *Appl. Environ. Microbiol.*, 58:3367.

Balkwill, D. L. 1989. Numbers, diversity, and morphological characteristics of aerobic, chemo-heterotrophic bacteria in deep subsurface sediments from a site in South Carolina. *Geomicrobiol. J.*, 7:33.

Balkwill, D. L. and W. C. Ghiorse. 1985. Characterization of subsurface bacteria associated with two shallow aquifers in Oklahoma. *Appl. Environ. Microbiol*, 50:580.

Barbaro, S. E., H.-J. Albrechtsen, B. K. Jensen, C. I. Mayfield, and J. F. Barker. 1994. Relationships between aquifer properties and microbial populations in the Borden aquifer. *Geomicrobiol. J.*, 12:203.

Beeman, R. E. and J. M. Suflita. 1990. Environmental factors influencing methanogenesis in a shallow anoxic aquifer: a field and laboratory study. *J. Ind. Microbiol.*, 5:45.

Beloin, R. M., J. L. Sinclair, and W. C. Ghiorse. 1988. Distribution and activity of microorganisms in subsurface sediments of a pristine study site in Oklahoma. *Microb. Ecol.*, 16:85.

Bjerg, P. L., K. Rugge, J. K. Pedersen, and T. H. Christensen.1995. Distribution of redox-sensitive groundwater quality parameters downgradient of a landfill (Grindsted, Denmark). *Environ. Sci. Technol.*, 29:1387.

Bjornstad, B. N., J. P. McKinley, T. O. Stevens, S. A. Rawson, J. K. Fredrickson, and P. E. Long. 1994. Generation of hydrogen gas as a result of drilling within the saturated zone. *Ground Water Monitoring Remediation*, 14:140.

Boone, D. R., Y. Liu, Z. Zhong-Ju, D. L. Balkwill, G. R. Drake, T. O. Stevens, and H. C. Aldrich. 1995. *Bacillus infernus* sp. nov., an Fe (III)- and Mn (IV)-reducing anaerobe from a deep terrestrial subsurface. *Int. J. Syst. Bacteriol.*, 45:441.

Bradley, P. M., M. Fernandez, and F. H. Chapelle. 1992. Carbon limitation of denitrification rates in an anaerobic groundwater system. *Environ. Sci. Technol.*, 26:2377.

Brockman, F. J., T. L. Kieft, J. K. Fredrickson, B. N. Bjornstad, S. W. Li, W. Spangenburg, and P. E. Long. 1992. Microbiology of vadose zone paleosols in south-central Washington State. *Microb. Ecol.*, 23:279.

Brockman, F. J., D. B. Ringelberg, D. C. White, and J. K. Fredrickson. Comparison of direct counts, viable biomass, and culturable microorganisms in sediments from six deep boreholes (in preparation).

Buswell, A. M. and T. E. Larson. 1937. Methane in ground waters. *J. Am. Water Works Assoc.*, 29:1978.

Chapatwala, K. D., G. V. Babu, E. Armstead, A. V. Palumbo, C. Zhang, and T. J. Phelps. Effects of micronutrients on microbial respiration in a shallow coastal subsurface and vadose zone. *Appl. Biochem. Biotechnol.*, in press.

Chapelle, F. H. 1993. *Ground-Water Microbiology and Geochemistry.* John Wiley & Sons, New York, chap. 9.

Chapelle, F. H. and D. R. Lovley. 1990. Rates of microbial activity in deep coastal plain aquifers. *Appl. Environ. Microbiol.*, 56:1865.

Chapelle, F. H. and P. B. McMahon. 1991. Geochemistry of dissolved inorganic carbon in a coastal plain aquifer. I. Sulfate from confining beds as an oxidant in microbial CO_2 production. *J. Hydrol.*, 127:85.

Chapelle, F. H., J. T. Morris, P. B. McMahon, and J. L. Zelibor. 1988. Bacterial metabolism and the $d^{13}C$ composition of groundwater, Floridan aquifer system, South Carolina. *Geology*, 16:117.

Chapelle, F. H., J. L. Zelibor, D. J. Grimes, and L. L. Knobel. 1987. Bacteria in deep coastal plain sediments of Maryland: a possible source of CO_2 to groundwater. *Water Resour. Res.*, 23:1625.

Colwell, F. S. 1989. Microbiological comparison of surface soil and unsaturated subsurface soil from a semiarid high desert. *Appl. Environ. Microbiol.*, 55:2420.

Conrad, R., T. J. Phelps, and J. G. Zeikus. 1985. Gas metabolism evidence in support of the juxtaposition of hydrogen-producing and methanogenic bacteria in sewage sludge and lake sediments. *Appl. Environ. Microbiol.*, 50:595.

Cozzarelli, I. M., J. S. Herman, and M. J. Baedecker. 1995. Fate of microbial metabolites of hydrocarbons in a coastal plain aquifer: the role of electron acceptors. *Environ. Sci. Technol.*, 29:458.

Dobbins, D. C. and F. K. Pfaender. 1988. Methodology for assessing respiration and cellular incorporation of radiolabeled substrates by surface and subsurface soil microbial communities. *Microb. Ecol.*, 15:257.

Dockins, W. S., G. J. Olson, G. A. McFeters, and S. C. Turbak. 1980. Dissimilatory bacterial sulfate reduction in Montana groundwaters. *Geomicrobiol. J.*, 2:83.

Ekendahl, S. and K. Pedersen. 1994. Carbon transformations by attached bacterial populations in granitic groundwater from deep crystalline bed-rock of the Stripa research mine. *Microbiology*, 140:1565.

Findlay, R. H., P. C. Pollard, D. J. W. Moriarty, and D. C. White. 1985. Quantitative determination of microbial activity and community nutritional status in estuarine sediments: evidence for a disturbance artifact. *Can. J. Microbiol.*, 31:493.

Francis, A. J., J. M. Slater, and C. J. Dodge. 1989. Denitrification in deep subsurface sediments. *Geomicrobiol. J.*, 7:103.

Fredrickson, J. K., D. L. Balkwill, J. M. Zachara, S. W. Li, F. J. Brockman, and M. A. Simmons. 1991. Physiological diversity and distributions of heterotrophic bacteria in deep Cretaceous sediments of the Atlantic coastal plain. *Appl. Environ. Microbiol.*, 57:402.

Fredrickson, J. K., T. T. Garland, R. J. Hicks, J. M. Thomas, S. W. Li, and S. M. McFadden. 1989. Lithotrophic and heterotrophic bacteria in deep subsurface sediments and their relation to sediment properties. *Geomicrobiol. J.*, 7:53.

Fredrickson, J. K., S. W. Li, F. J. Brockman, D. L. Haldeman, P. S. Amy, and D. L. Balkwill. 1995. Time-dependent changes in viable numbers and activities of aerobic heterotrophic bacteria in subsurface samples. *J. Microbiol. Methods*, 21:253.

Ghiorse, W. C. and D. L. Balkwill. 1983. Enumeration and morphological characterization of bacteria indigenous to subsurface sediments. *Dev. Ind. Microbiol.*, 24:213.

Ghiorse, W. C. and J. T. Wilson. 1988. Microbial ecology of the terrestrial subsurface. *Adv. Appl. Microbiol.*, 33:107.

Gillham, R. W., R. C. Starr, and D. J. Miller. 1990. A device for *in situ* determination of geochemical transport parameters. II. Biochemical reactions. *Ground Water*, 28:858.

Haldeman, D. L., P. S. Amy, D. Ringelberg, and D. C. White. 1993. Characterization of microbiology within a 21 m^3 section of rock from the deep subsurface. *Microb. Ecol.*, 26:145.

Haldeman, D. L., P. S. Amy, D. C. White, and D. B. Ringelberg. 1994. Changes in bacteria recoverable from subsurface volcanic rock samples during storage at 4°C. *Appl. Environ. Microbiol.*, 60:2697.

Haldeman, D. L., P. S. Amy, T. Kieft, W. Spangenberg, D. Ringelberg and D. C. White. 1995a. The microbiology of a chronosequence of buried paleosols from the channeled scabland of arid eastern Washington. Abstr. (number 1172) Am. Soc. Microbiol. Meetings, Washington, D.C.

Haldeman, D. L., P. S. Amy, D. Ringelberg, D. C. White, R. E. Garen, and W. C. Ghiorse. 1995b. Microbial growth and resuscitation alter community structure after perturbation. *FEMS Microb. Ecol.*, 17:27-38.

Hazen, T. C., L. Jimenez, G. Lopez de Victoria, and C. B. Fliermans. 1991. Comparison of bacteria from deep subsurface sediment and adjacent groundwater. *Microb. Ecol.*, 22:293.

Hazen, T. C., K. H. Lombard, B. B. Looney, M. V. Enzien, J. M. Dougherty, C. B. Fliermans, J. Wear, and C. A. Eddy-Dilek. 1994. Summary of *in situ* bioremediation demonstration (methane biostimulation) via horizontal wells at the Savannah River Site integrated demonstration project, in In Situ *Remediation: Scientific Basis for Current and Future Technologies*, G. W. Gee, and R. R. Wing, Eds., Battelle Press, Richland, WA, pp. 137.

Hicks, R. J. and J. K. Fredrickson. 1989. Aerobic metabolic potential of microbial populations indigenous to deep subsurface environments. *Geomicrobiol. J.,* 7:67.

Hiebert, F. K. and P. C. Bennett. 1992. Microbial control of silicate weathering in organic-rich groundwater. *Science,* 258:278.

Hirsch, P. and E. Rades-Rohkohl. 1988. Some special problems in the determination of viable counts of groundwater microorganisms. *Microb. Ecol.,* 16:99.

Hirsch, P. and E. Rades-Rohkohl. 1990. Microbial colonization of aquifer sediment exposed in a groundwater well in northern Germany. *Appl. Environ. Microbiol.,* 56:2963.

Insam, H. and K. H. Domsch. 1988. Relationship between soil organic carbon and microbial biomass on chronosequences of reclamation sites. *Microb. Ecol.,* 15:177.

Jahnke, R. A., S. R. Emerson, and J. W. Murray. 1982. A model of oxygen reduction, denitrification, and organic matter mineralization in marine sediments. *Limnol. Oceanogr.,* 27:610.

Johnson, A. C. and M. Wood. 1992. Microbial potential of sandy aquifer material in the London basin. *Geomicrobiol. J.,* 10:1.

Johnson, A. C. and M. Wood. 1993. Sulfate-reducing bacteria in deep aquifer sediments of the London Basin: their role in anaerobic mineralization of organic matter. *J. Appl. Bacteriol.,* 75:190.

Jones, R. E., R. E. Beeman, and J. M. Suflita. 1989. Anaerobic metabolic processes in the deep terrestrial subsurface. *Geomicrobiol. J.,* 7:117.

Jorgensen, B. B. 1982. Mineralization of organic matter in the sea bed: the role of sulphate reduction. *Nature,* 296:643.

Kazumi, J. and D. G. Capone. 1994. Heterotrophic microbial activity in shallow aquifer sediments of Long Island, New York. *Microb. Ecol.,* 28:19.

Kieft, T. L. 1995. Dwarf cells in soil and subsurface terrestrial environments, in *Non-Culturable Microorganisms in the Environment,* R. R. Colwell and D. J. Grimes, Eds., Chapman & Hall, New York. In press.

Kieft, T. L., P. S. Amy, F. J. Brockman, J. K. Fredrickson, B. N. Bjornstad, and L. L. Rosacker. 1993. Microbial abundance and activities in relation to water potential in the vadose zones of arid and semiarid sites. *Microb. Ecol.,* 26:59.

Kieft, T. L., J. K. Fredrickson, J. P. McKinley, B. N. Bjornstad, S. A. Rawson, T. J. Phelps, F. J. Brockman, and S. M. Pfiffner. 1995. Microbiological comparisons within and across contiguous lacustrine, paleosol, and fluvial subsurface sediments. *Appl. Environ. Microbiol.,* 61:749.

Kieft, T. L., D. B. Ringelberg, and D. C. White. 1994. Changes in ester-linked phospholipid fatty-acid profiles of subsurface bacteria during starvation and desiccation in porous media. *Appl. Environ. Microbiol.,* 60:3292.

Kieft, T. L. and L. L. Rosacker. 1991. Application of respiration- and adenylate-based soil microbiological assays to deep subsurface terrestrial sediments. *Soil Biol. Biochem.,* 23:563.

Kogure, K., U. Simidu, and N. Taga. 1979. A tentative direct microscopic method for counting living marine bacteria. *Can. J. Microbiol.,* 25:415.

Kolbel-Boelke, J., E.-M. Anders, and A. Nehrkorn. 1988. Microbial communities in the saturated groundwater environment. II. Diversity of bacterial communities in a Pleistocene sand aquifer and their *in vitro* activities. *Microb. Ecol.,* 16:31.

Konopka, A. and R. Turco. 1991. Biodegradation of organic compounds in vadose zone and aquifer sediments. *Appl. Environ. Microbiol.,* 57:2260.

Lind, A.-M. and F. Eiland. 1989. Microbiological characterization and nitrate reduction in subsurface soils. *Biol. Fertil. Soils,* 8:197.

Lovley, D. R. and F. H. Chapelle. 1995. Subsurface Microbial Processes. *Rev. Geophys.,* 33:365.

Lovley, D. R., F. H. Chapelle, and E. J. P. Phillips. 1990. Fe(III)-reducing bacteria in deeply buried sediments of the atlantic coastal plain. *Geology,* 18:954.

Lovley, D. R., F. H. Chapelle, and J. C. Woodward. 1994. Use of dissolved H_2 concentrations to determine distribution of microbially catalyzed redox reactions in anoxic groundwater. *Environ. Sci. Technol.,* 28:1205.

Madsen, E. L. and J.-M. Bollag. 1989. Aerobic and anaerobic microbial activity in deep subsurface sediments from the Savannah River Plant. *Geomicrobiol. J.,* 7:93.

Madsen, E. L. and W. C. Ghiorse. 1993. Ground water microbiology: subsurface ecosystems processes. In *Aquatic Microbiology: An Ecological Approach,* T. Ford, Ed., Blackwell Scientific, Cambridge, MA.

Marxsen, J. 1988. Investigations into the number of respiring bacteria in groundwater from sandy and gravelly deposits. *Microb. Ecol.,* 16:65.

McMahon, P. B. and F. H. Chapelle. 1991. Microbial production of organic acids in aquitard sediments and its role in aquifer geochemistry. *Nature,* 349:233.

McMahon, P. B., D. F. Williams, and J. T. Morris. 1990. Production and carbon isotopic composition of bacterial CO_2 in deep coastal plain sediments of South Carolina. *Ground Water,* 28:693.

Morris, J. T., G. J. Whiting, and F. H. Chapelle. 1988. Potential denitrification rates in deep sediments from the southeastern coastal plain, *Environ. Sci. Technol.,* 22:832.

Murphy, E. M., J. A. Schramke, J. K. Fredrickson, H. W. Bledsoe, A. J. Francis, D. S. Sklarew, and J. C. Linehan. 1992. The influence of microbial activity and sediment organic carbon on the isotope geochemistry of the Middendorf Aquifer. *Water Resour. Res.,* 28:723.

Obenhuber, D. C. and R. Lowrance. 1991. Reduction of nitrate in aquifer microcosms by carbon additions. *J. Environ. Qual.,* 20:255.

Olson, G. J., W. S. Dockins, G. A. McFeters, and W. P. Iverson. 1981. Sulfate-reducing and methanogenic bacteria from deep aquifers in Montana. *Geomicrobiol. J.,* 2:327.

Palumbo, A. V., S. P. Scarborough, C. Zhang, S. M. Pfiffner, and T. J. Phelps. 1995. Influence of nitrogen and phosphorus on the *in situ* bioremediation of trichloroethylene. *Appl. Biochem. Biotechnol.,* 51/52:637.

Pedersen, K. and S. Ekendahl. 1990. Distribution and activity of bacteria in deep granitic groundwaters of southeastern Sweden. *Microb. Ecol.,* 20:37.

Pfiffner, S. M., D. B. Ringelberg, D. B. Hedrick, T. J. Phelps, and A. V. Palumbo. 1995. Subsurface microbial communities and degradative capacities during trichloroethylene degradation. In *Bioremediation of Chlorinated Solvents,* R. E. Hinchee, A. Leeson, and L. Semprini, Eds., Battelle Press, Richland, WA, pp. 263.

Phelps, T. J. 1991. Similarity between transformation rates and turnover rates of organic matter biodegradation in anaerobic environments. *J. Microbiol. Methods,* 13:243.

Phelps, T. J., D. B. Hedrick, D. Ringelberg, C. B. Fliermans, and D. C. White. 1989a. Utility of radiotracer activity measurements for subsurface microbiology studies. *J. Microbiol. Methods,* 9:15.

Phelps, T. J., E. M. Murphy, S. M. Pfiffner, and D. C. White. 1994a. Comparison between geochemical and biological estimates of subsurface microbial activities. *Microb. Ecol.,* 28:335.

Phelps, T. J., S. M. Pfiffner, K. A. Sargent, and D. C. White. 1994b. Factors influencing the abundance and metabolic capabilities of microorganisms in eastern coastal plain sediments. *Microb. Ecol.,* 28:351.

Phelps, T. J., E. G. Raione, D. C. White, and C. B. Fliermans. 1989b. Microbial activities in deep subsurface environments. *Geomicrobiol. J.,* 7:79.

Phelps, T. J., D. Ringelberg, D. Hedrick, J. Davis, C. B. Fliermans, and D. C. White. 1988. Microbial biomass and activities associated with subsurface environments contaminated with chlorinated hydrocarbons. *Geomicrobiol. J.,* 6:157.

Phelps, T. J., R. L. Siegrist, N. E. Korte, D. A. Pickering, J. M. Strong-Gunderson, A. V. Palumbo, J. F. Walker, C. M. Morrissey, and R. Mackowski. 1994c. Bioremediation of petroleum hydrocarbons in soil column lysimeters from Kwajalein Island. *Appl. Biochem. Biotechnol.,* 45/46:835.

Phelps, T. J. and J. G. Zeikus. 1984. Influence of pH on terminal carbon metabolism in anoxic sediments from a mildly acidic lake. *Appl. Environ. Microbiol.,* 48:1088.

Phelps, T. J. and J. G. Zeikus. 1985. Effect of fall turnover on terminal carbon metabolism in Lake Mendota sediments. *Appl. Environ. Microbiol.,* 50:1285.

Plummer, L. N., J. F. Busby, R. W. Lee, and B. B. Hanshaw. 1990. Geochemical modeling of the Madison aquifer in parts of Montana, Wyoming, and South Dakota. *Water Resour. Res.,* 26:1981.

Providenti, M. A., H. Lee, and J. T. Trevors. 1993. Selected factors limiting the microbial degradation of recalcitrant compounds. *J. Ind. Microbiol.,* 12:379.

Rice, D. D. 1992. Controls, habitat, and resource potential of ancient bacterial gas. In: R. Vilay, Ed., *Bacterial Gas,* Editions Technip, Paris, pp. 91.

Rodriguez, G. G., D. Phipps, K. Ishiguro, and H. F. Ridgway. 1992. Use of a fluorescent redox probe for direct visualization of actively respiring bacteria. *Appl. Environ. Microbiol.,* 58:1801.

Roszak, D. B. and R. R. Colwell. 1987. Survival strategies of bacteria in the natural environment. *Microbiol. Rev.,* 51:363.

Russell, C. E., R. Jacobson, D. L. Haldeman, and P. S. Amy. 1994. Heterogeneity of deep subsurface microorganisms and correlations to hydrogeological and geochemical parameters. *Geomicrobiol. J.,* 12:37.

Rusterholtz, K. J. and L. M. Mallory. 1994. Density, activity, and diversity of bacteria indigenous to a karstic aquifer. *Microb. Ecol.,* 28:79.

Sarbu, S. M., B. K. Kinkle, L. Vlasceanu, and T. C. Kane. 1994. Microbiological characterization of a sulfide-rich groundwater ecosystem. *Geomicrobiol. J.,* 12:175.

Severson, K. J., D. L. Johnstone, C. K. Keller, and B. D. Wood. 1991. Hydrogeologic parameters affecting vadose-zone microbial distributions. *Geomicrobiol. J.,* 9:197.

Shanker, R., J.-P. Kiser, and J.-M. Bollag. 1991. Microbial transformation of heterocyclic molecules in deep subsurface sediments. *Microb. Ecol.,* 22:305.

Siegrist, R. L., T. J. Phelps, N. E. Korte, D. A. Pickering, R. Mackowski, and L. W. Cooper. 1994. Characterization and biotreatability of petroleum contaminated soils in a coral atoll in the Pacific Ocean. *Appl. Biochem. Biotechnol.,* 45/46:757.

Sinclair, J. L., and W. C. Ghiorse. 1989. Distribution of aerobic bacteria, protozoa, algae, and fungi in deep subsurface sediments. *Geomicrobiol. J.,* 7:15.

Smith, R. L. and J. H. Duff. 1988. Denitrification in a sand and gravel aquifer. *Appl. Environ. Microbiol.* 54:1071.

Smith, R. L. and R. W. Harvey. 1991. Development of sampling techniques to measure *in situ* rates of microbial processes in a sand and gravel aquifer. In *Proc. First Int. Symp. Microbiology of the Deep Subsurface,* C. B. Fliermans, and T. C. Hazen, Eds., WSRC Information Services, Aiken, South Carolina, pp. 2-19 to 2-33.

Stevens, T. O. and B. S. Holbert. 1995. Variability and density dependence of bacteria in terrestrial subsurface samples: implications for enumeration. *J. Microbiol. Methods,* 21:283.

Stevens, T. O. and J. P. McKinley. 1995. Geochemically produced hydrogen supports microbial ecosystems in deep basalt aquifers. *Science,* 270:450-454.

Stevens, T. O., J. P. McKinley, and J. K. Fredrickson. 1993. Bacteria associated with deep, alkaline, anaerobic groundwaters in southeast Washington. *Microb. Ecol.,* 25:35.

Swindoll, C. M., C. M. Aelion, D. C. Dobbins, O. Jiang, S. C. Long, and F. K. Pfaender. 1988. Aerobic biodegradation of natural and xenobiotic organic compounds by subsurface microbial communities. *Environ. Toxicol. Chem.,* 7:291.

Thorn, P. M. and R. M. Ventullo. 1988. Measurement of bacterial growth rates in subsurface sediments using incorporation of tritiated thymidine into DN. *Microb. Ecol.,* 16:3.

Trudell, M. R., R. W. Gillham, and J. A. Cherry. 1986. An *in situ* study of the occurrence and rate of denitrification in a shallow unconfined sandy aquifer. *J. Hydrol.,* 83:251.

Webb, O. F., B. H. Davison, T. C. Scott, and C. D. Scot. 1995. Design and demonstration of an immobilized-cell fluidized-bed reactor for the efficient production of ethanol. *Appl. Biochem. Biotechnol.,* 51/52:559.

White, D. C., R. J. Bobbie, J. D. King, J. Nickels, and P. Amoe. 1979a. Lipid analysis of sediments for microbial biomass and community structure. in: *Methodology for Biomass Determinations and Microbial Activities in Sediments,* (ASTM STP 673), C. D. Litchfield and P. L. Seyfried, Eds., American Society for Testing Materials, Philadelphia, pp. 87.

White, D. C., W. M. Davis, J. S. Nickels, J. D. King, and R. J. Bobbie. 1979b. Determination of sedimentary microbial biomass by extractable lipid phosphate. *Oecologia,* 40:51.

White, D. C., D. B. Ringelberg, J. B. Guckert, and T. J. Phelps. 1991. Biochemical markers for *in situ* microbial community structure. in *Proc. First Int. Symp Microbiology of the Deep Subsurface*, C. B. Fliermans, and T. C. Hazen, Eds., WSRC Information Services, Aiken, South Carolina, pp. 4-45 to 4-56.

White, D. C., G. A. Smith, M. H. Gehron, J. H. Parker, R. H. Findlay, R. F. Martz, and H. F. Fredrickson. 1983. The groundwater aquifer microbiota: biomass, community structure, and nutritional status. *Dev. Ind. Microbiol.*, 24:201.

Williams, P. M. and F. Carlucci. 1976. Bacterial utilization of organic matter in the deep sea. *Nature*, 262:810.

Wilson, J. T., J. F. McNabb, D. L. Balkwill, and W. C. Ghiorse. 1983. Enumeration and characterization of bacteria indigenous to a shallow water table aquifer. *Ground Water*, 21:134.

Wood, B. D., C. K. Keller, and D. L. Johnstone. 1993. In situ measurement of microbial activity and controls on microbial CO_2 production in the unsaturated zone. *Water Resour. Res.*, 29:645.

Zhang, C., S. Liu, J. Logan, R. Mazumder, and T. J. Phelps. Enhancement of Fe (III), Co (III), and Cr (VI) reduction at elevated temperatures and by a thermophilic bacterium. *Appl. Biochem. Biotechnol.*, In Press.

Zimmerman, R., R. M. Iturriaga, and J. Becker-Birck. 1978. Simultaneous determination of the number of aquatic bacteria and the number thereof involved in respiration. *Appl. Environ. Microbiol.*, 36:926.

10

Phylogenetic Analysis and Implications for Subsurface Microbiology

Robert H. Reeves

CONTENTS

KEY WORDS: *taxonomy, sequencing, DNA hybridization, DNA, Southern analysis, restriction endonuclease analysis, PCR, polymerase chain reaction, 16S rRNA, 5S rRNA, 23S rRNA, ribosomal RNA, DNA probes (probing), primers, fingerprinting, maximum parsimony, distance analysis, maximum likelihood, genome (analysis), phylogeny, phylogenetic analysis, classification.*

10.1   Introduction

The phylogenetic analysis of bacteria from the subsurface has been underway for only about five years. The classification of bacteria from subsurface environments has been analyzed for some time, but most of the analyses have relied on phenetic traits and taxonomic schemes that often have little congruence with phylogenetic relationships. Most of the recent efforts to organize the bacteria by phylogenetic criteria come from analysis of genomic DNA by DNA-DNA reassociation or rRNA-DNA hybridization techniques (Johnson, 1994a, 1994b), by restriction endonuclease fragment analysis combined with hybridization techniques (Southern analysis) (Grimont and Grimont, 1986; Kroczek, 1993; Southern, 1975), or by sequence analysis of specific genes such as the 5S rRNA or 16S rRNA genes (De Wachter et al., 1985; Olsen, 1988; Olsen et al., 1986; Woese, 1987). Several different methods have been used to obtain rRNA sequences (Brosius et al., 1978; Fox et al., 1977; Hoffman and Brian, 1991; Johnson, 1994b; Lane et al., 1985; Weller and Ward, 1989), and phylogenetic analysis of the aligned sequences has caused a major upheaval in the organization of the eubacteria (Bacteria) and the archebacteria (Archaea) (Winker and Woese, 1991; Woese and Fox, 1977; Woese et al., 1990).

Historically, the classification of bacteria has been based on phenetic traits that can be easily determined in the standard microbiological laboratory — traits such as bacterial morphology, staining characteristics, utilization of various carbon sources, types of energy metabolism, and the presence of specific enzymatic activities (Smibert and Krieg, 1994). Bacteria were classified by some combination of shared phenetic traits. In most classification schemes certain traits have priority, and bacteria sharing one of these traits are placed in a taxonomic group *a priori*. Examples of these high-priority traits are unusual morphology (spirochetes, vibrios, stalked forms), photosynthetic activity (cyanobacteria, purple and green bacteria), and obligate parasitism (the chlamydia and rickettsia). Our standard for bacterial classification has been *Bergy's Manual* since its first edition in 1923 (Bergey, 1923), and the latest edition is still based primarily on the phenetic rather than phylogenetic characterization of the bacteria (Holt, 1989).

The classification of bacteria based on phylogenetic methods began about 30 years ago with the introduction of batch DNA-DNA reassociation and DNA-RNA hybridization experiments (Denhardt, 1966; Gillespie and Spiegelman, 1966). With the introduction of rapid DNA sequencing methods, phylogenetic methods for the classification of bacteria have become commonplace. Today it is often easier to "identify" a new bacterial isolate by phylogenetic techniques than by phenotypic analysis. From an isolated bacterial colony it takes only a day or two to amplify a fragment of its 16S (or small subunit [SSU] rRNA gene) (Hoffman and Brian, 1991) to sequence a portion of this DNA fragment and to compare the resulting sequence to the very large database of SSU rRNA sequences (Maidak et al., 1994). Even a relatively short sequence (350 to 500 bases) is usually enough to identify a bacterial isolate at the genus-family level (Reeves et al., 1995).

In recent years a great deal of time and experimentation has dealt with comparisons of the results of phylogenetic analyses of bacteria to earlier taxonomic groupings. One of the earliest studies of this kind was the analysis of the *Pseudomonas* group by rRNA-DNA hybridization (Palleroni et al., 1973). Taxonomically the *Pseudomonas* group contains aerobic, Gram-negative, rod-shaped bacteria with polar flagella, able to catabolize a wide variety of organic compounds (Palleroni, 1978). Initial DNA-DNA hybridization results indicated that the group was probably not phylogenetically homogeneous, and subsequent rRNA-DNA hybridization results indicated that there were at least five distinct groups of pseudomonads (Palleroni, 1978). Sequencing results have confirmed the differences in the five rRNA-DNA groups (Woese, 1987). Clearly this taxonomic group is a phylogenetic mixture which now consists of at least six or seven very different clades (Maidak et al., 1994). These results create a horrendous nomenclature problem. Typical is the original *Pseudomonas maltophilia* group renamed *Xanthomonas maltophilia* in 1979, and again renamed *Stenotrophomonas maltophilia* in 1993 (Palleroni and Bradbury, 1993).

The methods used to classify subsurface bacteria have included a mixture of phenetic and phylogenetic techniques (Amy et al., 1992; Balkwill, 1989; Balkwill and Ghiorse, 1985; Braun-Howland et al., 1992; Haldeman and Amy, 1993; Reeves et al., 1995; Stim, 1995). It is relatively easy to obtain hundreds or thousands of isolates from a single subsurface environment, and rapid techniques are now available for placing these isolates in either taxonomic or phylogenetic schemes. The phylogenetic methods offer the advantage of quickly placing an unknown isolate into a stable group (clade) that share not only genotypic properties, but because of the similar genotypes share phenotypic properties as well. The knowledge of the phylogenetic makeup of the microbial population in subsurface habitats can give insights into the types of bacteria suited to these extreme environments and into the collective physiology and biochemistry of the population. For more detail, Woese (1985) has discussed at length the rationale for examining the phylogenetic relationships of the bacteria.

10.2 Phenotypic Methods

Historically, bacteria have been classified primarily by a combination of shared phenotypic traits (Smibert and Krieg, 1994). Individual traits are obviously an end product of the genotype of an organism. However, in simple organisms there are usually only a few traits unique to an individual species, and many individual traits are the result of a combination of genetic elements. In addition, the problem of convergent evolution of traits in simple organisms is extreme, and the transfer of genetic material among distantly related species, although rare, does occur, especially by the distribution of extrachromosomal elements (Mazodier and Davies, 1991). For these and probably many other reasons, the reliance on phenotypic characteristics to classify bacteria has resulted in taxonomic schemes that although useful for identification have in many cases given incongruent phylogenetic relationships (Olsen, 1988; Woese, 1985).

Habitat and natural history have been useful in clustering bacteria and in some cases has been at least partially congruent with phylogeny. For example, the obligate intracellular growth of certain parasitic bacteria has been used as a taxonomic discriminator, and this habitat defines a limited number of phylogenetic groups. The *Rickettsia* and the *Chlamydia* fall together in this taxonomic group, but are obviously very phylogenetically distant from each other (Weisburg et al., 1986). However, the chlamydial isolates fall together as a monophyletic group, and most of the rickettsial-like organisms are also monophyletic. Although, even within the *Rickettsia*, defined broadly as bacteria that grow intracellularly in both vertebrate and arthropod hosts, there are genera that are not closely related, i.e., the *Rickettsia* in the α-Proteobacteria and the *Coxiella* and *Wolbachia* in the γ-Proteobacteria (Weisburg et al., 1989).

The morphology of bacteria is rather limited, but it has been one of the major tools for taxonomic classification. Staining properties are included under morphology since most of the differences seen in the staining of bacteria are due to differences in morphological structures such as the cell envelope or the presence of specific structures (spores, capsules, or flagella). In many cases, morphology has been useful in predicting relatedness (e.g., the spirochetes), but reliance on morphological traits has also resulted in clustering many organisms with little or no phylogenetic relationships. Perhaps the prime example of the latter is the well-known case of the pseudomonads, mentioned above, which are taxonomically clustered primarily on the basis of morphology, but are phylogenetically very heterogeneous.

Biotyping is the characterization of bacteria by their collective biochemical traits, and it is probably the most widely used phenotypic method for classifying bacteria (Smibert and Krieg, 1994). The number of examined traits can be very large, and most commonly include the fermentation or utilization of sugars and other metabolites, the production of gases, the oxidation or reduction of specific dyes or metabolites, the presence of specific enzyme activities such as catalases or phosphatases, and the presence of specific metabolites (e.g., ubiqinones) or pigments. Biotyping has probably been carried to its extreme with the characterization of members of the Enterobacteriaciae. Many of the rapid biochemical test that have now been commercialized were first developed to characterize these bacteria, and specific tests have been developed for very specific purposes, e.g., the classic IMViC tests to distinguish coliforms. Biotyping has not been very useful as a general or preliminary method for classification of bacteria. Its forte is its application in distinguishing species or strains of bacteria after they have been placed within a genus (or family) by some other criteria.

Several commercially available kits and methods have been developed for identifying bacteria based on biotyping. These methods have different purposes and very different scopes. Most have been developed to rapidly identify human pathogens, and thus are not very useful for environmental or subsurface isolates. Three examples of these commercial kits with widely different scopes are the Enterotubes from Roche Diagnostics used to differentiate pathogenic enteric bacteria, the API Rapid NFT strips from Analytab Products for Gram-negative bacteria, and the Biolog MicroPlate kits and Analyzer from Biolog, Inc. which claims to taxonomically place over 800 bacteria based on the results of a battery of nutritional tests (Smibert and Krieg, 1994).

Other phenotypic methods used to cluster bacteria include such techniques as serotyping, phage typing, antibiotic susceptibility testing, pathogenicity, and the presence of specific extrachromosomal elements. These methods have been developed primarily for the classification of clinical isolates at the strain level, and have not been very useful for classifying isolates from the subsurface.

One phenotypic method that deserves special attention and has been useful in environmental studies is the analysis of the fatty acid composition of bacterial isolates (Bobbie and White, 1980). Fatty acids are easily extracted from a bacterial colony or culture, and small amounts can be quickly analyzed by GC-mass spectrometry. The composition and relative abundance of fatty acids (and other lipids) gives a pattern, a fatty acid profile or fingerprint, which is distinctive for a bacterium or a whole group (family) of bacteria. This method has been compared to phylogenetic methods and in general is congruent (Kohring et al., 1994). (For more information on fatty acid analysis, see Chapter 8.)

Finally, the results of a phenotypic characterization of bacteria are usually analyzed by clustering methods, which group taxa by their shared phenotypic traits. Numerical taxonomy has become the accepted method for analyzing phenotypic traits, and it has been used for many years to classify bacteria (Colwell and Austin, 1981). A large set of unweighted phenotypic (phenetic) characters are used to define related clusters, and dendrograms or trees based on phenetic traits are constructed. One statistical package developed for computer analysis of data by numerical taxonomy is NTSYS (Rohlf, 1987).

10.3 Genotypic Methods

By definition genotypic methods of classification rely on the sequence of nucleotides in the genome of an organism, and since the evolution of the genome is responsible for biological diversity, comparative analysis of bacterial genomes is the most direct and reliable indicator of phylogenetic or evolutionary relationships. In the past the problem with genomic analysis has been the size and seemingly repetitive nature of DNA. Bacterial genomes range in size from about 6×10^5 to 8×10^6 nucleotide base pairs (Krawiec and Riley, 1990), and until the restriction enzymes became available, obtaining specific DNA fragments of even the simplest bacterium was in the realm of science fiction. The first analyses of bacterial DNA were directed toward DNA base composition (GC content) and genome size. DNA base composition was one of the earliest phylogenetic tools, used mainly (and successfully) to group (or separate) clusters of taxonomically related bacteria (Silvestri and Hill, 1965). Genome size has not been used very successfully as a phylogenetic tool. The early estimates of size were not very accurate, and many bacteria have similar sized genomes. Recent genome size determinations based on large restriction fragment analysis have been more informative, but the paucity of data precludes useful phylogenetic analysis (Krawiec and Riley, 1990). Neither genome size nor base composition, however, are very useful phylogenetic tools except in grouping very closely related taxa.

10.3.1 Hybridization Methods

The first comprehensive phylogenetic analyses (DNA-based) utilized DNA-DNA reassociation methods. Shortly after DNA-DNA reassociation experiments were described (Schildkraut et al., 1961), the method was applied to the reassociation of labeled DNA from one species to immobilized DNA from another (Denhardt, 1966). Thus DNA-DNA "hybridization" experiments were begun, and over the years many variations of this basic technique have been described (Johnson, 1994a).

It did not take long to realize that DNA-DNA reassociation methods had serious limitations. DNA-DNA reassociation could only be detected among very closely related bacteria. Probably the best example of the use of this method has been the study of the Enterobacteriaceae, where DNA-DNA reassociation results have been compared to other phylogenetic (and taxonomic) methods (Sanderson, 1976). The method confirmed that *Shigella* spp. were actually within the *Escherichia coli* group, but at the same time could barely establish a relationship between *Escherichia* and *Proteus*. Within this family of bacteria DNA-DNA reassociation experiments usually confirmed close "species" relationships, but could not be used to determine family boundaries since distantly related members within the family gave background levels of reassociation. Relationships between distant species-pairs within a family, however, can be established by hybridization studies by using labeled DNA from several species that are positioned phylogenetically across the family, thus establishing relationships among several key members of the family. This type of analysis has been accomplished with the Enterobacteriaceae and with many other groups of bacteria (Grimont, 1988; Schleifer and Stackebrandt, 1983).

The hybridization of ribosomal RNA to DNA has been used to establish more distant relationships, and rRNA-DNA hybridization methods have been applied to a number of bacterial families, often with very surprising results (Schleifer and Stackebrandt,

1983). In this method, labeled rRNA from one bacterial species is hybridized to the DNA of the test species which is immobilized on a membrane (Denhardt, 1966). The membranes are treated with a ribonuclease solution and the extent of hybridization estimated from label retained on the membrane. Other methods involving rRNA-DNA hybridization include determining a thermal denaturation profile and competition hybridization (Johnson, 1994b). The utility of this method relies on the fact that rRNA genes are more highly conserved than genomes as a whole, and therefore, rRNA-DNA hybridization can be measured between distantly related bacteria. This method has been applied to several large taxonomic groups of bacteria (Schleifer and Stackebrandt, 1983). As stated above, the pseudomonads were one of the first groups to be analyzed by this method, and the method quite accurately predicted the phylogenetic heterogeneity of this taxonomic cluster (Palleroni, 1978).

10.3.2   Restriction Endonuclease Analysis

The discovery of enzymes that specifically cut both strands of DNA at a specific nucleotide sequence led the way to a variety of new methods for phylogenetic analysis. The restriction endonucleases isolated from a variety of bacteria were first used to construct physical maps of small viral genomes (Danna and Nathans, 1971; Smith, 1979). The use of restriction endonucleases in analyzing bacteria, however, is much more complex because of the very much larger genome size. A simple calculation reveals that a six-base-specific restriction endonuclease should cut an average sized bacterial genome into about 1000 fragments. These fragments, separated on a simple agarose gel, give a reproducible pattern for a given bacterial strain, and can be used to differentiate strains within a species. However, the same limitation applies here as it does with DNA-DNA reassociation experiments. Genome restriction endonuclease patterns are only useful in grouping very closely related bacteria. Genomic restriction endonuclease patterns can vary widely even among strains of the same species of bacteria.

A more recent approach with a somewhat broader scope is the use of a set of endonucleases with more restrictive cutting (Krawiec and Riley, 1990; Smith and Condemine, 1990). Digestion of a bacterial genome with these enzymes gives a limited set of large restriction fragments (often less than ten) that are separated by pulse-field electrophoresis (Schwartz and Cantor, 1984). These limited genome digests have been used to give reproducible patterns for bacterial strain identification and phylogenetic analysis (Krawiec and Riley, 1990; Romling et al., 1992). These patterns have also been used to more accurately determine the size of bacterial genomes (by summing the fragment sizes) (Krawiec and Riley, 1990). Several enzymes have been used for this purpose. Perhaps the most interesting is I-*Ceu*I, which cuts at a specific site within the 23S rRNA gene (Liu et al., 1993). This type of analysis reveals not only genome size, but also the number of rRNA gene clusters, and should be useful in grouping phylogenetically related bacteria at least at the family level. Other restriction endonucleases useful for this type of genome analysis include the eight-base-specific endonucleases and those that recognize cutting sites with unusual base compositions.

Restriction endonuclease analysis combined with hybridization techniques is now the most common form of DNA fragment analysis, sometimes called Southern hybridization analysis (or just Southern analysis) or RFLP (restriction-fragment length polymorphism) analysis (Southern, 1975). Genomic DNA is digested with a single restriction endonuclease and the fragments are separated on an agarose gel. The DNA pattern is transferred from the gel to a membrane either by Southern blotting or by one of several other transfer techniques. One or more of the DNA fragments on this membrane are detected by hybridization with a labeled DNA or RNA probe. The pattern of labeled fragments can be used for identifying bacteria and can be used to group closely related strains and species. The first probes were labeled rRNA (ribotyping), followed by cloned specific DNA fragments, and finally highly specific synthesized oligonucleotides (Grimont, 1988; Grimont and Grimont, 1986; Stull et al., 1988). Again, the limitation for phylogenetic analysis is that even with rRNA probes the patterns are useful only for closely related bacteria.

10.3.3   PCR Methods

In its simplest form the polymerase chain reaction (PCR) is the selective amplification of a double-stranded fragment of DNA between two input primers that determine the boundaries of the target DNA fragment (Bej et al., 1991). The method became widely available for DNA fragment amplification after the commercial production of the *Taq* polymerase and thermal cyclers. DNA primers of exact sequence had been available for several years before PCR was developed because of their use in DNA sequencing protocols. The main limitation for specific DNA fragment production is the knowledge of the sequence of the target DNA, or at least a knowledge of conserved sequences within or flanking the target DNA. This knowledge is necessary in order to synthesize the specific primers used in the PCR reaction.

For phylogenetic analysis PCR techniques have been used in several different ways. As discussed below, PCR is often used to obtain specific DNA fragments for sequencing, particularly the comparative sequencing of the rRNA genes. Highly conserved sequences within these genes has led to the synthesis of very useful PCR primer sets (Greisen et al., 1994; Lane et al., 1985).

A second way that PCR has been used in phylogenetic studies is in the differentiation of closely related bacterial strains and species by PCR pattern analysis or DNA fingerprinting using PCR (van Belkum, 1994). Very simply, DNA fragments from a PCR reaction using a specific primer set are separated on an agarose or polyacrylamide gel and the resulting patterns generated from different bacteria are compared. This method is appealing because no radioactive or other labeling is required; the DNA fragment patterns can be visualized on the gel by simple staining. One ideal site on bacterial DNA for this type of analysis is the 16S–23S rRNA intergene region. Nearly universal eubacterial primers for the amplification of this region can be made at the 3'-end of the 16S rRNA gene and the 5'-end of the 23S rRNA gene, so that almost any eubacterial species can be analyzed with the same primer set. Many bacteria have multiple copies of this intergene region, and they differ in length even within the same organism, usually from 300 to 500 base pairs. The pattern of DNA fragments resulting from the amplification of this intergene region is like a fingerprint for the bacterium. Like many of the methods discussed above, phylogenetic analysis based on these patterns is limited to the genus/species level, but the method has proved useful for rapid bacterial identification. Human DNA fingerprinting that relies on VNTRs (variable number of tandem repeats) uses the same overall PCR strategy — two conserved sequences where primers bind that flank a target region of DNA of variable length (van Belkum, 1994).

Finally, PCR amplification of a variable region can be combined with restriction endonuclease digestion of the resulting PCR fragment. In this method the DNA fragment from a PCR reaction is digested, usually with a four-base-specific restriction endonuclease, and the resulting pattern of smaller DNA fragments is resolved on an agarose or polyacrylamide gel. This method is rapid, requires no special equipment, and no labeling of DNA probes. It is useful for strain and species differentiation and "bacterial fingerprinting", but has limited phylogenetic usefulness.

10.3.4   Sequencing Methods

Rapid advances in the sequencing of proteins and nucleic acids has been primarily responsible for the development of phylogenetic methods for classifying bacteria. The sequencing of entire bacterial genomes is now a reality, and the rapid amplification and sequencing of individual genes is now routine (Fleischman et al., 1995). DNA methods have developed so rapidly that most techniques being used today were not even in the planning stages three decades ago, and in fact were not even thought possible. As indicated above, techniques such as PCR and automated DNA sequencing allow the phylogenetic identification of a bacterial isolate within a few days, and DNA probing techniques can potentially identify phylogenetically similar organisms *in situ* within a few hours.

10.3.4.1   RNA Sequencing

The application of RNA sequencing as a phylogenetic tool has for the most part been replaced by more rapid and reliable DNA sequencing methods. However, RNA sequencing played a key role in directing attention to the selection the stable RNAs as the molecules of choice for phylogenetic analysis, and provided us with some of the earliest comprehensive data on general bacterial phylogeny (Brosius et al., 1978; Brownlee and Cartwright, 1977; Fox et al., 1977; Lane et al., 1985). RNA sequencing is now over 30 years of age, with the publication of the yeast alanine tRNA sequence in 1965 (Holley et al., 1965). Sequences of tRNA, the 5S rRNA, and the SSU and LSU rRNA generated by RNA sequencing methods have all been used for phylogenetic analysis. The stability of these RNAs meant that they could be easily isolated and purified and thus were available for sequencing.

The first complete 16S rRNA sequence was reported in 1978, and was sequenced from the cloned *Escherichia coli rrn*B gene cluster using DNA sequencing techniques (Brosius et al., 1978). A much more rapid method for RNA sequencing of the 16S rRNA was developed by Lane et al. (1985). Reverse transcriptase was used to sequence purified 16S rRNAs by a modification of the Sanger dideoxy-terminator method using a set of synthetic primers complementary to three regions spaced along the 16S rRNA template. This method provided a way to obtain extended nucleotide sequences for bacterial 16S rRNAs, and for the first time provided a collection of nearly complete sequences for phylogenetic studies based on this molecule. For a short period of time the method was very useful because even though DNA sequencing was faster and more reliable, DNA cloning of the rRNA genes from different bacteria was slow and tedious, and PCR techniques were not yet generally available.

10.3.4.2   DNA Sequencing

The development of DNA sequencing techniques over the last 20 years has been truly remarkable. At the present time semiautomatic Sanger dideoxynucleotide sequencing can provide a single laboratory with thousands of base pairs of sequence in just a few days (McBride et al., 1989; Sanger et al., 1977). Thus, comparative DNA sequencing has become a reality, and DNA sequences are accumulating faster than most individual research laboratories can handle them.

Twenty years ago recombinant DNA cloning of specific genes provided the raw material (template DNA) for sequencing, and since then a wide variety of cloning vectors (and hosts) have been developed specifically for sequencing the inserted DNA. Both double-stranded and single-stranded DNA sequencing methods and cloning procedures have been developed. However, for the sequencing of rRNA genes from bacteria, cloning techniques which usually take several days and require purified DNA are no longer necessary. In most cases PCR can generate a double-stranded fragment of DNA from a rRNA gene (rDNA) suitable for sequencing in less than a day even without DNA isolation (Hoffman and Brian, 1991). In our laboratory DNA sequencing of rRNA genes is routine, and sets of primers (and their complements) spaced about every 200 nucleotides along the gene have been synthesized. One of the first truly universal primers used in sequencing both prokaryotic and eukaryotic rRNA, primer C, located at position 1392-1406 of the *E. coli* 16S rRNA, was also one of the first rRNA sequencing primers developed (Lane et al., 1985). This primer is also used as a universal biological probe (see below). The procedure that is currently used in our laboratory for obtaining a 16S rDNA sequence takes only a few days, and 10 to 20 isolates can be processed at the same time. Starting with an axenic culture, a colony PCR procedure is used to isolate template DNA for sequencing (Hoffman and Brian, 1991). Primers located near the 5'- and 3'-ends of the gene yield a double-stranded fragment of about 1500 nucleotides. One or two PCR reactions can provide enough DNA for several sequencing reactions. After a brief purification step, the DNA fragment is ready for sequencing, and a complete sequence can usually be obtained with three sequencing reactions using well-spaced primers. Opposite-strand sequences can also be quickly done. We are currently using *Taq* dideoxy cycle sequencing and an Applied Biosystems automated DNA sequencer (McBride et al., 1989).

At this time, most phylogenetic analyses in both the prokaryotic and eukaryotic kingdoms are being done with sequences generated from the SSU (16S and 18S) rRNA (Olsen, 1988; Olsen et al., 1986; Woese, 1988). The molecule is about the right size for a robust analysis and there are both conserved sequences for alignment (and primer synthesis) and enough variation that phylogenetic inferences can be made. Another reason that this gene is an excellent candidate for analysis is that a great deal is known about the secondary structure of the 16S rRNA, which allows unambiguous alignments of homologous sequence (see below). Other candidates for DNA sequencing and phylogenetic analysis include the 5S rRNA gene (De Wachter et al., 1985), the 23S (LSU) rRNA gene (Trust et al., 1994), and other genes involved universally with DNA replication or gene expression, such as a subunit of RNA polymerase or one of the elongation factors. Genes with protein products usually are much more variable than the RNA-product genes making both DNA template production and primer synthesis a more difficult task. Alignments of the resulting sequence are also more difficult. The one advantage that these latter genes have over the RNA genes is that their variability is desirable for phylogenetic analysis of closely related bacteria.

10.3.4.3 DNA Probes

One of the many byproducts of comparative DNA sequencing is the knowledge necessary to make DNA probes, and many specific probes have been developed for the rapid identification (and classification) of bacterial isolates (Braun-Howland et al., 1992; DeLong et al., 1989; Falkingham, 1994; Giovannoni et al., 1988; Holben et al., 1988; Schmidt et al., 1991). The 16S and 23S rRNA sequence databases are extensive, and thus a variety of species-specific probes have been synthesized. These are usually oligomers of about 20 nucleotides that hybridize to a portion of the 16S or 23S rRNA gene that is common only to the strains within a species. These probes are end-labeled with [^{32}P] phosphate or with a fluorescent dye, and DNA from either whole cells or cell extracts can be used as the hybridization target. Species- or strain-specific probes can also be developed from sequences of more specific DNA sequences, such as toxin genes and genes involved in specific types of metabolism. Probes specific for extrachromosomal elements have also been developed (Falkingham, 1994).

There are two advantages of the 16S and 23S rRNA probes that immediately come to mind. One is the natural amplification of these genes as rRNAs of the cell. In a rapidly growing *E. coli* cell there are about 30,000 ribosomes, and thus the same number of potential targets for complementary rRNA probes. Hybridization of a fluorescent rRNA probe to whole cells requires no further amplification step that would normally be necessary for DNA-binding probes. Target cells probed in this way often glow against a dark background when viewed under a fluorescence microscope. This technique allows the detection of target cells *in situ*, in some cases even without culturing or isolation. A second advantage for using rRNA probes is that the database for these sequences is extensive and allows the creation of probes at all phylogenetic levels. Indeed, there are universal probes, kingdom probes, and probes for specific groups of bacteria down to the species level (Braun-Howland et al., 1992; DeLong et al., 1989; Giovannoni et al., 1988).

10.4 Data Analysis

Nucleotide sequence data are readily available for several bacterial genes, but by far the largest database for a single gene is that for the 16S rRNA gene (Maidak et al., 1994). This section on data analysis will concentrate on this gene, but in general the discussion applies to any DNA or protein sequence data. Assuming that the sequence data are accurate, the next step in a phylogenetic analysis is to use some method for comparing the sequences, and there are a multitude of methods available for doing this (Felsenstein, 1988; Nei, 1987; Swofford and Olsen, 1990). However, for the analysis to be meaningful, all methods depend first on an accurate alignment of the sequences. After a "satisfactory" alignment is obtained, the three most common methods for inferring relationships among organisms are distance (similarity), maximum likelihood, and maximum parsimony methods. In each case the robustness of the inferred phylogenetic relationships depend on an accurate alignment of the sequence data. For a comprehensive list of computer-based phylogenetic methods used for sequence analysis visit the home page of PHYLIP at http://evolution.genetics.washington.edu/phylip.html (Felsenstein, 1991).

10.4.1 Alignments

The ideal alignment of two or more sequences for phylogenetic analysis results in comparing characters (individual nucleotide positions) that are derived from common ancestral nucleotides. Such an alignment is said to be homologous. Analogous alignments refer to alignments of characters that are related usually by gene products of similar function or by the position of a sequence. For example, the intergene regions between the 16S and 23S rRNA genes in the eubacteria are analogous by position, but are not necessarily homologous. These intergene region sequences are found in the same relative gene location, but they may not be derived by simple base substitution from a common source. Even within an homologous gene or large nucleotide sequence there may be smaller regions that are not truly homologous among all taxa. In the 16S rRNA gene with its many conserved regions it is often difficult to obtain an alignment that is truly homologous across the entire sequence. This is usually due to deletions or additions of nucleotides within the gene, which usually become evident when the alignment is being constructed. These sequences of nonhomology should be omitted from the analysis. However, even with their omission, the borders of the addition/deletion are not always clear, and truly homologous alignments of nucleotides near the addition/deletion site may not be possible. In the case of the 16S rRNA gene alignments, these problems have become less acute as more sequences have become available.

The use of secondary structure data for the 16S rRNA has been a great benefit in aligning multiple sequences of this gene. Gutell and collaborators have been examining this aspect of 16S rRNA sequence alignment in some detail (Gutell, 1994; Gutell et al., 1994). They have shown that compensatory base changes observed in many bacterial 16S rRNA sequences agree well with secondary structure models built with the results of a variety of experimental studies. Primarily from comparative sequence data, they have put together secondary structure diagrams for a variety of prokaryotes, thus providing a scaffold, as it were, for the alignment of homologous sequence. It is worthwhile to examine these diagrams when attempting the alignment of 16S rRNA sequences within a domain or a group of bacteria. One easy way to access these diagrams is through the internet site "16S rRNA Comparative Structure Database" at http://pundit.colorado.edu:8080/RNA/16S/16s.html.

The usual procedure for aligning multiple sequences is to start with the two most similar sequences (determined by sequence identity) and perform a pair-wise alignment. The next most similar sequence is then added to the pair of aligned sequences, followed by the next most similar, and so forth. There are several alignment programs available for computer analysis of both protein and nucleic acid alignments. FASTA is often used to perform both rapid sequence identities and pair-wise alignments (Pearson and Lipman, 1988).

Multiple alignment programs are becoming available, but most still suffer from problems whenever a large addition/deletion is encountered within one or more of the sequences. No alignment program is completely successful in obtaining an homologous alignment, and a certain amount of manual "tweaking" is usually required. With the 16S and 23S rRNA sequences the secondary structure diagrams of Gutell (1994) come in very handy.

10.4.2   Distance (Similarity) Methods

There are a variety of distance methods that have been described, and computer programs are abundant for their application in inferring relationships among taxa. The most widely used programs containing distance analysis packages are PHYLIP and MEGA (Felsenstein, 1991; Kumar et al., 1993). In essence, all distance methods rely on constructing a pair-wise set of distance parameters between all taxa (e.g., bacterial isolates). From sequence data, the simplest distance is merely the percentage of nucleotide differences observed in an alignment of two taxa (the overall alignment of all taxa should be considered to evaluate homologous positions). A matrix of distances among all taxa is then used to build (or infer) a tree relating all the taxa. Differences between the various distance methods involve the exact way the distance parameters are calculated, how various nucleotide changes (or amino acid changes) are weighted, how possible multiple changes during evolution are treated, and the protocol for tree construction. The simplicity of distance methods is that only one tree results from the analysis, but the weakness of these methods is that much of the information contained within the sequences is lost in the analysis. The tree is inferred only from pair-wise distances between each of the taxa, and these distance parameters, although relating similar overall sequences, say little about how the actual sequences vary. As such, distance methods are said to produce phenetic rather than cladistic trees. The inferred trees are the result of the similarity of overall sequences with no attempt to establish an ancestral relationships among the sequences upon which the tree is built.

Still, distance methods are very popular and even when other methods such as maximum parsimony or maximum likelihood are used to infer phylogeny, it is common to compare the results to those obtained by distance analysis.

10.4.3   Maximum Parsimony

Two aligned sequences that are the most similar in a multiple alignment are also said to be the most parsimonious, and the two taxa from which these sequences are derived are the most closely related. The two sequences (taxa) are joined, and at the node that joins them to the rest of the tree is a common or ancestral sequence. The changes in going from this nodal sequence to the terminal sequence (or taxa) is a branch length. The most parsimonious tree is that which requires the least number of sequence changes from one taxa or node to another in the construction of the overall tree, also said to be the tree with the shortest overall branch length. There are a variety of ways to construct a tree by maximum parsimony, and these methods have been reviewed and evaluated (Felsenstein, 1988; Nei, 1987; Swofford and Olsen, 1990). The most common method now used to infer trees by maximum parsimony is that described by Swofford (1983): PAUP (Phylogenetic Analysis Using Parsimony). From the multiple alignment of sequences a reduced data set of informative sites is used to construct the tree. Each possible tree topology (branching order) is evaluated, and the one with the shortest overall length is chosen as the resultant tree. Unfortunately, for trees with more than eight taxa evaluating the length of every possible tree takes a great deal of computer time. Heuristic methods (quicker, approximate methods with defined rules) for searching for the most parsimonious tree are generally used. The most common is a branch-swapping method to move rationally toward the tree of shortest overall length.

Maximum parsimony methods are generally said to be cladistic in that an attempt is made to build a tree based on ancestral sequences (found at the internal nodes of the tree). The problem with maximum parsimony methods is that much more computer time is needed to evaluate trees, and trees with large numbers of taxa can only be evaluated by inexact (heuristic) methods.

10.4.4   Maximum Likelihood

The maximum likelihood method of inferring phylogeny is based on the selection of the overall tree that is most likely based on the evaluation of all possible trees. It is a statistical method where evolutionary change in nucleotide sequence is followed by a probability model. Felsenstein developed an early model for tree construction based on maximum likelihood, but the amount of computer time necessary to evaluate large trees (with many taxa) made the method impractical (Felsensein, 1988). The latest versions of PHYLIP contains an updated version of Felsenstein's maximum likelihood method (DNAml), and another version used by the Ribosomal Database Project called fastDNAml is also available (Olsen et al., 1994). The latter version runs best on minicomputers or work stations, and thus makes it somewhat limited for general use. However, the SSU rRNAs of nearly 500 taxa have been analyzed by fastDNAml and Overbeek (Olsen et al., 1994) claims that 1600 taxa have recently been analyzed by this method.

10.5 Phylogenetic Relationships of Subsurface Bacteria

Any kind of comprehensive phylogenetic analysis of the bacteria is only a few years old. The first large phylogenetic trees were constructed just ten years ago, when enough sequence data became available to assemble the first multiple alignments. During the last decade, computer methods for the analysis of large collections of sequences were also developed at a rapid rate. These efforts have now culminated in large databases of DNA sequence, some devoted just to particular genes such as tRNAs or the ribosomal RNAs. If a new bacterium is isolated from any source it takes only a few days to "place" it phylogenetically, and techniques are available to make this placement without an axenic culture or even without culturing.

In the 1980s the search for microorganisms in subsurface soils and aquifers began in earnest (Amy et al., 1992; Balkwill, 1989; Balkwill and Ghiorse, 1985; Haldeman et al., 1993; Hirsch and Rades-Rohkohl, 1983; Kieft et al., 1993; Russell et al., 1994; White et al., 1993), even though there were many sporadic earlier reports of such isolates (Ghiorse and Wilson, 1988). In deep subsurface environments to depths of over 300 m and in habitats as inhospitable as solid rock, microorganisms are being isolated and characterized (Balkwill, 1989; Haldeman et al., 1993). Most of the characterizations of subsurface bacteria have involved cataloging phenotypic traits. Thus far only a few reports of phylogenetic analysis of such isolates have appeared (Fredrickson et al., 1995; Kohring et al., 1994; Reeves et al., 1995; Stim, 1995). For the most part these have described a limited number of isolates from deep subsurface aquifers. It is probably much too early to make generalizations about the abundance and relationships of subsurface bacteria, but at least in one study in which nearly 200 isolates from one deep subsurface well were examined, a limited number of related groups emerged. These were the groups related to the *Arthrobacter* (Reeves et al., 1992), *Acinetobacter* (Baldwin et al., 1993), *Commamonas* (LoCascio et al., 1993), *Pseudomonas* (fluorescent group), and the *Sphingomonas* (Fredrickson et al., 1995). One of these groups, the *Arthrobacter*, seems to be common to several subsurface environments, including the Siberian permafrost (Ting, unpublished data).

At this early stage, the phylogenetic analyses of subsurface bacteria have not been rigorous. In fact these studies are more accurately described as bacterial identifications using phylogenetic tools. In most cases the intent of the analysis has been bacterial identification, but with DNA sequencing becoming so rapid and the 16S rRNA database so large, a phylogenetic identification (classification) is now easier and more informative than phenotypic typing. It is relatively easy to classify new isolates to the genus-family level by sequencing and analyzing one 400-base region of 16S rRNA gene, and it is likely that the phylogenetic characterization of new bacterial isolates (by 16S rRNA sequencing) will become standard procedure for reasons that have already been adequately explained (Woese, 1985, 1987).

10.6 Conclusions

Our understanding of the phylogeny of the bacteria (and archea) has definitely changed our way of thinking about the prokaryotic world. The new three-domain evolutionary tree, based on SSU rRNA sequencing, is only ten years old and is already well accepted as evidenced by its appearance in textbooks of general

TABLE 10.1

Phylogenetic Methods and Their Ability to Resolve Bacterial Phylogenetic Relationships

Method	Strain	Species	Genus	Family	Division	Phylum	Domain
DNA-DNA reassociation	+++++++++++++++++						
DNA-rRNA hybridization	++++++++++++++++++++++++++++++++++						
Genomic restriction patterns	+++						
Genomic restriction patterns (restrictive cutting)		+++++++++++++++++					
Southern analysis with rRNA[a]		+++++++++++++++++		+	+	+	+
PCR fingerprinting[a]	++++++++	+	+	+	+	+	+
DNA probing[a]		+	+	+	+	+	+	+	+	+
5S rRNA sequencing		+++					
16S/23S rRNA sequencing		+++					

[a] DNA probing, and to a lesser extent ribotyping and PCR fingerprinting, can detect bacteria within a species, genus, family, division, or domain, but they are not very useful for establishing relationships among different bacteria within these groups. Ribotyping and PCR fingerprinting give patterns which can be compared at the species-genus level.

microbiology (Brock et al., 1994). The methods used to construct these trees are now commonplace, and in the future will be used to not only analyze bacteria for their phylogenetic relationships, but will be routinely used to "identify" new bacterial isolates, placing them within a framework of bacterial species or type stains by nature of their 16S rRNA sequence similarities. Since the SSU rRNA database has grown so rapidly, it appears that at least for the next decade or longer phylogenetic relationships will use these aligned sequences as the "gold standard" for phylogenetic analysis. It is possible that as more refinements are needed a larger sequence will be necessary; the LSU rRNA database is also growing rapidly (Maidak et al., 1994). However, it appears that the SSU rRNA gene has provided us with a very acceptable molecular clock, at least to establish overall evolutionary relationships.

The various methods used for phylogenetic analysis described above have different scopes or resolving powers. Most of those that result in a pattern of DNA fragments with or without hybridization techniques are best at classifying bacteria at the strain or species level, although ribotyping can relate bacteria that are more distant. DNA-DNA reassociation methods that give a index of similarity, like a percent reassociation or a DNA similarity value, can relate isolates from different genera. The sequencing methods can be used to relate bacteria at the species level to the kingdom or domain level. Table 10.1, modified from a similar table compiled by Schleifer and Stackebrandt (1983), gives an indication of what phylogenetic level a given method can be used to establish relationships.

Thus far in our analyses of subsurface bacteria the choice of sequence analysis method has made little difference. Distance methods are the easiest to use and for the most part have given us the same overall results as either maximum parsimony or maximum likelihood methods. As mentioned above, fine-tuning the sequence alignment itself may be more important than the method used to construct the phylogenetic trees. However, differences between these methods become more obvious when fine structure analyses are examined among a large number of very closely related bacteria, and trees with very different topologies can and do result. Analyses of this type based on the 16S rRNA gene, however, rely on only a few base changes among the different sequences (taxa), and it should not be surprising that different trees can result depending on how the aligned sequences are analyzed.

One final comment: we are inferring phylogenies of living bacteria from the sequence of a single gene. Actually we are inferring relationships among 16S rRNA genes that reside within these bacteria and, as yet, we know little of how this gene (with its multiple copies per genome) is inherited or to what extent there is lateral transfer (with genetic recombination) of this DNA sequence among the bacteria. We already know that recombination occurs among the *rrn* clusters within a genome. It will be interesting to compare the phylogenetic relationships of the bacteria based on the SSU rRNA to those constructed from other genes or DNA sequences when they become available.

References

Amy, P. S., D. L. Haldeman, D. Ringleberg, D. H. Hall, and C. Russell. 1992. Comparison of identification systems for classification of bacteria isolated from water and endolithic habitats within the deep subsurface. *Appl. Environ. Microbiol.*, 58:3367.

Baldwin, M. A., J. Y. Reeves, D. L. Balkwill, and R. H. Reeves. 1993. The Phylogenetic Characterization of *Acinetobacter* Isolates from the Deep Subsurface, 1993 Int. Symp. Subsurface Microbiology, Bath, U.K., September, 1993.

Balkwill, D. L., and W. C. Ghiorse. 1985. Characterization of subsurface bacteria associated with two shallow aquifers in Oklahoma. *Appl. Environ. Microbiol.*, 50:580.

Balkwill, D. L. 1989. Numbers, diversity, and morphological characteristics of aerobic, chemoheterotrophic bacteria in deep subsurface sediments from a site in South Carolina. *Geomicrobiology J.*, 7:33.

Bej, A. K., M. H. Mahbubani,and R. M. Atlas. 1991. Amplification of nucleic acids by polymerase chain reaction and other methods and their application, *Crit. Rev. Biochem. Mol. Biol.*, 26:301.

Bergey, D. A. 1923. *Bergey's Manual of Determinative Bacteriology;* a key for the identification of organisms of the class Schizomycetes, arranged by a committee of the Society of American Bacteriologists, American Society for Microbiology, Williams and Wilkins, Baltimore.

Bobbie, R. J. and D. C. White. 1980. Characterization of benthic microbial community structure by high resolution gas chromatography of fatty acid methyl esters. *Appl. Environ. Microbiol.*, 39:1212.

Braun-Howland, E. B., S. A. Danielsen, and S. A. Nierzwicki-Bauer. 1992. Development of a rapid method for detecting bacterial cells *in situ* using 16S rRNA-targeted probes. *BioTechniques*, 13:928.

Brock, T. D., M. T. Madigan, J. M. Martinko, and J. Parker. 1994. *The Biology of Microorganisms*, Prentice-Hall, Englewood Cliffs, NJ, pp. 909.

Brosius, J., M. L. Palmer, P. J. Kennedy, and H. R. Noller. 1978. Complete nucleotide sequence of a 16S ribosomal RNA gene from *Escherichia coli. Proc. Natl. Acad. Sci. U.S.A.*, 75:4801.

Brownlee, G. G. and E. M. Cartwright. 1977. Rapid gel sequencing of RNA by primed synthesis with reverse transcriptase. *J. Mol. Biol.*, 114:93.

Colwell, R. R., and B. Austin. 1981. Numerical taxonomy, in *Manual of Methods for General Bacteriology*, P. Gerhardt, R. G. E. Murray, R. N. Costilow, E. W. Nester, W. A. Wood, N. R. Krieg, and G. B. Phillips, Eds., American Society for Microbiology, Washington D.C.

Danna, K. and D. Nathans. 1971. Sequence specific cleavage of simian virus 40 DNA by restriction endonucleases of *Hemophilus influenzae. Proc. Natl. Acad. Sci. U.S.A.*, 68:2913.

DeLong, E. F., G. S. Wickham, and N. R. Pace. 1989. Phylogenetic stains: Ribosomal RNA-based probes for the identification of single cells. *Science*, 243:1360.

Denhardt, D. T. 1966. A membrane-filter technique for the detection of complementary DNA. *Biochem. Biophys. Res. Commun.*, 23:641.

De Wachter, R., E. Huysmans, and A. Vandenberghe. 1985. 5S ribosomal RNA as a tool for studying evolution, in *Evolution of Prokaryotes*, K. H. Schleifer and E. Stackebrandt, Eds., Academic Press, New York, pp. 115-141.

Falkingham, J. O. 1994. Nucleic acid probes, in *Methods for General and Molecular Bacteriology*, P. Gerhardt, R. G. E. Murray, A. W. Wood, and N. R. Krieg, Eds., American Society for Microbiology, Washington D.C., chap. 28.

Felsenstein, J. 1988. Phylogenies from molecular sequences: Inference and reliability. *Annu. Rev. Genet.*, 22:521.

Felsenstein, J. 1991. PHYLIP (Phylogeny Inference Package), version 3.4, University of Washington, Seattle, WA.

Fleischman, R. D., M. D. Adams, O. White, R. A. Clayton, E. F. Kirkness, A. R. Kerlavage, C. J. Bult, J. F. Tomb, B. A. Dougherty, J. M. Merrick, K. McKenney, G. Sutton, W. Fitzhugh, C. Fields, J. D. Gocayne, J. Scott, R. Shirley, L. I. Liu, A. Glodek, J. M. Kelley, J. F. Weidman, C. A. Phillips, T. Spriggs, E. Hedblom, M. D. Cotton, T. R. Utterback, M. C. Hanna, D. T. Nguyen, D. M. Saudek, R. C. Brandon, L. D. Fine, J. L. Fritchman, J. L. Fuhrmann, N. S. M. Geoghagen, C. L. Gnehm, L. A. McDonald, K. V. Small, C. M. Fraser, H. O. Smith, and J. C. Venter. 1995. Whole-genome random sequencing and assembly of *Haemophilus influenzae* Rd. *Science*, 269:496.

Fox, G. E., K. R. Pechman, and C. R. Woese. 1977. Comparative cataloging of 16S ribosomal ribonucleic acid: molecular approach to procaryotic systematics. *Int. J. Syst. Bacteriol.*, 27:44.

Fredrickson, J. K., D. L. Balkwill, G. A. Drake, M. F. Romine, D. B. Ringleberg, and D. C. White. 1995. Aromatic-degrading sphingomonas isolates from the deep subsurface. *Appl. Environ. Microbiol.*, 61:1917.

Ghiorse, W. C. and J. T. Wilson. 1988. Microbial ecology of the terrestrial subsurface. *Adv. Appl. Microbiol.*, 33:107.

Gillespie, D. and S. Spiegelman. 1966. A quantitative assay for DNA-RNA hybrids with DNA immobilized on a membrane. *J. Mol. Biol.*, 12:829.

Giovannoni, S. J., E. F. DeLong, G. J. Olsen, and N. R. Pace. 1988. Phylogenetic group-specific oligonucleotide probes for the identification of single microbial cells. *J. Bacteriol.*, 170:720.

Greisen, K., M. Loeffelholz, A. Purohit, and D. Leong. 1994. PCR primers and probes for the 16S rRNA gene of most species of pathogenic bacteria, including bacteria found in cerebrospinal fluid. *J. Clin. Microbiol.*, 32:335.

Grimont, P. A. D. 1988. Use of DNA reassociation in bacterial classification. *Can. J. Microbiol.*, 34:541.

Grimont, F. and P. A. D. Grimont. 1986. Ribosomal ribonucleic acid gene restriction patterns as potential taxonomic tools. *Ann. Inst. Pasteur Microbiol.*, 137B:165.

Gutell, R. R. 1994. Collection of small subunit (16S- and 16S-like) ribosomal RNA structures: 1994. *Nucleic Acids Res.*, 22:3502.

Gutell, R. R., N. Larsen, and C. R. Woese. 1994. Lessons from an evolving rRNA: 16S and 23S rRNA structures from a comparative perspective. *Microbiol. Rev.*, 58:10.

Haldeman, D. L., and P. S. Amy. 1993. Bacterial heterogeneity in deep subsurface tunnels at Rainier Mesa, Nevada Test Site. *Microb. Ecol.*, 25:183.

Haldeman, D. L., P. S. Amy, D. Ringelberg, and D. C. White. 1993. Characterization of the microbiology within a 21 m³ section of rock from the deep subsurface. *Microb. Ecol.*, 26:145.

Hirsch, P. and E. Rades-Rohkohl. 1983. Microbial diversity in a groundwater aquifer in northern Germany. *Dev. Ind. Microbiol.*, 24:183.

Hoffman, M. A. and D. A. Brian. 1991. Sequencing PCR DNA amplified directly from a bacterial colony. *BioTechniques*, 11:30.

Holben, W. E., J. K. Jansson, B. K. Chelm, and J. M. Tiedje. 1988. DNA probe method for the detection of specific microorganisms in the soil bacterial community. *Appl. Environ. Microbiol.*, 54:703.

Holley, R. W., J. Apgar, G. A. Everett, J. T. Madison, M. Marquisee, S. H. Merrill, J. R. Penswick, and A. Zamir. 1965. Structure of a ribonucleic acid. *Science*, 147:1462.

Holt, J. G. 1989. Editor-in-chief, *Bergey's Manual of Systematic Bacteriology*, Vol. I, 1984; Vol. II, 1986; Vol. III and IV, 1989, Williams & Wilkens, Baltimore.

Johnson, J. L. 1994a. Similarity analysis of DNAs, in *Methods for General and Molecular Bacteriology*, P. Gerhardt, R. G. E. Murray, A. W. Wood, and N. R. Krieg, Eds., American Society for Microbiology, Washington, D.C., chap. 26.

Johnson, J. L. 1994b. Similarity analysis of rRNAs, in *Methods for General and Molecular Bacteriology,* P. Gerhardt, R. G. E. Murray, A. W. Wood, and N. R. Krieg, Eds., American Society for Microbiology, Washington, D.C., chap. 27.

Kieft, T. L., P. S. Amy, F. J. Brockman, J. K. Fredrickson, B. N. Bjornstad, and L. L. Rosacker. 1993. Microbial abundance and activities in relation to water potential in the vadose zones of arid and semiarid sites. *Microb. Ecol.,* 26:59.

Kohring, L. L., D. B. Ringelberg, R. Devereux, D. A. Stahl, M. W. Mittelman, and D. C. White. 1994. Comparisons of phylogenetic relationships based on phospholipid fatty acid profiles and ribosomal RNA sequence similarities among dissimilatory sulfate-reducing bacteria. *FEMS Microbiol. Lett.,* 119:303.

Krawiec, S. and M. Riley. 1990. Organization of the bacterial chromosome. *Microbiol. Rev.,* 54:502.

Kroczek, R. A. 1993. Southern and Northern analysis, *J. Chromatogr.,* 618:133.

Kumar, S., K. Tamura, and M. Nei. 1993. MEGA: Molecular Evolutionary Genetics Analysis, version 1.01, The Pennsylvania State University, University Park, PA.

Lane, D. J., G. Pace, G. J. Olsen, D. A. Stahl, M. L. Sogin, and N. R. Pace. 1985. Rapid determination of 16S ribosomal RNA sequences for phylogenetic analyses. *Proc. Natl. Acad. Sci. U.S.A.,* 82:6955.

Liu, S. L., A. Hessel, and K. E. Sanderson. 1993. Genomic mapping with I-*Ceu*I, an intron-encoded endonuclease specific for genes for ribosomal RNA, in *Salmonella* spp., *Escherichia coli*, and other bacteria. *Proc. Natl. Acad. Sci. U.S.A.,* 90:6874.

LoCascio, M., D. L. Balkwill, and R. H. Reeves. 1993. Characterization of Ribosomal RNA Gene Clusters from *Commamonas testosteroni* and related isolates from the Deep Subsurface, abstr. H-23, Abstr. 93rd Annu. Meet. Am. Soc. Microbiol., Atlanta, May, 1993, pp. 194.

Maidak, B. L., N. Larsen, M. J. McCaughey, R. Overbeek, G. J. Olsen, K. Fogel, J. Blandy, and C. R. Woese. 1994. The ribosomal database project. *Nucleic Acids Res.,* 22:3485.

Mazodier, P., and J. Davies. 1991. Gene transfer between distantly related bacteria. *Annu. Rev. Genet.,* 25:147.

McBride, L. J., S. M. Koepf, R. A. Gibbs, W. Salser, P. E. Mayrand, M. W. Hunkapiller, and M. N. Kronick. 1989. Automated DNA sequencing methods involving polymerase chain reaction. *Clin. Chem.,* 35:2196.

Nei, M. 1987. Phylogenetic trees, in *Molecular Evolutionary Genetics*, Columbia University Press, New York, chap. 11.

Olsen, G. J. 1988. Phylogenetic analysis using ribosomal RNA. *Methods Enzymol.,* 164:793.

Olsen, G. J., D. J. Lane, S. J. Giovannoni, N. R. Pace, and D. A. Stahl. 1986. Microbial ecology and evolution: a ribosomal RNA approach. *Annu. Rev. Microbiol.,* 40:337.

Olsen, G. J., H. Matsuda, R. Hagstrom, and R. Overbeek. 1994. FastDNAmL: a tool for construction of phylogenetic trees of DNA sequence using maximum likelihood. *Comput. Appl. Biosci.,* 10:41.

Palleroni, N. J. 1978. *The Pseudomonas Group,* Meadowfield Press, Durham, England.

Palleroni, N. J. and J. F. Bradbury. 1993. *Stenotrophomonas,* a new bacterial genus for *Xanthomonas maltophilia* (Hugh 1980) Swings et al., 1983, *Int. J. Syst. Bacteriol.,* 43:606.

Palleroni, N. J., R. Kunizawa, R. Contopoulou, and M. Douderroff. 1973. Nucleic acid homologies in the genus *Pseudomonas. J. Syst. Bacteriol.,* 23:333.

Pearson, W. R. and D. J. Lipman. 1988. Improved tools for biological sequence comparison. *Proc. Natl. Acad. Sci. U.S.A.,* 85:2444.

Reeves, J. R., D. A. Balkwill, and R. H. Reeves. 1992. Phylogenetic Analysis of "Arthrobacter-like" Isolates from the Deep Subsurface, abstr. N-18, Abstr. 92rd Annu. Meet. Am. Soc. Microbiol., New Orleans, May, 1992, pp. 295.

Reeves, R. H., J. Y. Reeves, and D. L. Balkwill. 1995. Strategies for phylogenetic characterization of subsurface bacteria. *J. Microbiol. Methods,* 21:235.

Rohlf, F. 1987. NTSYSpc: Numerical Taxonomy and Multivariate Analysis System for the IBM PC Microcomputer, version 1.30, Applied Biostatistics, Inc., Setauket, NY.

Romling, U., D. Grothues, T. Heuer, and B. Tummler. 1992. Physical genome analysis of bacteria. *Electrophoresis,* 13:626.

Russell, C. E., R. Jacobson, D. L. Haldeman, and P. S. Amy. 1994. Heterogeneity of deep subsurface microorganisms and correlations to hydrogeological and geochemical parameters. *Geomicrobiol. J.,* 12:37.

Sanderson, K. E. 1976. Genetic relatedness in the family Enterobacteriaceae. *Annu. Rev. Microbiol.,* 30:327.

Sanger, F., S. Nicklen, and A. R. Coulson. 1977. DNA sequencing with chain terminating inhibitors. *Proc. Natl. Acad. Sci. U.S.A.,* 74:5463.

Schildkraut, C. L., J. Marmur, and P. Doty. 1961. The formation of hybrid DNA molecules and their use in studies of DNA homologies. *J. Mol. Biol.,* 3:595.

Schleifer, K. H. and E. Stackebrandt. 1983. Molecular systematics of prokaryotes. *Annu. Rev. Microbiol.,* 37:143.

Schmidt, T. M., E. F. Delong, and N. R. Pace. 1991. Phylogenetic identification of uncultivated microorganisms in natural habitats, in *Rapid Methods and Automation in Microbiology and Immunology,* A. Vaheri, R. C. Tilton, and A. Balows, Eds., Springer-Verlag, New York, pp. 37-46.

Schwartz, D. C. and C. R. Cantor. 1984. Separation of yeast chromosome-sized DNAs by pulsed field gradient gel electrophoresis. *Cell,* 37:67.

Silvestri, L. G. and L. R. Hill. 1965. Agreement between deoxyribonucleic acid base composition and taxonomic classification of Gram-positive cocci. *J. Bacteriol.,* 90:136.

Smibert, R. M. and N. R. Krieg. 1994. Phenotypic characterization, in *Methods for General and Molecular Bacteriology,* P. Gerhardt, R. G. E. Murray, A. W. Wood, and N. R. Krieg, Eds., American Society for Microbiology, Washington, D.C., chap. 25.

Smith, H. 1979. Nucleotide sequence specificity of restriction endonucleases. *Science,* 205:455.

Smith C. L. and G. Condemine. 1990. New approaches for physical mapping of small genomes. *J. Bacteriol.,* 172:1167.

Southern, E. M. 1975. Detection of specific sequences among DNA fragments separated by gel electrophoresis. *J. Mol. Biol.,* 98:503.

Stim, K. P. 1995. A phylogenetic analysis of microorganisms isolated from subsurface environments. *Mol. Ecol.,* 4:1.

Stull, T. L., J. J. LiPuma, and T. D. Edlind. 1988. A broad-spectrum probe for molecular epidemiology of bacteria: ribosomal RNA. *J. Infect. Dis.,* 157:280.

Swofford, D. L. 1993. PAUP: Phylogenetic Analysis Using Parsimony, version 3.1, Illinois Natural History Survey, Champaign, Ill.

Swofford, D. L. and G. J. Olsen. 1990. Phylogeny reconstruction, in *Molecular Systematics,* D. M. Hillis and C. Moritz, Eds., Sinauer Associates, Inc., Sunderland, MA, chap. 11.

Ting, S., R. H. Reeves, E. I. Friedmann, and D. A. Gilichinsky. Unpublished.

Trust, T. J., S. M. Logan, C. E. Gustafson, P. J. Romaniuk, N. W. Kim, V. L. Chan, M. A. Ragan, P. Guerry, and R. R. Gutell. 1994. Phylogenetic and molecular characterization of a 23S rRNA gene positions the genus *Campylobacter* in the epsilon subdivision of the *Proteobacteria* and shows that the presence of transcribed spacers is common in *Campylobacter* spp. *J. Bacteriol.,* 176:4597.

van Belkum, A. 1994. DNA fingerprinting of medically important microorganisms by use of PCR. *Clin. Microbiol. Rev.,* 7:174.

Weisburg, W. G., M. E. Dobson, J. E Samuel, G. A. Dasch, L. P. Mallavia, O. Baca, L. Mandelco, J. E. Sechrest, E. Weiss, and C. R. Woese. 1989. Phylogenetic diversity of the Rickettsiae. *J. Bacteriol.,* 171:4202.

Weisburg, W. G., T. P. Hatch, and C. R. Woese. 1986. Eubacterial origin of chlamydiae. *J. Bacteriol.,* 167:570.

Weller, R. and D. M. Ward. 1989. Selective recovery of 16S ribosomal RNA sequences from natural microbial communities in the form of cDNA. *Appl. Environ. Microbiol.,* 55:1818.

White, D. C., G. A. Smith, M. J. Gehron, R. H. Parker, R. F. Findlay, Martz, and H. L. Fredrickson. 1983. The groundwater aquifer microbiota: biomass, community structure, and nutritional status. *Dev. Ind. Microbiol.,* 24:201.

Winker, S. and C. R. Woese. 1991. A definition of the domains *Archaea Bacteria* and *Eucarya* in terms of small subunit ribosomal RNA characteristics. *Syst. Appl. Microbiol.,* 14:305.

Woese, C. R. 1985. Why study evolutionary relationships among bacteria?, in *Evolution of Prokaryotes*, K. H. Schleifer and E. Stackebrandt, Eds., Academic Press, New York, pp. 1-30.

Woese, C. R. 1987. Bacterial evolution. *Microbiol. Rev.*, 51:221.

Woese, C. R. and G. E. Fox. 1977. Phylogenetic structure of the prokaryotic domain. The primary kingdoms. *Proc. Natl. Acad. Sci. U.S.A.*, 74:5088.

Woese, C. R., O. Kandler, and M. L. Wheelis. 1990. Towards a natural system of organisms. Proposal for the domains *Archaea*, *Bacteria*, and *Eucarya*. *Proc. Natl. Acad. Sci. U.S.A.*, 87:4576.

11

Microbial Dormancy and Survival in the Subsurface

Penny S. Amy

CONTENTS

KEY WORDS: *survival, starvation, dormancy, VBNC, biogeochemical cycles, marine, soil, spores, resuscitation, bioremediation, starvation-survival phenomenon, endogenous metabolism, PLFA, DGFA, nucleic acid probes, heat shock proteins/stress proteins, radiation waste repository.*

11.1 Introduction

Topics of microbial survival and dormancy have been of interest to both microbiologists and others for some time. As early as the 1920s, those involved in oil recovery noticed that some wells became "soured" by bacterial activity (Ivanov, 1990; Kuznetsov et al., 1963). The question arose as to whether the undesirable microorganisms were previously present in the deep oil deposits or whether they were introduced by human activity (Ghiorse and Wilson, 1988). In the 1950s, scientists on mid-Pacific expeditions cultured bacteria from sediment cores retrieved from the deep ocean. The material was laid down over 1 million years ago, and scientists hypothesized that the microbes had been present since the time of sediment deposition (Morita and Zobell, 1955). However, skeptics thought that the microbes must have come from seawater contamination. The concept of long-term survival of bacteria was postulated but not accepted, and thus it was not introduced to the general scientific community. Advances in sampling technology (see Chapters 3 and 4) have demonstrated that microbes are indigenous to even the deepest subsurface environments. Because researchers have become interested in subsurface microorganisms for practical reasons such as bioremediation (Ghiorse and Wilson, 1988; Swindoll et al., 1988; Truax et al., 1992) and waste repository design (see the Proceedings from the High-Level Radioactive Waste Management Meetings, Las Vegas, 1996; and Chapters 15 and 16), subsurface microbial ecology has developed into an active field of investigation. Specifically, the concept of survival of microorganisms has become important to more members of the scientific community than ever before. Evidence will be presented that supports the concept that even vegetative microbes (mostly bacteria) can survive for millions of years in the subsurface.

11.2 Survival of Microbes in Natural Environments

11.2.1 Marine Environment

11.2.1.1 *Evidence for Long-Term Survival*

The impetus to study the long-term survival of microorganisms was based on investigations of marine water mass movements, and the necessity for microorganisms to survive in low-nutrient water masses from the time of convergence to the time of upwelling in nutrient-rich waters. Morita and others investigated the occurrence of bacteria in open ocean water masses and their survival for what was then thought to be phenomenally long periods of time. Water masses carrying bacteria for several hundred years were considered relatively closed systems where the consistent culturable microbial counts at depth could not readily be accounted for by microbial transport from the surface. The physiology of deep marine microorganisms was investigated and a process of "starvation-survival" was described (Amy and Morita, 1983a,b; Amy et al., 1983; Kurath, 1980; Novitsky and Morita, 1977).

11.2.1.2 *The Starvation-Survival Phenomenon*

The slow-growing, often psychrophilic bacterial isolates from the oceans served for some time as the stereotypical bacteria capable of starvation-survival. The typical survival curves of marine heterotrophic bacteria (nearly all open-ocean bacteria are

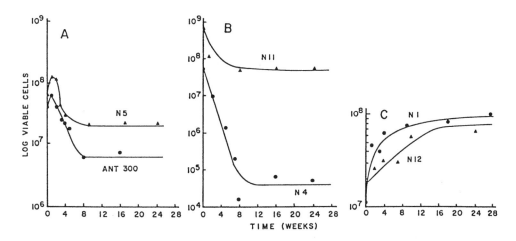

FIGURE 11.1

Viability patterns of 17 open-ocean isolates during nutrient starvation. The pattern of Ant-300 and 7 of the 16 isolates showed initial cell fragmentation followed by a decrease in cell viability (A); the pattern of initial rapid cell death was seen with 4 of the 16 isolates (B); and the unusual pattern of initial viable cell increase was seen in 5 of the 16 isolates (C). (From Amy, P. S. and Morita, R. Y., *Appl. Environ. Microbiol.*, 45, 1109, 1983. With permission.)

Gram negative) are provided in Figures 11.1A and B with autotrophic bacteria most likely described by Figure 11.1C (Amy and Morita, 1983a). As the study of starvation proceeded, and other strains were tested, basic principles for survival of Gram-negative cells under conditions of nutrient deprivation were described. Some of those principles included a series of structural and physiological changes leading to a resistant state (Albertson et al., 1990; Geesey and Morita, 1979; Hood et al., 1986; Matin, 1991; Novitsky and Morita, 1977; Nystrom et al., 1990).

Starvation-survival characteristics include nearly total metabolic arrest in some cell types (Boylen and Ensign, 1970b; Dawes, 1976), miniaturization of cells (Kieft, 1996; MacDonnell and Hood, 1982), changes in macromolecular quantities and cellular density (Amy et al., 1983; Novitsky and Morita, 1977), and resistance to starvation and other hardships (Novitsky and Morita, 1978; Rockabrand et al., 1995; Siegele et al., 1993). Cell membrane transport mechanisms may be enhanced in low-nutrient environments (Geesey and Morita, 1979), and plasmids and chromosomal DNA may change in quantity and the ability to be expressed (Moyer and Morita, 1989; Caldwell et al., 1989). Specific starvation-related proteins were first described by Amy and Morita (1983b) for the marine vibrio, Ant-300. These proteins were synthesized under conditions of nutrient deprivation, and were either upregulated constitutive proteins or *de novo* synthetic products. Recent genetic techniques have allowed researchers to further investigate many of the important proteins synthesized during the starvation process (Hengge-Aronis, 1993; Spector and Foster, 1993). For example, the sigma factor product of *dnaK* is expressed as cells enter the stationary phase and during successful starvation-survival (Spence et al., 1990). Additionally, specific stress proteins are expressed as a result of various types of nutrient deprivation. For example, some proteins are expressed as a result of nitrogen deprivation, others in response to phosphate or carbon deprivation, and some are expressed as a response to general nutrient deprivation (Groat et al., 1986; Matin et al., 1989).

11.2.1.3 Techniques for the Study of Starvation-Survival

Investigation of the physiology of the starvation phenomenon continues to be problematic because by the time a batch culture of starving cells reaches its stable culturable number (days to months, depending on the bacterial type, see Figures 11.1A and B), the population is morphologically, physiologically, and probably genetically mixed. Moyer and Morita (1989) made use of a chemostat to prepare uniform cultures of starved cells. The cells appeared to go through the starvation-survival phenomenon and to reach the starvation-survival plateau, however, nutrient addition was necessary to prevent washout of cells and no evidence was presented that demonstrated that the population was indeed physiologically uniform.

Attempts to separate or distinguish members of the mixed population have not been very successful. Centrifugation and filtration have been used in soil environments to separate populations by size with some success (Bakken, 1985). Results suggested that larger-sized cells were culturable, whereas in the marine environment it is thought that miniaturized cells remain viable/culturable (Amy et al., 1983). Therefore, there is no guarantee that a small cell is viable or that it is physiologically the same as the rest of the cells of its size (or size range). In fact, many cells that miniaturize will die or become nonculturable during the process, with only 0.001 to 10% of the cells remaining culturable after extended starvation (Figure 11.1) (Amy and Morita, 1983a; Amy et al., 1983; Kurath, 1980; Novitsky and Morita, 1977). Rodriguez et al. (1993) developed a method for the visualization of viable cells which reduced a fluorescent dye through respiratory activity. This method has served as an alternative to the older, less sensitive nonfluorescent methods that made use of the same mechanism. Both methods allow viable cells to be visualized and compared to the total number of cells present. However, difficulties with these techniques have included underestimation of some cell types (Smith et al., 1994), difficulty in visualizing formazan crystals in small cells (more problematic with nonfluorescent dyes), and variability of performance depending upon sample type and preparation (unpublished data).

Other techniques based on macromolecular extraction, quantification, and characterization do not resolve the problem of investigating cells within physiologically nonuniform populations because they are based on or normalized to the total population.

New techniques for separating cells of differing physiological properties have recently been employed. Flow cytometry, coupled with vital dyes, can separate viable (respiring) from nonviable cells (Kaprelyants and Kell, 1993; Kaprelyants et al., 1996). This will allow for the study of a cell class with a specific physiological status. However, adaptation of this technique to mixed populations, microbial communities, and soil samples has not yet been reported.

11.2.2 Soil Environment

Gram-negative cell starvation-survival studies were in full swing when some researchers asked how other cell types, in other environments, survived starvation. Soil was the logical place to investigate Gram-positive cell starvation-survival. The soil is an active and dynamic environment for microorganisms. It contains organic matter as part of its physical and chemical structure. Not all the organic matter is bioavailable due to the presence of complex and recalcitrant compounds such as lignin. Additionally, because the earth's surface is often exposed to extreme weather conditions, survival is based on more than just nutrient availability. Temperature, desiccation, and pollutants are other factors which are important in soil. It is clear that organisms in surface soils do survive because large numbers of soil microbes (10^8 to 10^{11} cells per gram) are found consistently.

Study of the survival of soil microorganisms has been based primarily on work of Boylen and Ensign (Boylen, 1973; Boylen and Ensign, 1970a,b; Boylen and Mulks,

1978). The metabolic shutdown of starving cells has been measured best by these authors. Their work with *Arthrobacter crystallopoites* has demonstrated that these Gram-positive cells behave in much the same manner that Gram-negative marine bacteria do during nutrient deprivation; many of the overall cellular protein changes are similar. Specific lipid changes, however, appear to be different in Gram-positive and -negative cells. White and Ringelberg have shown that although Gram-negative cells change lipid types, and demonstrate stress lipid ratios different from non-stressed cells, Gram-positive arthrobacters do not demonstrate such extreme reactions to starvation (Guckert et al., 1986; Kieft et al., 1994).

Survival of Gram-negative soil microbes has also been investigated (Buttner, 1989; Haldeman et al., 1995; Kieft et al., 1996). One of the most interesting studies was that involving pathovars of *Pseudomonas fluorescens* and other pseudomonad strains that were documented to survive in distilled water on a laboratory shelf for 24 years (Iacobellis and DeVay, 1986). Although this is a relatively short survival time compared to that suggested by many researchers for bacterial survival in natural environments, it is a clearly documented experiment and one which helps bring the scientific community closer to understanding the long-term nature of microbial survival.

The discussion thus far has included representatives from both Gram-positive and -negative bacterial groups, however, they have been nonspore-forming cell types. A number of kinds of eubacteria form spores and their survival is usually attributed to the spore structure (Sneath, 1962). Spore-forming bacteria, fungi, algae, and virus survival have been reviewed by Kennedy et al. (1994) and Henis (1987). Small acid-soluble proteins have been shown to help protect spore DNA from damage by dry heat and other environmental stresses (Setlow and Setlow, 1995; Setlow, 1995), and similar molecules may be important in the survival of vegetative cells.

11.2.3   Subsurface Environment

Both soil and aquatic microbial communities may have served as the inoculum for subsurface environments in the past. At the Nevada Test Site, repeated volcanic ash deposition and subsequent colonization of the ash during rudimentary soil formation gave rise to tuff (rock from volcanic ash) laced with what were once soil microorganisms. The White Bluffs, Hanford, WA is an ancient lacustrine environment which contains culturable and active bacteria, some of which were once probably of aquatic origin (Brockman et al., 1992; Kieft et al., 1993; Kieft et al., 1995). At Cerro Negro, ancient marine sediments have developed into sedimentary rock, from which marine bacteria have been isolated (Zlatkin and Forney, 1996). Each subsurface environment is unique because of the geological/physical/biological events that have impacted its history.

Within the U.S., several sites have been investigated for the presence of microorganisms which have been geologically and hydrologically isolated for long periods of time. The basalts of the Idaho National Engineering Laboratory, the tuffs of the Nevada Test Site, the sedimentary deposits at Cerro Negro, New Mexico, the Hanford Reservation, Washington, and the very deep laminated shale deposits of the Taylorsville Triassic Basin in Virginia have been locations of experimental work by interdisciplinary teams of geologists, microbiologists, chemists, and others to define the probable time of microbial isolation and degree of microbial transport from the surface (Balkwill et al., 1994). Investigations of recharge water from the surface, fractures and rock porosity, age of rock material, and biology have left no particular site without some uncertainty regarding the age of the subsurface microorganisms that have been recovered. However, the sum of these studies and others has laid the basis for an increasing belief that subsurface microorganisms are not only present, but viable and culturable, from deep subsurface environments which are millions of years old (Amy et al., 1992; Colwell, 1989; Fredrickson et al., 1995a; Haldeman et al., 1993a,b; Phelps et al., 1994).

At several sites, environmental conditions appear to have defined the types of microbiota observed, and suggest that the organisms were not the result of surface contamination, but indigenous subsurface microbes. Piceance Basin, CO and Triassic Basin, VA are sites where geothermal heating and anaerobic conditions appear to have selected for specific bacterial types (Boone et al., 1995; Colwell et al., 1996; DOE, 1995). For example, from the anaerobic, CH_4-pressurized environment of Piceance Basin, anaerobic, metal-reducing bacteria have been cultured (Colwell et al., 1996). In an earlier study by Lovley et al. (1990), bacteria which could oxidize acetate and reduce iron (III) were found in subsurface aquifers. A new species, *Bacillus infernus,* has been characterized from the unusual environment in the deep subsurface at the Triassic Basin (Boone et al., 1995). The unique physiological properties of this organism further suggest its long-term physical separation from the surface; it is the first anaerobic *Bacillus* sp. have been described. Other interesting organisms have come from deep samples taken from Columbia River Basin basalts. A chemical reaction in the deep subsurface between basalt and water appears to produce hydrogen gas by abiotic means, and hydrogen-utilizing bacteria were prevalent (Stevens and McKinley, 1995). At Cerro Negro, NM, researchers have found higher bacterial activity at sandstone:shale interfaces as compared to other sedimentary rock layers. Increased sulfate-reducing activity was observed in sandstone layers adjacent to shale deposits which presumably supply electron donors to the sulfate-reducing bacteria (Krumholz and Suflita, 1996). In several other subsurface environments where aerobic/facultative heterotrophic bacteria predominate, the cells appear to have gone through the miniaturization process (by direct observation, the cells are very small) yet many of the cells can be recultured even though they are thought to be very old (Amy et al., 1993; Haldeman et al., 1995; unpublished observations). In these subsurface environments, the good match between the highly selective physical/chemical conditions of the environment and the physiological abilities of the microbes suggest that subsurface microorganisms from these sites have been isolated from the surface.

At the Nevada Test Site, tuff material, altered by zeolitization, is found at 400 m depth within Rainier Mesa. Experiments in our laboratory and at the Idaho National Engineering Laboratory have demonstrated the nonpermeability of altered tuff to water, and therefore nutrients and bacteria. Likewise, tuff which has a high clay content has been shown to be impermeable to water. Therefore, the microorganisms trapped in the sealed pore spaces of either of these rock materials have been isolated, minimally since the time of rock alteration (Amy et al., 1992) and probably longer in zeolitized tuffs, where the age of the pore water is greater than the minimum time since zeolitization (250,000 years). Culturable bacterial counts of 10^2 to 10^5 cfu/g dry weight have been isolated from unfractured zeolitized tuff (Amy et al., 1992; Haldeman et al., 1994b).

11.2.3.1 *Microbial Communities*

Microbial burial, isolation, and reculture implies microbial survival. Do we have indications that subsurface microbes are different in any way from surface counterparts? The answer is "yes" from several lines of evidence. Research in our laboratory has included the investigation of microbial communities associated with fracture water, fracture faces and rubble, clay, and isolated zeolitized rock from the tunnel systems mined into tuffaceous rock at Rainier Mesa, on the Nevada Test Site (Amy et al., 1992;

Haldeman, 1994; Story, 1994) and an elevational study of Rainier Mesa (Haldeman et al., 1994b). In each of these studies, the microbial community found inside the isolated, zeolitized rock of the mesa contained bacterial numbers and types not like those of samples more recently impacted by the surface or by recharge water. Work done by Colwell and colleagues with samples from the Taylorsville Triassic Basin site demonstrated that metabolic patterns, as displayed by BIOLOG plates, were quite different between organisms present in the drilling fluid (surface organisms) and those from uncontaminated subsurface rock (Lehman et al., 1995).

Subsurface samples from the Savannah River Plant, South Carolina were obtained from an aquifer which "fed" the bacteria at a slow but steady rate. Unlike more nutrient-deprived environments, numbers of culturable bacteria were nearly equal to the total bacterial count (Balkwill, 1989; Fredrickson et al., 1991). The close correlation in numbers implies that the cells were active and most of them were culturable; however, the turnover time for microbes in that environment was estimated to be decades to centuries (Phelps et al., 1994).

11.2.3.2 Molecular Analyses

Most subsurface environments contain low numbers of cells, and therefore the study of the microbial communities in the subsurface has had to await techniques sensitive enough to describe them. Cell culture has recently met with opposition due to the inability to culture all cell types, and molecular techniques like lipid analysis and nucleic acid probes have proved useful in supplementing traditional methods, e.g., culturing and metabolic activities.

The use of extraction, derivitization, and analysis of lipids by gas chromatography and mass spectrometry (see Chapter 8) has been invaluable in understanding the amount of biomass, and its nature, in the subsurface. This method was pioneered and has been refined over the last several years by White and colleagues for the analysis of subsurface samples. The analysis allows for measurement of living biomass by phospholipid fatty acids (PLFA) and the measurement of "dead" biomass by diglyceride fatty acids (DGFA) (Haldeman et al., 1995; Kieft et al., 1994). Some specific PLFA types can be assigned to particular species (signature lipids), others are indicative of general microbial groups such as Gram-negative and -positive bacteria, and changes in specific lipid ratios provide an index of stress or nutrient status.

Ribosomal RNA probes have been developed by Nierzwicki-Bauer and associates. These fluorescently labeled probes bind to cellular ribosomes and can be used to determine the number of metabolically active cells in environmental samples using *in situ* hybridization methods. It is believed that a detectable fluorescent signal is produced only if there are sufficient numbers of ribosomes present in the cell, thus allowing for the ability to distinguish between "active" (positive ribosome signal) from "nonactive" cells (negative ribosome signal but positive for protein counterstain). These probes have been used to enumerate metabolically active bacteria in the deep subsurface in a paleosol chronosequence in eastern Washington state (Haldeman et al., 1995; DOE, 1995).

11.2.3.3 Application of Molecular Techniques to the Subsurface

The microbial ecology of a paleosol chronosequence in eastern Washington state was investigated in the fall of 1994 (Haldeman et al., 1995). The soil in this region was deposited by aeolian forces (wind) over several million years in layers. Soil horizons developed over time were periodically buried, and ultimately became paleosols. A series of these paleosols (a chronosequence of ancient soil horizons) were exposed by roadcuts in the 1970s. Buried ash layers from nearby volcanos and magnetic polarity measurements allowed age boundaries to be set for the paleosol chronosequence: from the present to 1 million years old. Although the paleosols may have been impacted by water during glacial retreat and concomitant flooding, they represent a chronological sequence of buried layers of nearly identical soil makeup. The paleosol series, from surface to depth, can be compared to a very long-term starvation-survival experiment. When a comparison of culturable cell number was graphed against depth (time of burial), the curve looks very much like that shown in Figure 11.1B. Parameters such as active cells (as measured by the rRNA probe method described above) and PLFA (living biomass) decrease with depth (soil age); while DGFA and "dead cells" (total cells minus active cells) increase with soil age. This implies, as one would predict, that living cells die over time and the result is a decrease in PLFA and an increase in DGFA. In this case, lipid analysis and rRNA probes have been used to investigate microbial communities in paleosols ranging in age from recent time to over 1 million years, and have augmented more traditional plating, metabolic activity and enrichment techniques.

11.3 Molecular Biology of Survival

Molecules which have been evaluated during starvation-survival of pure cultures in liquid menstruum include proteins, nucleic acids, lipids, and storage products. More recent techniques that have been used to study pure, mixed, and native microbial communities (including investigations of survival in geological materials) are (1) RNA probes for the presence of specific microbial types and activities; (2) lipid analyses for the determination of microbial biomass, specific cell classes/types and stress markers; and (3) DNA sequencing to examine the rate of change of specific genes with time (molecular clocks).

11.3.1 Heat-Shock or Stress Proteins

Since the early 1980s, heat-shock proteins (now called stress proteins) have been found in all types of organisms. Much of the recent work on bacterial stress proteins has been conducted in the laboratory of Abdul Matin, where the individual as well as the common nature of stress proteins produced as the result of starvation have been elucidated (Matin et al., 1989; Matin, 1991). In some cases, the lack of an individual nutrient, e.g., nitrogen or phosphate, triggered the synthesis of a specific set of proteins, whereas some of these proteins in each of these sets are commonly synthesized by the absence of any necessary nutrient.

Heat-shock or stress proteins have been best elucidated by acrylamide gel electrophoresis (O'Farrell, 1975). The first dimension separates the proteins by their isoelectric point (on a pH gradient) and the second dimension separates them by molecular weight (in a sodium dodecyl sulfate, polyacrylamide gel). The two-dimensional pattern of protein spots that results is specific for each organism and its culture conditions, and thus specific stressors can be tested to evaluate the protein patterns that result as a response of the specific organism. Most of the extensive investigations of the stress response in bacteria have been examined in *Escherichia coli*. In this organism, the *rpoS/katF* gene product, a sigma factor, was found to be necessary for transcription of a series of proteins synthesized after activation by nutrient deprivation and other stressors (van Elsas and van Overbeek, 1993). In this way, the sigma factor is believed to control the expression of stress proteins in *E. coli.*

11.3.2 Stress-Induced Resistance

Unicellular as well as multicellular organisms become resistant to heat stress and other environmental stresses after experiencing an initial stressor (thus the name stress proteins). Stressors include elevated temperature, lack of oxygen, oxidizing conditions, and toxicants (Matin, 1991). Often, exposure to one stress can provide cross protection against another stress (Rockabrand et al., 1995). The enhanced survival of starved bacteria compared to actively growing cells in the presence of heat and hydrogen peroxide stressors is shown in Figure 11.2. Starved Gram-negative cells demonstrated a much greater resistance to disinfecting agents and toxicants than those which were actively growing (Nystrom et al., 1992; Hengge-Aronis, 1992, Rockabrand et al., 1995). With *Micrococcus luteus,* a nonsporulating Gram positive bacterium, cells became resistant to mechanical breakage or enzyme lysis after heat-shock exposure (Amy and Dupre, unpublished data). Compounds like cAMP and gene products such as katF appear to be important in triggering the stress response and controlling protein expression (Blum et al., 1990; Schultz et al., 1988), but exact mechanisms of how these proteins allow organisms to become more resistant to environmental stresses are as yet unknown.

11.3.3 Lipid Analysis

In addition to the brief description given above, lipid analyses can be performed to reveal much about the subsurface microbial community and its activity. General PLFA and DGFA levels can be used to describe the abundance of living and "dead" biomass, respectively. Additionally, specific classes of fatty acid molecules can be separated and analyzed to describe representative microbial groups, such as Gram-negative/positive bacteria, or more specific groups, such as sulfate-reducing bacteria. The ratio of a particular saturated/unsaturated fatty acid product, *trans/cis* fatty acids, or cyclopropyl saturated/monoenoic precursors can describe the stress response of bacteria to starvation or desiccation (Guckert et al., 1986; Kieft et al., 1994). These ratios and the presence of storage products such as polyhydroxy alkanoates can provide an indication of the state of nutrient deprivation, carbon starvation, oligotrophy, and metabolic status (see Chapter 8). Lipid analyses are invaluable in environments where biomass is low, i.e., many subsurface environments, because of the sensitivity afforded by these techniques. Additionally, the DGFA values can be used to describe dead biomass, something that other techniques often do not address.

11.4 Microbial Dormancy

To this point, the discussion has centered around survival of microbes for long periods of time, but are those microbes dormant? Logic tells us that an organism that has been trapped for millennia in a rock matrix containing little or no organic carbon (heterotrophs) or other oxidizable substrates (autotrophs), and is subsequently cultured, must have been in some state of metabolic arrest or dormancy. The more we know

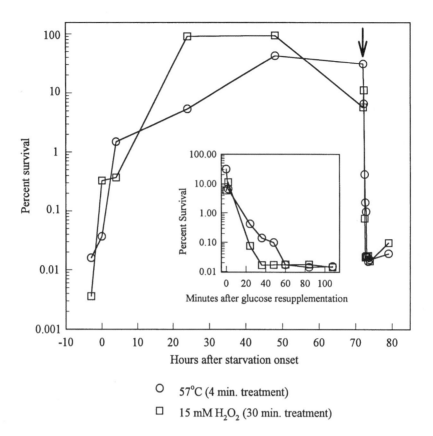

FIGURE 11.2
Thermotolerance and H_2O_2 resistance in glucose-starved and resupplemented *E. coli*. Thermotolerance of
PBL500 (wild-type) cell samples was determined at 57°C after 4 min of heat exposure (circles). H_2O_2
resistance was determined by exposure to 15 mM H_2O_2 for 30 min (squares). The number of surviving
cells was normalized to the number of viable cells in an untreated but otherwise identical sample. The
values shown are the averages of duplicate samples. The onset of glucose starvation is indicated as time
zero, and the time at which cultures were resupplemented with glucose is indicated by the arrow. The
insert shows cell survival frequencies for glucose-resupplemented cells on an expanded time scale. The
insert x-axis value for time zero corresponds to 72 h of glucose starvation. (From Rockabrand, D., Arthur,
T., Korinek, G., Lievers, K., and Blum, P., *J. Bacteriol.*, 177, 3695, 1995. With permission.)

about the subsurface, the more it appears that many cells must be in a dormant state
or surviving by maintaining extremely low (unmeasurable) endogenous metabolism.
But how do they maintain viability? Is water excluded, as is the case with spore formation? Are the cell contents/membranes crystallized? Do chaperone proteins coat
the DNA to promote stability in vegetative cells as has recently been shown with
small, acid-soluble proteins in spores (Setlow and Setlow, 1995)? The answers to these
questions await advances in technology and techniques and continued interest in
cells which survive long-term starvation.

11.4.1   Spores

Although spores have mastered the art of survival under extreme conditions, it now
appears that vegetative cells rival, if not exceed, the ability of spores to survive in non-
extreme environments where nutrients are limiting, like many deep subsurface environments. The special ability of spores to remain viable is not related only to the
length of time in the spore state, but also to the ability of the spore to withstand

extremes in drying, heat exposure, and toxic materials. Within the "meso" ranges of environmental parameters, starving vegetative cells survive for very long periods of time without specialized spore coat structures or many other gross morphological changes other than miniaturization and loss of cytoplasmic density. Thus, it appears as if vegetative cells have adapted to long-term survival by a different mechanism that also includes enhanced protection from environmental stresses. This protective mechanism may be triggered by starvation and other environmental stresses.

The existence of spores in the subsurface has been inferred primarily through culture of fungi, actinomycetes, and spore-forming bacteria (Amy et al., 1992; Balkwill, 1989; Ghiorse and Balkwill, 1982). However, it is difficult to know when cells have germinated from spores or have proliferated from hyphal fragments or individual cells. Culture of deep subsurface tuff from Rainier Mesa has yielded large numbers of actinomycete colonies from some sites and few in others. Traditional methods for spore detection (exposure of samples to elevated heat for short time periods) has resulted in recovery of few isolates. Fungal colonies were seldom observed except in association with exposed surfaces of tunnel walls (Haldeman and Amy, 1993; see Chapter 4).

In a study at Yucca Mountain, the proposed site for the long-term high-level nuclear waste repository, vadose rock from a mined area was ground, packed into microcosms, and was allowed to support cell growth at room temperature for several weeks (Pitonzo et al., unpublished data; see Chapter 5 for cell proliferation during storage). After this treatment, many more spore-forming cell types were recovered than in the initial plating of the rock material, and the nonirradiated control microcosms did not show this reaction. After extensive radiation exposure, even spore-forming fungi and actinomycetes became nonculturable. It is not known whether spore-forming bacteria and fungi were present in the original rock as hyphae or as spores, or what conditions might have "resuscitated" fungal spores or hyphae, although these experiments suggest that low levels of radiation might play a role.

Although Gram-positive bacteria are prevalent in the subsurface, they are not commonly spore-forming species, i.e., *Arthrobacter, Micrococcus* and *Staphylococcus* (see Chapter 7). The impact of spores as a major factor in the microbial ecology of the subsurface is unknown. If resuscitation is needed for significant spore germination or hyphal growth, perhaps conditions have not been provided for the germination of fungal, actinomycete, or bacterial spores in previous investigations. More work will have to been done to determine the ecological significance of spores in the subsurface.

11.4.2 Vegetative Cells

Vegetative cells have been the most commonly cultured and microscopically observed cell forms from subsurface terrestrial material. Until recent anaerobic and geothermally heated sites were analyzed, the most commonly recovered cultured microbes were aerobic/facultative mesophilic heterotrophic bacteria (Amy et al., 1992; Balkwill, 1989; Fredrickson et al., 1991, 1993; Ghiorse and Balkwill, 1982). As evidence for the ancient nature of cells in many subsurface environments has mounted, it has been necessary to imagine vegetative cells surviving in rock pore spaces, with or without water, for thousands to millions of years. The term "dormancy" has been used by some to describe vegetative cells that have been in conditions of long-term starvation for millennia.

The exact nature of dormancy has been difficult to define. Through the study of pure cultures in various phases of growth, a somewhat continuous series of cell states has been described. These include actively growing, stationary phase, starving, respiring, viable but nonculturable, dormant, and dead cells. Much of the distinction at this point is based on the growth rate, metabolic activity, and culturability of the cells in question.

11.4.2.1 Metabolism: Evidence for Dormancy

Cells in natural settings, by the nature of their surroundings, eventually reach a state of starvation-survival under most conditions (Poindexter, 1981). This appears to hold true in the subsurface because total counts generally exceed culturable counts by several orders of magnitude. If lack of culturability is an indicator of dormancy, this would suggest that most subsurface environments contain dormant or dead organisms. An exception has been the subsurface at Savannah River, SC where an aquifer fed the immobilized cells by bathing them periodically with low levels of nutrients. Here, total bacterial counts approached culturable bacterial counts (Balkwill, 1989).

Under starvation conditions, cellular metabolism is greatly arrested (Lappin-Scott and Costerton, 1990) and a low level of "endogenous" metabolism, which cannot often be measured, but which is assumed to be needed for membrane and nucleic acid integrity, is achieved (Dawes, 1976). During the normal course of events in starvation of heterotrophs, stored products are metabolized, cellular components including proteins and nucleic acids (mostly ribosomal material) are utilized, and some lipid structures or ratios are changed (Amy et al., 1983; Guckert et al., 1986; Moyer and Morita, 1989; Novitsky and Morita, 1977; also see Chapter 8). But what happens after this plethora of available carbon sources is exhausted? Is there a state of suspended animation for bacteria? We often think of spores somewhat along these lines, but it now seems that vegetative cells have as good or better ability to survive in the subsurface.

In a study of tuff from Rainier Mesa, NV, correlations between geochemical, geological and microbiological parameters were made for 19 samples from a 3-D section of rock (Russell et al., 1994). One of the strongest correlations was that of the presence of nitrate ion and nitrate-reducing bacteria. If the cells were metabolically active, nitrate would have been reduced to nitrite, or further reduced compounds. However, since metabolism (allowing for nitrate reduction) was arrested, an electron donor or other factor must have limited microbial metabolism, rendering the bacteria in those samples dormant. Other researchers have described sites where either electron donors (as in the case described) or electron acceptors limited metabolism but associated bacteria were viable, and thus provide evidence for bacterial dormancy (Fredrickson et al., 1989; Whitelaw and Reese, 1980; and see Chapter 6).

11.4.2.2 Viable but Nonculturable Cells

Viable but nonculturable cells (VBNC) were described by Kogure et al. (1979) when they observed cellular elongation of Gram-negative marine bacteria in the presence of nalidixic acid, a cell division inhibitor. Later, Colwell et al. (1985) coined the term "viable but nonculturable" to describe cells that retain viability but cannot be cultured on media that would normally support their growth. This definition can be extended to include cells which are viable but which cannot grow on the media or under the conditions used in a particular experiment. For example, nitrifying bacteria in a subsurface sample would be considered as viable but nonculturable if plated on heterotrophic media. The large discrepancy between total and culturable cell counts in nearly all natural environments can now partly be explained by VBNC.

The VBNC state has been studied extensively in *Vibrio* species and other Gram-negative cells (Colwell et al., 1985; Kogure et al., 1979; Oliver, 1993; Roszak et al., 1984). This cell state is very important to those who study pathogens because living bacteria that are found in natural environments (e.g., natural waters and foods such as oysters) can escape detection, and host populations are then at risk. Microorganisms can become VBNC in drinking water (Byrd et al., 1991), and can potentially cause illness when they are resuscitated in the host organism. This cell state is probably important in all native environments. It is in fact probably one of the most common states of cells on the planet!

11.4.2.3 Microbial Resuscitation

Resuscitation is the process by which VBNC become culturable. Colwell and co-workers have described the process of resuscitation for *Vibrio cholerae* and other pathogens from the VBNC state (Colwell et al., 1985; Rozak and Colwell, 1987). This was accomplished by passing them through rabbit ileal loops to produce culturable cells on standard media. Haldeman et al. (1994a, 1995) and Fredrickson et al. (1995b) have described subsurface bacterial communities which change in numbers and types of culturable cells over time at 4°C, due partly to growth of some cell types, and partly to resuscitation of originally nonculturable cells. Pitonzo (1996) have shown that subsurface rock communities at Yucca Mountain can be resuscitated from a non-culturable state, due to extended radiation exposure that rendered them VBNC, to a culturable state after storage at 4°C.

Kaprelyants et al. (1993) define dormancy as "a reversible state of low metabolic activity, in which cells can persist for extended periods without division." They further suggest that many or most of these cells require a resuscitation step to become culturable again. In other words, they are VBNC during dormancy and require resuscitation. The resuscitation process can be accomplished by a number of previously described means, including low temperature (Haldeman et al., 1995), warming (Nilsson et al., 1991), a short growth period on nonselective media followed by replica-plating on selective media (Clesceri et al., 1989), ionic shock (Murno et al., 1989), and distilled water. Most importantly, resuscitation can probably be accomplished in many ways that have never been tested or observed as of yet. Resuscitation will be an interesting research challenge as more people want or need to resuscitate populations and communities of subsurface bacteria.

11.5 Implications for Microbial Ecology

At many sites, the subsurface is teeming with life, primarily bacteria. At some sites the subsurface is a dynamic environment due to nutrient and microbial influx. However, many subsurface environments are isolated from surface impact, and thus contain microorganisms which have been in place for long periods of time. Even in subsurface environments dominated by dormant bacteria, the potential for resuscitation of indigenous bacteria exists. This potential has great implications for microbial ecology, and for applications of microorganisms to the bioremediation of deep subsurface contamination.

11.5.1 Biogeochemical Cycles of Elements

With the knowledge that living bacteria exist to at least 9180 ft in the subsurface, Gould hypothesized that microorganisms must comprise the majority of the biomass on earth (Gould, 1996). Additionally, because microorganisms display the greatest metabolic diversity of all living things, they have the potential for greatly impacting the biogeochemical cycles of elements. In fact, some chemical transformations within natural elemental cycles are known to be accomplished, primarily, if not exclusively, by microbes (e.g., nitrogen fixation, nitrification, sulfate reduction).

The recent discovery of hydrogen-utilizing bacteria, fed by the chemical reaction of basalt and water (Stevens and McKinley, 1995), has also suggested that biochemical transformations may be occurring in the extended biosphere, perhaps to very great depths. New species (e.g., *Bacillus infernus*, Boone et al., 1995) and unusual biogeochemical transformations have been reported for microbes in both aquifer and geothermally heated, gas pressurized, anaerobic shale and sandstone environments (Colwell et al., 1996; Lovely et al., 1990). By extending the biosphere to a minimum of 9000 ft below the surface in terrestrial environments, and to unknown depths in ocean sediments, questions regarding the potential impact of indigenous microorganisms on biogeochemistry become apparent. It is clear that the potential for subsurface microbial impact is significant and that further study is likely to provide answers to questions regarding the importance of indigenous subsurface microbes to biogeochemical cycles, as well as revealing undiscovered phylogenetic types and metabolic capabilities.

11.5.2   Bioremediation/Waste Storage

Two driving forces have been at the core of subsurface science support in the U.S., (1) the need to clean up severely contaminated sites, including deep subsurface environments, and (2) the need to know about subsurface microbial ecology for long-term geological storage of toxic/radioactive materials.

The fact that living microbes (actively metabolizing or potentially capable of resuscitation) can be found in virtually every subsurface environment yet studied, means that there is a natural inoculum for bioremediation efforts. However, many conditions may need to be met before the existing microbes can actively degrade target toxic materials (see Chapter 14). This is intuitively true, otherwise the toxic materials in those environments would have already been degraded. The stimulation of (resuscitation?), or the addition of limiting nutrients to indigenous microorganisms, remains a challenge. Likewise, injection of specific microbes to very deep subsurface environments also remains challenging (for an example of how this challenge is being met, see Chapter 17).

For long-term storage of toxic/radioactive wastes, lack of microbial activity is desired. Ideally, the organisms remain dormant, leaving materials placed underground unaltered. However, the decomposer role of microbes in nature is not consistent with this anthropocentric view. There is currently much concern about the long-term integrity of the containers used for high-level nuclear waste storage because nearly every material is subject to microbial deterioration (see Chapter 15). How to keep the microbes in a dormant state while perturbing them with heat, radiation, and desiccation is as challenging as trying to resuscitate them and stimulate them to degrade toxic materials.

Acknowledgments

I wish to thank Dr. R. Y. Morita for his inspiration to study starvation-survival and dormancy of microorganisms in natural environments and Dana Haldeman for scientific partnership during seven years of subsurface research. Also, thanks to Dr. Frank Wobber for seven continuous years of funding in the USDOE Subsurface Science Program to support our investigations of the deep subsurface at the Nevada Test Site and paleosols in Washington State.

References

Albertson, N. H., T. Nystom, and St. Kjelleberg. 1990. Exoprotease activity of two marine bacteria during starvation. *Appl. Environ. Microbiol.*, 56:218-223.

Amy, P. S. and R. Y. Morita. 1983a. Starvation survival patterns of sixteen freshly isolated open-ocean bacteria. *Appl. Environ. Microbiol.*, 45:1109-1115.

Amy, P. S. and R. Y. Morita. 1983b. Protein patterns of growing and starved cells of a marine Vibrio sp. *Appl. Environ. Microbiol.*, 45:1748-1752.

Amy, P. S., C. Pauling, and R. Y. Morita. 1983. Starvation-survival processes of a marine vibrio. *Appl. Environ. Microbiol.*, 45:1041-1048.

Amy, P. S., D. L. Haldeman, D. Ringelberg, D. H. Hall, and C. Russell. 1992. Comparison of identification systems for the classification of bacteria isolated from water and endolithic habitats within the deep subsurface. *Appl. Environ. Microbiol.*, 58:3367-3373.

Amy, P. S., C. Durham, D. Hall, and D. Haldeman. 1993. Starvation-survival of deep subsurface isolates. *Appl. Environ. Microbiol.*, 26:345-352.

Bakken, L. 1985. Separation and purification of bacteria from soil. *Appl. Environ. Microbiol.*, 49:1482-1487.

Balkwill, D. L., 1989. Numbers, diversity, and morphological characteristics of aerobic, chemo-heterotrophic bacteria in deep subsurface sediments from a site in South Carolina. *Geomicrobiol. J.*, 7:33-51.

Balkwill, D. L., D. R. Boone, F. S. Colwell, T. Griffin, T. L. Kieft, R. M. Lehman, J. P. McKinley, S. Nierzwicki-Bauer, T. C. Onstott, H. Y. Tseng, G. Gao, T. J. Phelps, D. Ringelberg, B. Russell, T. O. Stevens, D. C. White, and F. J. Wobber. 1994. D.O.E. seeks origin of deep subsurface bacteria. *EOS*, 75:395-396.

Balkwill, D. L., J. K. Fredrickson, and J. M. Thomas. 1989. Vertical and horizontal variations in the physiological diversity of the aerobic chemoheterotrophic bacterial microflora in deep southeastern coastal plain subsurface sediments. *Appl. Environ. Microbiol.*, 55:1058-1065.

Blum, P. H., S. B. Jovanovich, M. McCann, J. Schultz, L. Lesley, R. Bergess, and A. Matin. 1990. Cloning and *in vivo* and *in vitro* regulation of cyclic AMP-dependent carbon starvation genes from *E. coli*. *J. Bacteriol.*, 172:3813-3820.

Boone, D. R., Y. Lui, Z. Zhao, D. L. Balkwill, G. R. Drake, T. O. Stevens, and H. C. Aldrich. 1995. *Bacillus infernus* sp. Nov., an Fe(III)- and Mn(IV)-reducing anaerobe from the deep subsurface. *Int. J. Syst. Bacteriol.*, 45:441-448.

Boylen, C. W. 1973. Survival of *Arthrobacter crystallopoietes* during prolonged periods of extreme desiccation. *J. Bacteriol.*, 133:33-37.

Boylen, C. W. and J. C. Ensign. 1970a. Long-term starvation survival of rod and spherical cells of *Arthrobacter crystallopoietes*. *J. Bacteriol.*, 103:569-577.

Boylen, C. W. and J. C. Ensign. 1970b. Intracellular substrates for endogenous metabolism during long-term starvation of rod and spherical cells of *Arthrobacter crystallopoietes*. *J. Bacteriol.*, 102:578-587.

Boylen, C. W. and M. H. Mulks. 1978. The survival of coryneform bacteria during periods of prolonged nutrient starvation. *J. Gen. Microbiol.*, 105:323-334.

Brockman, F. J., T. L. Kieft, J. K. Fredrickson, B. N. Bjornstad, S. W. Li, W. Spangenburg, and P. E. Long. 1992. Microbiology of vadose zone paleosols in south-central Washington State. *Microb. Ecol.*, 23:279-301.

Buttner, M. P. 1989. Survival of Ice Nucleation-Active and Genetically Engineered Inactive Strains of *Pseudomonas syringae*. M.S. thesis, University of Nevada, Las Vegas.

Byrd, J. J., H. S. Xu, and R. R. Colwell. 1991. Viable but nonculturable bacteria in drinking water. *Appl. Environ. Microbiol.*, 57:875-878.

Caldwell, B. A., C. Ye, R. P. Griffiths, C. L. Moyer, and R. Y. Morita. 1989. Plasmid expression and maintenance during long-term starvation-survival of bacteria in well water. *Appl. Environ. Microbiol.*, 55:1860-1864.

Clesceri, L. S., A. E. Greenberg, R. R. Trussell. 1989. *Standard Methods for the Examination of Water and Wastewater.* American Public Health Association, Washington, D.C.

Colwell, F., M. Delwiche, T. C. Ontsott, Q.-J. Yao, R. Griffiths, D. Ringelberg, D. White, J. K. Fredrickson, and R. Lehman. 1996. Microbial communities from deep natural gas-bearing rocks. Am. Soc. Microbiology Ann. Meet. Abstr., American Society for Microbiology, Washington, D.C. #N58.

Colwell, F. S. 1989. Microbiological comparison of surface soil and unsaturated subsurface soil from a semiarid high desert. *Appl. Environ. Microbiol.*, 55:2420-2423.

Colwell, R. R., B. R. Brayton, D. J. Grimes, D. B. Roszak, S. A. Huq, and L. M. Palmer. 1985. Viable but non-culturable *Vibrio cholerae* and related pathogens in the environment. *Bio/Technology*, 3:817-820.

DOE. 1995. Subsurface Science Program. Deep subsurface microbiology principal investigator's annual meeting, July, 1995. Lake Oswego, Oregon. Department of Energy, Washington, D.C.

Dawes, E. A. 1976. Endogenous metabolism and the survival of starved prokaryotes. *Symp. Soc. Gen. Microbiol.*, 26:19-53.

Fredrickson, J. K., T. R. Garland, R. J. Hicks, J. M. Thomas, L. W. Li, and K. M. McFadden. 1989. Lithotrophic and heterotrophic bacteria in deep subsurface sediments and their relation to sediment propperties. *Geomicrobiol. J.*, 7:53-66.

Fredrickson, J. K., D. L. Balkwill, J. M. Zachary, S. W. Li, F. J. Brockman, and M. A. Simmons. 1991. Physiological diversity and distributions of heterotrophic bacteria in deep Cretaceous sediments of the Atlantic coastal plain. *Appl. Environ. Microbiol.*, 57:402-411.

Fredrickson, J. K., F. J. Brockman, B. N. Bjornstad, P. E. Long, S. W. Li, J. P. McKinley, J. V. Wright, J. L. Conca, T. L. Kieft, and D. L. Balkwill. 1993. Microbiological characteristics of pristine and contaminated deep vadose zone sediments from an arid region. *Geomicrobiol. J.*, 11:95-107.

Fredrickson, J. K., J. P. McKinley, S. A. Nierzwicki-Bauer, D. C. White, D. B. Ringelberg, S. A. Rawson, S. Li, F. J. Brockman, and B. N. Bjornstad. 1995a. Microbial community structure and biogeochemistry of Miocene subsurface sediments: implications for long-term microbial survival. *Mol. Ecol.*, 4:619-626.

Fredrickson, J. K., S. W. Li, F. J. Brockman, D. L. Haldeman, P. S. Amy, and S. L. Balkwill. 1995b. Time-dependent changes in viable count and activities of heterotrophic bacteria in subsurface samples. *J. Microbiol. Methods*, 21:305-316.

Geesey, G. G. and R. Y. Morita. 1979. Capture of arginine at low concentrations by a marine psychrophilic marine bacterium. *Appl. Environ. Microbiol.*, 38:1092-1097.

Ghiorse, W. C. and D. L. Balkwill. 1982. Microbiological characterization of subsurface environments. In Proc. 1st Int. Conf. on Groundwater Quality Research. Ward, C. H. and B. Giger, Eds., Wiley Interscience, New York, 887-401.

Ghiorse, W. C. and J. T. Wilson. 1988. Microbial ecology of the terrestrial subsurface. *Adv. Appl. Microbiol.*, 33:107-172.

Gould, S. J. 1996. Microcosms. *Natural History Magazine.* 3/96:21-68.

Groat, R. G., J. E. Schultz, E. Zychlinski, A. T. Bockman, and A. Matin. 1986. Starvation proteins in *Escherichia coli*: Kinetics of synthesis and role in starvation survival. *J. Bacteriol.*, 168:486-493.

Guckert, J. B., M. A. Hood, and D. C. White. 1986. Phospholipid ester-linked fatty acid profile changes during nutrient deprivation of *Escherichia coli*. *Appl. Environ. Microbiol.*, 52:794-801.

Haldeman, D. L. 1994. Recovery and Characterization of Microbiota from Rainier Mesa, Nevada Test Site. Doctoral dissertation, University of Nevada, Las Vegas.

Haldeman, D. L. and P. S. Amy. 1993a. Bacterial heterogeneity in deep subsurface tunnels at Rainier Mesa, Nevada Test Site. *Microb. Ecol.*, 25:183-194.

Haldeman, D. L., P. S. Amy, D. Ringelberg, and D. C. White. 1993b. Characterization of the microbiology within a 21 m^3 section of rock from the deep subsurface. *Microb. Ecol.*, 26:145-159.

Haldeman, D. L., P. S. Amy, D. C. White, and D. R. Ringelberg. 1994a. Changes in bacteria recoverable from subsurface volcanic rock samples during storage at 4°C. *Appl. Environ. Microbiol.*, 60:2697-2703.

Haldeman, D. L, B. Pitonzo, S. P. Story, and P. S. Amy. 1994b. Comparison of the microbiota recovered from surface and deep subsurface rock, water, and soil samples along an elevational gradient. *Geomicrobiol. J.*, 12:99-111.

Haldeman, D. L., P. S. Amy, D. Ringelberg, D. C. White, R. E. Garen, and W. C. Ghiorse. 1995. Microbial growth and resuscitation alter community structure after resuscitation. *FEMS Microb. Ecol.*, 17:27-38.

Haldeman, D. L., P. S. Amy, T. Kieft, W. Spangenberg, D. Ringelberg, and D. White. 1995. The microbiology of a chronosequence of paleosols from the Channeled Scablands of arid eastern Washington. Am. Soc. Microbiology Annu. Meet. Abstr., Amer. Soc. Microbiol., Washington, D.C. #N142.

Haldeman, D. L., L. E. Ragatz, and P. S. Amy. 1995. Starvation-survival of deep subsurface bacteria in liquid and sand microcosms. Am. Soc. Microbiology Annu. Meet. Abstr., American Society Microbiology, Washington, D.C. #Q157.

Hengge-Aronis, R. 1993. The role of rpoS in early stationary phase gene regulation in *Escherichia coli* K12. In *Starvation in Bacteria*, Kjelleberg, S., Ed., Plenum Press, New York, 171-200.

Henis, Y. 1987. Survival and dormancy of bacteria. In *Survival and Dormancy of Microorganisms.* Henis, Y., Ed., Wiley Interscience, New York.

Hood, M. A., J. B. Guckert, D. C. White, and F. Deck. 1986. Effect of nutrient deprivation on lipid, carbohydrate, DNA, RNA, and protein levels in *Vibrio cholerae. Appl. Environ. Microbiol.*, 52:788-793.

Iacobellis, N. S. and J. E. DeVay. 1986. Long-term storage of plant-pathogenic bacteria in sterile distilled water. *Appl. Environ. Microbiol.*, 52:388-389.

Ivanov, M. V. 1990. Subsurface microbiological research in the U.S.S.R. In Fliermans, C. B. and T. C. Hazen, Eds., *Proc. First Int. Symp. Microbiology of the Deep Subsurface*. WSRC Information Services, Aiken, SC. p. 1-7.

Kaprelyants, A. S. and D. B. Kell. 1993. Dormancy in stationary-phase cultures of *Micrococcus luteus*. Flow cytometric analysis of starvation and resuscitation. *Appl. Environ. Microbiol.*, 59:3187-3196.

Kaprelyants, A. S., J. C. Gottschall, and D. B. Kell. 1993. Dormancy in non-sporulating bacteria. *FEMS Microbiol. Rev.*, 104:271-286.

Kaprelyants, A. S., G. V. Mukamolova, H. M. Davey, and D. B. Kell. 1996. Quantitative analysis of the physiological heterogeneity within starved cultures of *Micrococcus luteus* by flow cytometry and cell sorting. *Appl. Environ. Microbiol.*, 62:1311-1316.

Kennedy, M. J., S. L. Reader, and L. M. Swierczynski. 1994. Preservation records of microorganisms: evidence of the tenacity of life. *Microbiology*, 140:2513-2529.

Kieft, T. L., P. S. Amy, F. J. Brockman, J. K. Fredrickson, B. N. Bjornstad, and L. L. Rosacker. 1993. Microbial abundance and activities in relation to water potential in the vadose zones of arid and semiarid sites. *Microb. Ecol.*, 26:59-78.

Kieft, T. L., D. B. Ringelberg, and D. C. White. 1994. Changes in ester-linked phospholipid fatty acid profiles of subsurface bacteria during starvation and desiccation in a porous medium. *Appl. Environ. Microbiol.*, 60:3292-3299.

Kieft, T. L., J. K. Fredrickson, J. P. McKinley, B. N. Bjornstad, S. A. Rawson, T. J. Phelps, F. J. Brockman, and S. M. Pfiffner. 1995. Microbial comparisons within and across contiguous lacustrine, paleosol, and fluvial subsurface sediments. *Appl. Environ. Microbiol.*, 61:749-757.

Kieft, T. L. 1996. Dwarf cells in soil and subsurface terrestrial environments, In *Non-culturable Microorganisms in the Environment*. R. R. Colwell and D. J. Grimes, Eds., Chapman and Hall, New York.

Kieft, T. L., K. O'Connor, E. Wilch, D. B. Ringelberg, and D. C. White. 1996. Long-term survival rates and phospholipid fatty acid signature lipids of subsurface microorgainsisms in sediment microcosms. Am. Soc. Microbiology Annu. Meet. Abstr., American Society for Microbiology, Washington, D.C. #Q105.

Kogure, K., U. Simidu, and N. Taga. 1979. A tentative direct microscopic method of counting living bacteria. *Can. J. Microbiol.,* 25:415-420.

Krumholz, L. R. and J. M. Suflita. 1996. Ecological interactions among sulfate-reducing and other bacteria living in subsurface Cretaceous rocks. Am. Soc. Microbiology Meet. Abstr., American Society for Microbiology, Washington, D.C. N56.

Kurath, G. 1980. Some Physiological Bases for Survival of a Marine Bacterium During Nutrient Starvation. M.S. thesis, Oregon State University, Corvallis.

Kuznetsov, S. I., M. V. Ivanov, and N. N. Lyalikova. 1963. *Introduction to Geological Microbiology.* (English translation). McGraw-Hill, New York.

Lapin-Scott, H. M. and J. W. Costerton. 1990. Starvation and penetration of bacteria in soils and rocks. *Experientia,* 46:807-812.

Lehman, R. M., F. S. Colwell, D. B. Ringelberg, and D. C. White. 1995. Combined microbial community-level analyses for quality assurance of terrestrial subsurface cores. *J. Microbiol. Methods,* 22:263-281.

Lovley, D. R., F. H. Chapelle, and E. J. P. Phillips. 1990. Fe(III)-reducing bacteria in deeply buried sediments of the Atlantic coastal plain. *Geology,* 18:954-957.

MacDonnell, M. T. and M. A. Hood. 1982. Isolation and characterization of ultramicrobacteria from a Gulf Coast estuary. *Appl. Environ. Microbiol.,* 43:566-571.

Matin, A. 1991. The molecular basis of carbon-starved-induced general resistance in *Escherichia coli. Mol. Microbiol.,* 5:3-10.

Matin, A., E. A. Auger, P. H. Blum, and J. E. Schultz. 1989. Genetic basis of starvation survival in nondifferentiating bacteria. *Annu. Rev. Microbiol.,* 43:293-316.

Morita, R. Y. and ZoBell, C. E. 1955. Occurrence of bacteria in pelagic sediment collected during the Mid-Pacific Expedition. *Deep Sea Res.,* 3:66-73.

Moyer, C. L. and Morita, R. Y. 1989. Effect of growth rate and starvation-survival on cellular DNA, RNA and protein of a psychrophilic marine bacterium. *Appl. Environ. Microbiol.,* 55:2710-2716.

Murno, P. M., M. J. Gauthier, V. A. Breittmayer, and J. Bongiovanni. 1989. Influence of osmo-regulation processes on starvation survival of *Escherichia coli* in seawater. *Appl. Environ. Microbiol.,* 55:2017-2024.

Nilsson, L., J. D. Oliver, and S. Kjelleberg. 1991. Resuscitation of *Vibrio vulnificus* from the viable but nonculturable state. *J. Bacteriol.,* 173:5054-5059.

Novitsky, J. A. and R. Y. Morita. 1977. Survival of a psychrophilic marine vibrio under long-term nutrient starvation. *Appl. Environ. Microbiol.,* 33:635-641.

Novitsky, J. A. and R. Y. Morita. 1978. Starvation induced barotolerance as a survival mecha-nism of a psychrophilic marine vibrio in the waters of the Antarctic Convergence. *Mar. Biol.,* 49:7-10.

Nystrom, T., K. Flardh, and S. Kjelleberg. 1990. Responses to multiple-nutrient starvation in marine *Vibrio* sp. Strain CCUG 15956. *J. Bacteriol.,* 172:7085-7097.

Nystrom, T., R. M. Olsson, and S. Kjelleberg. 1992. Survival, stress resistance and alterations in protein expression in the marine Vibrio sp. S14 during starvation for different individual nutrients. *Appl. Environ. Microbiol.,* 58:55-65.

O'Farrell, P. H. 1975. High resolution two-dimensional electrophoresis of proteins. *J. Biol. Chem.,* 250:4007-4021.

Oliver, J. D. 1993. Formation of viable but nonculturable cells. In *Starvation in Bacteria,* Kjelleberg, S., Ed., Plenum Press, New York. pp. 239-272.

Phelps, T. J., E. M. Murphy, S. M. Pfiffner, and D. C. White. 1994. Comparison between geochemical and biological estimates of subsurface microbial activities. *Microb. Ecol.,* 28:335-349.

Pitonzo, B. J. 1996. Characterization of microbes implicated in microbially-influenced corrosion from the proposed Yucca Mountain repository. PhD dissertation, U. Nevada, Las Vegas.

Poindexter, J. S. 1981. Oligotrophy: fast and famine existence. *Adv. Microb. Ecol.,* 5:63-90.

Rockabrand, D., T. Arthur, G. Korinek, K. Lievers, and P. Blum. 1995. An essential role for the *Escherichia coli* DnaK protein in starvation-induced thermotolerance, H_2O_2 resistance and reductive division. *J. Bacteriol.,* 177:3695-3703.

Rodriguez, G. G., D. Phipps, D. Ishiguro, and H. F. Ridgway. 1993. Use of a fluorescent redox probe for direct visualization of actively respiring bacteria. *Appl. Environ. Microbiol.,* 58:1801-1808.

Roszak, D. B. and R. R. Colwell. 1987. Survival strategies of bacteria in the natural environment. *Microbiol. Rev.,* 51:365-379.

Roszak, D. B., D. J. Grimes, and R. R. Colwell. 1984. Viable but nonrecoverable stage of *Salmonella enteritidis* in aquatic systems. *Can. J. Microbiol.,* 30:334-338.

Russell, C. E., R. Jacobson, D. L. Haldeman, and P. S. Amy. 1994. Heterogeneity of deep subsurface microorganisms and correlations to hydrogeological and geochemical parameters. *Geomicrobiol. J.,* 12:37-51.

Schultz, J. E., G. I. Latter, and A. Matin. 1988. Differential regulation by cyclic AMP of starvation protein synthesis in *E. coli. J. Bacteriol.,* 170:3903-3909.

Setlow, P. 1995. Mechanisms for the prevention of damage to the DNA in spores of *Bacillus* species. *Annu. Rev. Microbiol.,* 49:29-54.

Setlow, B. and P. Setlow. 1995. Small, acid-soluble proteins bound to DNA protect *Bacillus subtilis* spores from killing by dry heat. *Appl. Environ. Microbiol.,* 61:2787-2790.

Siegele, D. A., M. Almiron, and R. Kolter. 1993. Approaches to the study of survival and death in stationary-phase *Escherichia coli.* In *Starvation in Bacteria,* Kjelleberg, S., Ed., Plenum Press, New York. 201-224.

Smith, J. J., J. P. Howington, and G. A. McFeters. 1994. Survival, physiological response, and recovery of enteric bacteria exposed to a polar marine environment. *Appl. Environ. Microbiol.,* 60:2977-2984.

Sneath, P. H. A. 1962. Longevity of micro-organisms. *Nature,* 4842:643-646.

Spector, M. P. and J. W. Foster. 1993. Starvation-stress response (SSR) of *Salmonella typhimurium.* Gene expression and survival during nutrient starvation. In *Starvation in Bacteria,* Kjelleberg, S., Ed., Plenum Press, New York. 201-224.

Spence, J., A. Cegielska, and C. Georgopoulos. 1990. Role of *Escherichia coli* heat shock proteins Dnak and HtpG (C62.5) in response to nutritional deprivation. *J. Bacteriol.,* 172:7157-7166.

Stevens, T. O. and J. P. McKinley. 1995. Lithoautotrophic microbial ecosystems in deep basaltic aquifers. *Science,* 270:450-454.

Story, S. P. 1994. Microbial Transport in Volcanic Tuff, Rainier Mesa, Nevada Test Site. M.S. thesis, University of Nevada, Las Vegas.

Swindoll, C. M., C. M. Aelion, D. C. Dobbins, O. Jiang, S. C. Long, and F. K. Pfaender. 1988. Aerobic biodegradation of natural and xenobiotic organic compounds by subsurface microbial communities. *Environ. Toxicol. Chem.,* 7:291-299.

Truax, M. J., F. J. Brockman, D. L. Johnstone, and J. K. Fredrickson. 1992. Effect of starvation on induction of quinoline degradation for a subsurface bacterium in a continuous-flow column. *Appl. Environ. Microbiol.,* 58:2386-2392.

van Elsas, J. D. and L. S. van Overbeek. 1993. Bacterial responses to soil stimuli In *Starvation in Bacteria,* Kjelleberg, S., Ed., Plenum Press, New York. 55-80.

Whitelaw, K. and J. F. Reese. 1980. Nitrate-reducing and ammonium-oxidizing bacteria in the vadose zone of the chalk aquifer of England. *Geomicrobiol. J.,* 2:179-187.

Zlatkin, I. V. and L. J. Forney. 1996. Marine bacteria cultivated from a terrestrial deep subsurface sediment. Am. Soc. Microbiology Annu. Meet. Abstr. American Society for Microbiology, Washington, D.C. #N61.

12

Subsurface Microbiology and the Evolution of the Biosphere

Todd O. Stevens

CONTENTS

KEY WORDS: *subsurface, bacteria, microbiology, chemolithotrophic, origin of life, Precambrian, Archean, Hadean, impact frustration, rock cycle, exobiology.*

12.1 Introduction

As related in previous chapters, we now know that the biosphere extends down into the upper portion of the earth's crust. As knowledge of the subsurface biosphere grows, we can begin to explore what significance it has to the global environment. What are the main functions and consequences of subsurface ecosystems on a global scale? How might subsurface ecosystems have contributed to the formation of the biosphere as we see it today?

Life on Earth is primarily bacteriological, and has been throughout the planet's history (Schopf, 1983). Bacterially mediated processes control the composition of the atmosphere and the balance of oxidants at the earth's surface. Earth's biomass is predominantly microbiological, and it now appears that a large portion of this bio-mass may exist below the ground surface (e.g., Gold, 1992; Parkes et al., 1994). Although the extent to which bacteria influence observable geochemical phenomena is uncertain, it is evident that cycling of many major elements is controlled by bacter-ial physiology. Because of the impalpability of microorganisms and the slow rates of global biogeochemical cycles, much of the complex interplay between geochemistry and microbial physiology remains to be investigated. A consequence of this is that intriguing discoveries and synergistic melding of diverse inquiries are almost com-mon occurrences in microbial ecology.

In this chapter, we explore potential intersections between studies of subsurface microbiology and various studies of planetary-scale ecology. In doing so, we first attempt to outline some important functional links between subsurface microorgan-isms and larger ecological processes. Using this context, we then discuss how recent results in subsurface microbiology may contribute to resolving several intriguing questions about the evolution of the earth's biosphere. These emerging lines of inquiry may show that microorganisms in the deep terrestrial subsurface are not merely a scientific curiosity, but an important functioning component of the global ecological system.

12.2   Interactions Between the Surface and the Subsurface Worlds

Recent advances in subsurface science have only begun to illuminate the ecological interactions of life in the earth's crust with its chemical and physical milieu. The quantitative significance of microbially mediated interactions between the subsurface biosphere and the surface biosphere — the familiar world of the crust/atmosphere or crust/hydrosphere interface — remains unknown. However, evidence for several important functions of the subsurface is growing. While large blank spots remain in our knowledge, several broad outlines can be drawn.

12.2.1   The Microbial Subsurface Cycle

Although there is continuing inquiry and debate concerning the conditions under which microorganisms are transported in the subsurface, it is clear that such phenom-ena take place on a variety of time scales. Field evidence indicates that microbial transport is affected by limitations of survivability imposed by physiology, and by the physical constraints of the transporting medium, e.g., groundwater velocity, sedi-ment permeability, and sediment pore-throat diameters (for more information, see Chapter 13). Photosynthetic microorganisms which cannot metabolize in the subsur-face have been detected in aquifers with modern groundwater ages, where transport times are very short (Sinclair and Ghiorse, 1989). In contrast, many reports have now shown that living bacteria are distributed throughout the upper part of the earth's crust in rocks ranging up to about 10^8 years old. Some evidence suggests that living bacteria have been entrapped within sediments for most of this time (DOE, 1994; White et al., 1993). On these longer time scales, microorganisms would be transported not only by groundwater movement, but also by the movement of rock over geolog-ical time. Thus, since microorganisms can travel with both water and rock, they are inevitably subject to the processes of both the hydrologic cycle and the rock cycle.

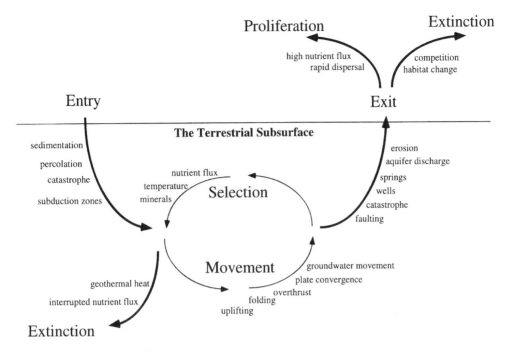

FIGURE 12.1

The subsurface microbial cycle. Geological and hydrological processes, operating on multiple time-scales, move microorganisms into the subsurface, transport them through the subsurface, and may return them to the surface environment. In the subsurface, selective pressures constrain the types of organisms that can survive there. A passage through the subsurface may be as short as a few weeks or as long as 100,000,000 years.

These overlapping processes, combined with selection by the subsurface environment, should result in the Subsurface Microbial Cycle (SMC, Figure 12.1).

There are three general phases that have been proposed for the SMC (Stevens and Long, 1995). The relative importance of these phases may depend on the specific time scales and mechanisms involved. Microorganisms enter the subsurface from their point of origin, which is presumably the crust/atmosphere interface or the crust/hydrosphere interface. In the subsurface, microorganisms are subjected to selective pressures which may result in their extinction or in the survival and adaptation of particular groups. Eventually, subsurface microorganisms may be transported to the surface environment, where they are again selected by the various ecological processes occurring there.

12.2.1.1   *Entering the Subsurface: Interment*

Two major hypotheses for the entry of bacteria into the subsurface are (1) percolation and subsequent advective transport to depth by ground water, or (2) burial by the same processes that bury the surrounding formation. Clearly both processes are operative to some extent, as exemplified by a brief survey of geologic provinces and microbial ecology investigations.

Algal and fungal cells have been reported in sediments hundreds of meters deep (Sinclair and Ghiorse, 1989) in ground water determined to be essentially modern (i.e., less than 1000 years from recharge). While there is no direct data describing how long algae and fungi can survive in the subsurface, data from dried soil specimens indicate that they survive less than 50 to 100 years away from their normal surface environment (Sneath, 1962). In karstic terrains, surface water enters the subsurface rapidly, and may contain considerable flora and fauna (Gilbert et al., 1994). In areas with highly fractured bedrock, such as the Snake River Plain of Idaho, entire rivers and streams disappear into the subsurface, carrying their microbial burden with them. Thus it appears that at least in some cases, microorganisms can enter the subsurface relatively rapidly by groundwater recharge.

Several rock cycle processes result in burial of rocks and any bacterial populations that may exist within them. Probably the most common such process is slow sedimentation. Sediments and their microbial communities are slowly buried by the deposition of new layers of sediment above them in aquatic habitats, and buried within crustal basins caused by subsidence. Most of the subsurface habitats that microbiologists have studied so far originated as marine or lacustrine sedimentary systems, sometimes intertwined with fluvial sediments (for more information, see Chapters 7 and 9). Because these sediments are typically rich in organic matter and other nutrients, bacteria should survive the burial process quite well, though nutrient fluxes probably decline as the sediments become compacted. In support of this hypothesis, bacteria have been detected in analogous deep-sea sediment from hundreds of meters below the sea floor, suggesting that microorganisms survive for millions of years in these sediments (Parkes et al., 1994). Observed numbers of cells declined from about 10^7/g at the sea floor to about 10^5/g at a few meters depth, then remained relatively constant with depth for hundreds of meters. Although most of these cores were disturbed by the sampling process and may have been contaminated by sea water (ODP, 1992), increasing methanogenesis and decreasing sulfate reduction with depth appeared to reflect a response to restricted diffusion of nutrients from the sea floor, and provided evidence that at least some organisms were indigenous to the sediments.

Subduction at the edge of tectonic plates also buries rocks over geological time. The subducting plate is forced deep beneath the overriding plate, eventually melting as it is forced deep into the crust. Sediments and other materials become mixed at the subduction zone and can be pushed on top of the overriding plate, where they can eventually become incorporated in mountain ranges (e.g., the coast ranges of western North America). No subsurface microbiology explorations have yet been made in these environments.

Catastrophic processes can bury soils and sediments containing microorganisms in relatively short time periods. Volcanic ash falls and pyroclastic flows can bury a landscape in just a few days or hours. A recent example in North America was provided by the 1980 eruption of Mount Saint Helens; however, much more dramatic examples can be found in the geological record. Other phenomena, such as mudslides, earthquakes, tsunamis, and lava flows all have the capability of rapidly burying surface organisms. No studies have yet been made of the changes that occur in subsurface microbial populations after such an event, although they may provide easily dated strata that could be used to study successional events after burial.

12.2.1.2 *Leaving the Subsurface: Exhumation*

Subsurface microorganisms (and rocks) must constantly be returned to the surface environment by geological processes. Primarily, these involve the reverse of processes discussed previously: discharge of ground water to the crustal surface and exposure at the surface of previously buried rocks.

Ground water discharges into rivers and oceans, or appears as springs, transporting bacteria and products of bacterial metabolism. For example, a subsurface methanogenic ecosystem beneath the Florida escarpment has been detected discharging into the Atlantic Ocean (Paull et al., 1985). Groundwater flux to the coastal zone off South Carolina was recently estimated to be 40% as large as the flux from rivers (Moore, 1996). Numerous springs, hot and cold, have been found to contain microorganisms distinct from the surrounding environment (e.g., Barns et al., 1994; Bath et al., 1987) although it has been difficult to distinguish whether the organisms are being emitted from the subsurface, or merely proliferate in the spring. In modern times, aquifers are artificially discharged to the surface by drinking water and irrigation wells. Many irrigation wells are hundreds of meters deep and irrigation waters may contain high numbers of subsurface bacteria (e.g., Olson et al., 1981; Stevens and McKinley, 1995) that are continuously dispersed onto the land and into the atmosphere as aerosols.

Deeply buried sediments return to the surface over geological time through uplift and erosion; hence marine sediments are found in high-altitude deserts and mountain ranges. Deeply buried strata are eventually exposed at the surface through removal of overlying material by wind and water. The most spectacular examples of subsurface exposure can be found in deep river canyons, though eroding bluffs along any river probably shed subsurface bacteria. Bacteria exposed in recently eroded surface outcrops have been studied as surrogates for subsurface bacteria (Brockman et al., 1992; Haldeman et al., 1994; Stevens and Holbert, 1995). Community-level studies show that outcrop organisms share properties of subsurface organisms from the same formation at depth, and are distinct from surface soil populations, though no taxonomic studies comparing the two have yet been done.

12.2.1.3 *Living and Dying in the Netherworld*

In the subsurface, numerous selective pressures act to constrain the types of microorganisms that can live there (also, see Chapter 11). For instance, one would not expect obligately photosynthetic organisms to survive for long, and in fact they have never been observed in subsurface samples more than a few hundred years removed from the surface. The slow nutrient fluxes in the subsurface probably prevent copiotrophs from thriving, and few fungi or actinomycetes have been found in subsurface habitats, except those with young ground waters. Similarly, no protozoans or other predators have been detected in the deep subsurface, though they are may be found in shallow ground waters and or karstic terrains. In more deeply buried formations, where temperature is elevated along a geothermal gradient, only thermophilic organisms have been detected (Stetter et al., 1993; DOE, 1994; Stevens and Boone, 1993; White et al., 1993).

It can again be seen that multiple time scales operate on these various processes. The passage of a microbial population through the subsurface may follow complex paths and might be as short as a few weeks or as long as 10^8 years. Thus, although microbial populations may be isolated in the subsurface for millions of years, there is no reason to suppose that they will be unknown organisms, or fundamentally different from all surface organisms. Since subsurface microorganisms are constantly recycled to the surface, they are already here.

This concept of the SMC is of importance to molecular biologists who may want to use subsurface microorganisms to calibrate "molecular clocks". There would probably be little point in comparing nucleic acid sequences of subsurface bacteria with those of surface bacteria, if they have not necessarily been continuously isolated from each other. One should be able to find the same sequence in bacteria shed from outcrops of the subsurface formation. However, two different populations of subsurface bacteria in geologically confined strata *would* have been continuously isolated from each other, and might show sequence divergence. For the same reason, calls to treat subsurface bacteria as potentially dangerous invaders of the surface biosphere (Kennedy et al., 1994) can safely be disregarded.

Of course, the mere fact that microorganisms can cycle through the subsurface and return to the surface does not mean that all microorganisms are capable of surviving the passage, or that the entire subsurface is amenable to microbial life. Shifting water tables can lead to desaturation of sediments and loss of nutrient flux. As sediments become more deeply buried, they compact and cement, resulting in decreased porosity and declining nutrient flux. In deeper formations, temperatures may rise to the point of sterilization and persist for millions of years. Certainly, molten lava flows, intrusions, and hot ash falls are sterile during emplacement and sterilize the sediments surrounding them. Some subducted sediment heats to the point of melting as it plunges into the crust. Indeed, the earth's geothermal activity ensures that most sediments buried more deeply than a few kilometers are heat sterilized (for an alternate view, see Deming and Baross, 1993).

On the other hand, the stability of the physical environment, slow but constant nutrient fluxes, and the presence of geochemical nutrient sources may make some subsurface habitats very comfortable homes for those organisms capable of exploiting them.

Some observers have proposed that subsurface organisms are "living fossils" that are somehow preserved in a state of dormancy in the subsurface (Kennedy et al., 1994). However, the presence of biogeochemical signatures in aquifers, and activity measurements along aquifer flow paths, suggest that this may not be the case (e.g., Chapelle et al., 1987; Chapelle and Lovley, 1990; Murphy et al., 1992; Stevens and McKinley, 1995). These observations indicate that microorganisms are metabolically active, albeit at extremely slow rates, in subsurface aquifers. Viable subsurface microorganisms have rarely been found in the absence of metabolizable nutrients, or nutrient flux. (Findings of microorganisms within fluid inclusions in salt deposits may provide a counterexample (Norton et al., 1993).) The role of dormancy in subsurface ecology has yet to be thoroughly examined (for more information, see Chapter 11). It seems logical that periods of dormancy would have to be punctuated by occasional periods of active metabolism with a frequency at least as great as the biochemical half-life of critical cell components. Such a model cannot be distinguished from constant infinitesimally low rates of metabolism by the entire population (enzyme turnover times have been estimated on the order of years!) with the data currently in hand.

12.2.2   A Source or a Sink?

Active subsurface microbial ecosystems can be divided into two main types of habitat, based on their trophic potentials. Microorganisms in detrital habitats are dependent on the metabolism of organic carbon deposited with the rock formation or possibly transported into the formation by groundwater or migration of petroleum fluids. In such a system, the flow of chemically reduced, energy-rich compounds is from the surface biosphere to the subsurface. Organisms in these habitats might essentially be considered exiles from the surface. Most subsurface ecosystems studied so far are detrital habitats, and they are mostly sedimentary systems. Recently, though, evidence has grown for lithotrophy-based habitats in the subsurface, in which inorganic geochemical nutrient sources support microbial populations. Lithoautotrophic metabolism can result in the production of biomass, energy gases, and organic acids. In these systems, the flow of reduced energy-rich compounds would be from the subsurface to the surface biosphere. Organisms in these habitats might be considered natives of the subsurface.

12.2.2.1 Heterotrophy-Based Ecosystems

Organic-rich sedimentary formations contain organic carbon that is ultimately derived from photosynthetic primary production at the earth's surface. These deposits may allow a diverse community of heterotrophic organisms to survive for as long as the organic carbon and other nutrients last. Diffusion and transport of electron acceptors through ground water appears to limit the microbial communities in such environments. Flow-path analysis of aquifers indicates that oxygen becomes depleted relatively rapidly, followed by nitrate, ferric iron, and finally sulfate (Champ et al., 1979; Murphy et al., 1992). Ferric iron may be the most abundant electron acceptor in many subsurface environments because it is an insoluble component of the sediments themselves, while other electron acceptors must be supplied by ground water through dissolution and transport. Eventually, nutrients enclosed in a detrital system become depleted, or nutrient flux is cut off by hydrologic confinement or desaturation. As resources decline, populations may also. Survival of microorganisms by punctuated metabolism might be described by stochastic population biology models. As the probability of encountering sufficient nutrients for a burst of metabolism declines, the probability that the population will become extinct increases. On the scale of a formation, over time, this should result in an increasingly patchy distribution of microorganisms, followed by extinction.

12.2.2.2 Lithotrophy-Based Ecosystems

There is growing evidence that a second major type of subsurface microbial ecosystem is based on lithotrophic food webs. Chemolithoautotrophy is primary production without photosynthesis. That is, inorganic carbon is converted to organic carbon and only inorganic compounds serve as electron acceptors and electron donors to provide energy. Many microorganisms are known that can grow chemolithoautotrophically in the laboratory, some of which are listed in Table 12.1. All of these metabolic processes should be possible in the subsurface, if the appropriate substrates are available. Those requiring oxygen, for instance, may be restricted to relatively near-surface environments. Ivanov, Gold, and others have noted the presence of reduced energy gases in the deep subsurface and speculated on the possibility that they might support microbial life (Boston et al., 1992; Gold, 1992). Recent subsurface investigations have provided evidence that lithoautotrophy does, in fact, occur in the earth's crust.

Several karst systems have been discovered in Europe and North America in which reduced fluids from deep anaerobic aquifers mix with oxygenated water and gases introduced through caves (Brigmon et al., 1994; Lauritzen and Bottrell, 1994; Sarbu et al., 1994). Chemolithoautotrophic microbial mats proliferate in these caves and are similar to those described at deep-sea hydrothermal vents (Jannasch, 1995). Reduced sulfur species from ground waters are oxidized with atmospheric oxygen to yield energy for primary production. There is some indication that endemic cave fauna live by consuming this bacterial biomass in a manner analogous to the oceanic vent communities (Sarbu et al., 1994). This sort of metabolism is probably limited to the shallow subsurface, where atmospheric oxygen is readily available. The ultimate source of reducing power in these karst environments has yet to be determined. Although these systems are chemolithoautotrophic, they are still indirectly dependent on photosynthesis, since O_2 is required as an electron acceptor.

TABLE 12.1

Some Chemolithoautotrophic Processes Carried Out by Bacteria

Process	Electron Donor	Electron Acceptor
Aerobic		
Nitrification	NH_4^+	O_2
Sulfide oxidation	SH^-	O_2
Sulfur oxidation	$S°$	O_2
Thiosulfate oxidation	$S_2O_3^=$	O_2
Ferrous iron oxidation	Fe^{++}	O_2
Manganese oxidation	Mn^{++}	O_2
Hydrogen oxidation	H_2	O_2
Methane oxidation	CH_4	O_2
Anaerobic		
Denitrification	H_2	NO_3^-
	Fe^{++}	NO_3^-
Ferric iron reduction	H_2	Fe^{+++}
Sulfate reduction	H_2	$SO_4^=$
	CH_4	$SO_4^{=b}$
Sulfur reduction	H_2	$S°$
Methanogenesis	H_2	CO_2
Acetogenesis	H_2	CO_2

[a] Recently reported by Straub et al. (1996).
[b] Geochemical evidence, but no isolates.

Recently, relatively high numbers of microorganisms have been reported in deep anaerobic aquifers within crystalline rocks, where no detrital carbon sources are expected (Pedersen and Ekendahl, 1990, 1992; Stevens et al., 1993; Stevens and McKinley, 1995). These observations were initially attributed to the presence of un-identified carbon sources seeping up from coal beds buried below the igneous rock formations, or percolating from the surface. However, closer examination of the ground waters in the Columbia River Basalt Group (CRB) revealed, in many cases, anomalous concentrations of reduced gases, including dissolved H_2, at high enough concentrations in ground waters, to support anaerobic lithoautotrophic microorganisms (Table 12.1; Stevens and McKinley, 1995).

In the complex food web of an ecosystem it is difficult to track electron flow; however, geochemical observations can provide clues to the major processes occurring there. A widely used method for determining the origin of energy gases is to measure the ratios of various isotopes of carbon and hydrogen and compare them with ratios known to be characteristic of various origins. For instance, enzyme systems involved in C-1 metabolism tend to have a kinetic preference for ^{12}C isotopes over ^{13}C isotopes; when microorganisms metabolize CO_2, they leave a metabolic signature in both the products and the remaining pool of reactants. Calculations based on this principle showed that in the CRB formation, concentrations of dissolved inorganic carbon (DIC) were largely controlled by microbial uptake in H_2-dependent methanogenesis (Figure 12.2). If microorganisms were significantly producing DIC by metabolizing infiltrating organic matter, this relationship could not have occurred. Thus, it appears that a **S**ubsurface **Li**thotrophic **M**icrobiological **E**cosystem (SLiME) exists in this igneous system (Stevens and McKinley, 1995).

Hydrogen gas features prominently in Table 12.1, and is also found widely in the earth's crust. H_2 is present in hot geothermal fluids that seep into microbially

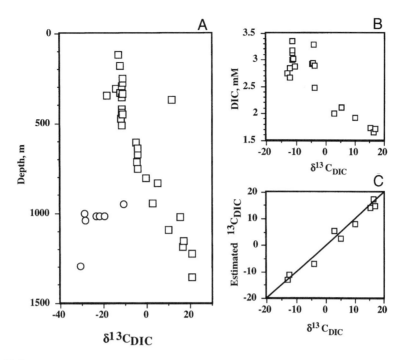

FIGURE 12.2

Carbon isotope evidence for SLiMEs in the Columbia River Basalt Group. A. With inceasing depth (and age) dissolved inorganic carbon (DIC) in methanogenic ground waters (squares) becomes enriched in ^{13}C, indicating carbon fixation by microorganisms. Bicarbonate in sulfate-reducing ground waters becomes enriched in ^{12}C, indicating biological oxidation of previously fixed carbon. B. This enrichment correlates with depletion of DIC, not production. C. Predictions of ^{13}C ratios, assuming that autotrophic methanogens consumed the DIC, correlate well with measured values — suggesting that primary production, not organic matter oxidation (which would increase DIC) is the carbon source for microorganisms. (From Stevens, T.O. and McKinley, J.P., *Science*, 270, 450, 1995. With permission.)

habitable areas near volcanic zones, but many observations of elevated H_2 concentration are not near any zone of active vulcanism. Geochemical literature suggests that most nonthermal H_2 occurrences are associated with ultramafic or mafic rocks. These rocks contain chemically reducing silicate minerals composed of magnesium and ferrous iron. They are formed from upwelling mantle magmas that penetrate the earth's crust, but may then be carried far from zones of vulcanism by rock cycle processes. The minerals are oxidized during the weathering process by atmospheric oxygen and water. In large igneous bodies they oxidize only slowly because of their crystalline nature, the very small water to rock ratio, and the limited diffusion or advection of oxygen into subsurface environments. Some thermodynamic calculations indicate that, in the absence of oxygen, these minerals should react with water to form hydrogen gas. In fact, laboratory experiments with CRB showed that, under appropriate conditions, H_2 was liberated rather rapidly (Stevens and McKinley, 1995). Oxidation of reduced sulfur-containing minerals also results in H_2 generation (Drobner et al., 1990).

Water-rock interactions may explain the presence of elevated H_2 concentrations in ground waters, and thus the presence of microbial populations. In microcosm experiments, groundwater bacteria and other anaerobic microorganisms have been shown to grow with basalt supplied as the sole electron donor (Stevens and McKinley, 1995; Stevens et al., 1996).

It appears that igneous rocks can provide the reducing power necessary for microbial metabolism and, with the right combination of geochemical and hydrologic conditions, can support lithoautotrophic microbial ecosystems that are independent of photosynthetic inputs. Thus, energy for life might be obtained from the difference in oxidation state between the earth's mantle and the crust rather than on incident radiation from the sun. Jannasch (1995) has called this "terrestrial energy", as opposed to "solar energy". Although such claims are often made for ocean-floor communities at hydrothermal vents, those communities depend on reduction of dissolved O_2 in seawater and at least partly on oxidation of pyrolyzed sedimentary organic matter that is cycled through the vent system. Dissolved O_2 and sedimentary organic carbon are both products of photosynthesis, and thus, like the karst systems described above, the vent communities are photosynthesis dependent. It is possible that SLiMEs exist in the basalt surrounding thermal vents (Deming and Baross, 1993), but this has not yet been determined.

Understanding SLiMEs may also shed light on the formation of natural gas in a number of regions where observations of gas composition and stable isotope ratios do not conform to conventional models of gas formation. Further geochemical and microbiological examination of large igneous terrains will be required to confirm the global significance of these systems. However, the fact that primary production occurs in the subsurface means that subsurface microbial ecosystems are not necessarily closed-ended decaying detrital systems. They may be a self-sustaining feature of the biosphere.

Given the patchwork nature of the earth's crust due to rock cycle processes, many subsurface environments probably contain a mixture of lithotrophy- and heterotrophy-based primary production. "End-member" communities, then, could arise mainly in relatively homogeneous terrains. Large igneous bodies such as flood basalts, and perhaps granite batholiths, might contain primarily lithoautotrophic ecosystems, while large sedimentary basins might contain primarily heterotrophic ecosystems. Detailed examination of microbial community structure from end-member environments should reveal the effects of these processes on selection and ecological succession of particular microorganisms.

12.3 Subsurface Roles in the Evolution of the Biosphere

The origin of life and the evolution of the biosphere is a fascinating multidisciplinary field of study to which subsurface science can make several contributions. Due to rock cycle processes, and to catastrophic events that occurred during the early history of the solar system, no materials are available on earth that are as old as the planet. Conclusions about events on the early earth are based on very few data points, extensive extrapolation, and mathematical modeling. Nevertheless, several lines of evidence indicate that life was present on Earth when the oldest known terrestrial rocks were formed 3.5 to 3.8 billion years ago, and may have been present before that time (Schidlowski, 1988; Schopf, 1983). Except where evidence points to specific catastrophic events or developmental changes, principles of uniformitarianism are used to explain past events in terms of present-day processes. Knowledge of how terrestrial subsurface systems work in the present may add an important factor that will help to determine how critical events happened in the distant past.

12.3.1 The Cradle Underground?

The nature and date of the event known as the Origin of Life (OOL) remain a mystery. Organic compounds that are important in biology can be formed by a variety of reactions between simple inorganic compounds. Organic molecules have been detected in deep space and as a component of some meteorites. Several groups of researchers have proposed partial schemes by which nonliving "prebiotic" chemicals might have been synthesized, self assembled, and somehow made the leap to becoming a self-replicating living organism. This stimulating body of work has been extensively reviewed elsewhere (Cairns-Smith, 1992; Oro et al., 1990; Schopf, 1983).

The minimal biochemical components necessary to define the first "living" cells remain unclear. Some researchers focus on abiotic production of nucleic acids as the enabling step, some on cyclic heterotrophic processes, and others on primitive intermediary metabolism. With the current state of knowledge, competition between these hypotheses remains at the level of chicken vs. egg arguments. Some OOL schemes require very specialized environments. Some require input of organic materials from meteorites. Energy sources proposed for these reactions include electrical discharge from lightning, UV irradiation, kinetic forcing by condensation of prebiotic "soups", and oxidation of pyrite at volcanic vents.

Could the OOL have occurred in the subsurface? Clearly, some prebiotic chemistry hypotheses would not work in the subsurface. No UV radiation or lightning strikes would be available, for instance. One family of OOL hypotheses, however, collectively known as the "iron-sulfur hypotheses" (Cairns-Smith, 1992; Wächtershauser, 1988, 1990) might be well-suited to the subsurface. Briefly stated, in this scheme chemical energy from reduced minerals provided the energy for prebiotic organic synthesis, and minerals provided catalytic surfaces. Membranes self-assembled on the mineral surfaces to form early containment of the prebiotic mixture. Finally, the first cells formed by budding off from these "two-dimensional organisms."

The many predictions of these theories should be testable, but so far few tests have been made. It is interesting, though, to note how many key enzymes contain reduced metal centers that may have been "stolen" from mineral surfaces by early protoenzymes. This scheme also has interesting parallels to the basalt SLiMEs described above. Clearly, such an OOL sequence could occur in the subsurface as well as on the surface, though there seems to be no way to differentiate between the two scenarios. One advantage of a subsurface location for fostering OOL reactions might be the stability of the environment: constant fluxes of reactants, protection from disruption by UV radiation, and constant temperature gradients might have aided in nurturing the earliest life forms. Also, the simple fact that, volumetrically, there is much more habitable subsurface environment than there is surface environment, may mean that life was statistically more likely to have originated in the subsurface.

12.3.2 The Subsurface as Refuge

Some planetologists have attempted to constrain the earliest date for the OOL by determining when the surface became continuously habitable (e.g., Maher and Stevenson, 1988; Sleep et al., 1989). As the solar system formed, large quantities of debris — dust, asteroids, planetesimals — were left over from the condensation of the planets. This debris was eventually swept up in collisions with the planets, some of which formed the familiar craters on the moon. The earth was subjected to similar asteroid bombardment (actually, 20-fold greater than the moon) at the same time, but the craters have been obliterated by rock cycle processes. Researchers have calculated that this early bombardment must have resulted in impacts large enough to flash-heat the surface of the earth and vaporize the oceans; the surface of the planet became a colossal autoclave for thousands of years after the strike. Using the crater age-distribution on the moon, they calculated the time at which surface-sterilizing impacts became improbable on earth. Since no known life could survive such an impact on the surface, these calculations were used to set the earliest OOL at about 3.8 billion years ago (Sleep et al., 1989).

The problem with this analysis is that rocks from that time appear to already contain highly evolved microbial life. The oldest samples of inorganic carbon (there are very few of them) appear to show isotopic fractionation by enzymatic biological carbon fixation (Schidlowski, 1988). Many of the oldest rocks contain structures that have been interpreted as fossils of microorganisms and microbial mats (Schopf, 1983). If these organisms evolved from abiotic precursors in the (relatively!) short time between the cessation of the bombardment and the formation of the rocks, then rates of evolution must have been much faster at that time than in all the subsequent ages of the earth.

Subsurface microbiology may provide a solution to this dilemma. Whether life originated on the surface, or in the subsurface, or even by panspermic processes, if subsurface ecosystems had been established at that time, organisms in them would have been protected from surface-sterilizing impacts. Surviving organisms could then migrate back to the surface after it had cooled, by the processes of the subsurface microbial cycle (Figure 12.1).

The question of the earliest OOL then becomes one of when the subsurface environment could have first hosted life. The entire planet probably would have been uninhabitable during the "iron catastrophe" — the event which differentiated the earth into core, mantle, and crust (Wetheril, 1990). Radioactive elements in the material that accreted to make up the earth slowly decayed, giving off heat. Various models agree that the heat was produced faster than it could dissipate, building up until the inner part of the planet melted. Heavier fluids (molten iron) sank and formed the core, while lighter elements (silicates) rose and became the crust. Impact of a Mars-sized body with the earth, which is a current favored theory for the formation of the moon, would also have contributed to forming this "Magma Ocean". The timing of these events remains uncertain, and estimates range from 4 to 5 billion years ago. Nevertheless, as the planet cooled, and volatiles such as water reaccumulated, the subsurface became habitable. The earliest continuously habitable potential biosphere was probably located in the subsurface, sandwiched between geothermal heat from below and asteroid impact heat from above (Figure 12.3).

Fossil records suggest that the tree of life is periodically pruned back by catastrophic asteroid impacts (McLaren and Goodfellow, 1990). Molecular records, in the form of nucleic acid sequence trees (Woese, 1987), have been interpreted to indicate similar extinction events during the early evolution of life (e.g., Forterre et al., 1995; Gogarten-Boekels et al., 1995). The survivors of these events appear to have been thermophilic microorganisms, from which all subsequent living organisms are believed to have descended. It has been suggested that sea-floor hot springs provided a refuge for thermophiles to survive surface-sterilizing impact events. Subsurface microorganisms might have been even better protected, since it would take more energy to sterilize the crust than the oceans. It is worth noting that the deepest-living, hence most protected, subsurface organisms are thermophiles (Stetter et al., 1993; Boone et al., 1995; DOE, 1994; Stevens and Boone, 1993; White et al., 1993). One can speculate that subsurface microorganisms may have formed the root from which the biosphere periodically resprouted.

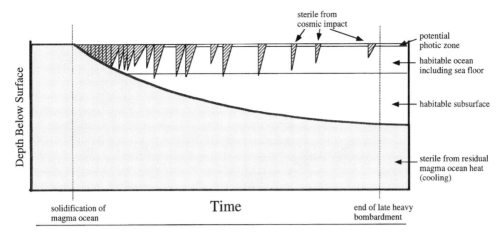

FIGURE 12.3

Possible locations of the earliest habitable biosphere on Earth as constrained by temperature. As the earth cooled, the surface and oceans were periodically resterilized by ocean-vaporizing asteroid impacts (Maher and Stevenson, 1988; Sleep et al., 1989). The deep subsurface would not have been greatly affected. The times scale is poorly constrained, but recent estimates indicate that solidification of the magma ocean was about 4.5 Gya and the end of the late heavy bombardment was about 3.8 Gya.

12.3.3 The Way Things Used to Be

At present, the ecology of the surface world is dominated by photosynthetic primary production. The atmosphere contains approximately 20% molecular oxygen as a consequence of photosynthesis. Oxygenic heterotrophs, such as aerobic bacteria and mammals, are wholly dependent on products of photosynthesis for life. To many people, including many scientists, this is a fundamental paradigm of ecology. Yet, as we have seen, it is not always true. Photosynthesis is a complex, highly evolved process that occurs only in organisms branching rather shallowly in phylogenetic trees. It was likely not present during the earliest evolution of life. At the time of the OOL, the earth was probably anaerobic (Holland, 1984) (but see Towe, 1994 for an alternate view). The deep subsurface may contain remnants of, or analogs to, ecosystems once present at the surface.

Anaerobic habitats are ubiquitous, from the intestines of animals, to sediments in mud puddles, to the subsurface. Yet the familiar anaerobic metabolic processes occurring there are usually assumed to merely accomplish decay of photosynthetically produced organic matter — a sideshow on the photosynthetic stage. The growing body of evidence supporting the existence of SLiMEs shows that they are a main-stage act in their own right.

SLiMEs may be an important model for studying how the biosphere evolved during the Archean Era. Studies of ancient soils show that the earth's atmosphere contained no significant amount of oxygen before about 2.8 billion years ago (Holland, 1984; Schopf, 1983). Soils older than that contained large amounts of ferrous iron and other reduced metals when they were formed. These soluble metals were then leached away by rainfall, leaving a residue depleted in iron. Similar soils have not formed since the end of the Archean Era because oxygen in the atmosphere rapidly oxidizes iron to the ferric form which is insoluble and therefore not leachable. At about the same time that oxygen built up in the atmosphere, there was a large increase in burial of organic carbon, suggesting a surge in primary productivity. (Though Des Marais et al. [1992, 1994] argue that an increase in the rate of organic carbon burial due to tectonic processes could have caused the increase in O_2.) Most life on the surface was probably wiped out by exposure to highly toxic O_2. This event is widely known as "the oxygen catastrophe". Although the mechanisms and sequence of this change are still the subject of vigorous debate, most investigators agree that it was coupled with the evolution of oxygenic photosynthesis.

It is difficult to determine how ecosystems functioned before the oxygen catastrophe. No one knows what a subareal microbial mat would look like on an anaerobic planet or what sorts of microorganisms would live in one, because there is no such environment today. In the absence of atmospheric oxygen, systems similar to the SLiMEs might even have existed on the surface and formed silicified mats. One can speculate that organisms in such an environment would have been well poised to evolve the earliest photosynthetic pathways. The combination of dense biomass, abundant ferrous iron, and incident radiation, which promotes photolysis of water (e.g., Braterman et al., 1993):

$$\left\{2\ Fe^{2+} + 3\ H_2O + hv \rightarrow Fe(OH)_3 + H_2 + H^+\right\} \tag{12.1}$$

could have promoted the biochemical capture of iron redox chemistry that allows photosynthesis. In such a scenario, modern photosynthetic stromatolites would be structural analogs to possibly ancestral nonphotosynthetic stromatolites.

In any case, there is every reason to think that the environment for SLiMEs — and possibly other subsurface systems — is pretty much the same today as it was 3 or 4 billion years ago. These anaerobic ecosystems may be representative of the surface biosphere between 3.8 and 2.8 billion years ago. As dissolved O_2 built up in surface waters and free O_2 permeated the atmosphere, anaerobic ecosystems would have been forced underground, or into protected diffusion-limited environments. It is possible that they function underground now in essentially the same way they did when the earth was young.

The difference between this Archean hypothesis (SLiME analogy) and the Hadean one (SMC refuge) presented above is that rocks from the late Archean Era still exist. In the future, it may become possible to determine whether lithoautotrophic microorganisms existed in these rocks. As more is learned about SLiMEs, it may prove rewarding to compare their properties with the observations of the oldest microfossils and rocks. In particular, it would be very useful to detect some biogeochemical signature, characteristic of SLiMEs, that could be used to determine whether they had ever been active in a particular rock formation. These sorts of studies have not yet been conducted, but could yield valuable information about the early evolution of the biosphere.

12.3.4   The Way Things Are Out There: Implications for Exobiology

Recent findings in subsurface microbiology can contribute to the search for extraterrestrial life. In the solar system, Earth is the only planet that provides clear evidence for life "as we know it", but the reasons for that are not entirely clear, and the search for extraterrestrial life continues. At the time of the OOL, conditions on Earth and on Mars, for instance, were relatively similar. Since that time, their histories and characteristics have diverged. However, if the OOL was a natural consequence of prebiotic chemistry, there is no known reason why it would have occurred on one planet rather than the other. In fact, evidence borne by the SNC group of meteorites (which are thought to have originated in the crust of Mars) indicates that the two planets probably could have cross-inoculated each other by ballistic exchange during the late heavy bombardment. If a piece of crust large enough to insulate microorganisms from the vacuum of space and the heat of impact were knocked from one planet to the other, panspermia may have occurred (Wallis and Wickramasinghe, 1995).

Several scientists have studied life in extreme environments on Earth to evaluate whether biological processes there could possibly function on Mars. Viking lander experiments conducted in the 1970s did not find evidence for life on the surface (Horowitz and Hubbard, 1976; Levin and Straat, 1976). In fact, McKay et al. (1992) have concluded that none of the earth's "extreme" biosystems could survive on the surface of modern Mars, but that several could have functioned there in the past. Boston et al. (1991) have speculated that the subsurface might have remained habitable on Mars after the surface was no longer habitable, and that lithoautotrophic systems might function there (Boston et al., 1992). Consistent with this work, the SLiMEs discussed above are an example of a terrestrial ecosystem that could conceivably function on Mars today (Stevens and McKinley, 1995).

Liquid water is not present on the surface of modern Mars, though ice is, and water vapor is present in the atmosphere. Surface features that clearly resulted from erosion have been interpreted to have resulted from short-lived outbursts of liquid water from subsurface aquifers, possibly caused by geothermal heating (Squyres and Kasting, 1994). Inorganic carbon is abundant in the Martian atmosphere, and probably is found in the crust (Griffith and Schock, 1995). Much of Mars is covered with volcanic features, including large basaltic terrains. Mars is not tectonically active, but is or has recently been volcanically active, so geothermal heat is present in the shallow crust. Speculative scenarios have been developed for a subsurface hydrothermal cycle on Mars (e.g., Clifford, 1991). Environments very similar to that of the CRB SLiMEs discussed above could very well exist in the Martian subsurface. Further studies of SLiMEs, especially detection of biogeochemical signatures, could yield tools that could be used to search for evidence of life (or extinct life) during future Martian lander missions.

Other planets in our solar system might also contain environments where self-sustaining subsurface microbial ecosystems could exist (Carle et al., 1992). Essentially, as pointed out by McKay (1992), the search for life is the search for liquid water. Liquid water is thought to exist in the subsurface of Europa, and possibly of Io, Ganymede, and Callisto in the Jupiter system. In the Saturn system, Titan is thought to possess subsurface water and has a nitrogen-methane atmosphere that is attractive to proponents of some prebiotic chemistry scenarios. While it may be many years before we know very much about these planets, the existence of subsurface microorganisms on Earth keeps the possibilities of extraterrestrial life an open question.

If life exists elsewhere in the solar system, it is most likely present in subsurface environments. Subsurface microbiology thus provides an important model for exobiology.

12.4 Subsurface Microbiology in the Big Picture

While a great deal has been learned about the microbial ecology of the terrestrial subsurface, the results to date are mostly tantalizing clues that "Great Events" may be happening underground. Progress is relatively slow, partly because of the great cost and difficulty of obtaining samples or otherwise making observations beneath kilometers of soil and rock. Laboratory microcosm simulations, discovery of biogeochemical signatures, and deployment of newly developed detection methods may help to accelerate the rate of discovery in the coming years. Nevertheless, one can paint an extremely speculative picture of the role of subsurface microbiology in the biosphere.

The biosphere extends from the lower atmosphere down into the earth's crust to the point at which local temperatures become too high for microbiological life. This thickness varies laterally from point to point around the planet, depending on the altitude of the terrain and the geothermal gradient. It appears, from elementary geometry, that the vast majority of the volume of the biosphere is found in the terrestrial subsurface (e.g., Gold, 1992; Gould, 1996). Living organisms, which are mainly bacteria, are busily involved in mediating the oxidation of the earth's crust, including the continually introduced reduced material from the mantle. Biomass concentrations and rates of metabolism are greatest at phase interfaces such as the land surface, the sea floor, and the sea surface. Because of its great bulk, though, the subsurface contains much of the biomass, albeit at low density.

Geological and hydrological processes cause a slow mixing, folding, and turning of the Earth's crust. Microorganisms — sometimes entire ecosystems — are carried along with this mixing. As they are carried further and further from interface zones, their community metabolism slows and "goes on geological time". Ecological selection eliminates some organisms, and given time, new adaptive traits may evolve. Along some paths, microorganisms are killed by increasing geothermal heat. Along others, nutrient flux declines to the point where microorganisms become completely dormant, and eventually, dead. Still other paths lead to new subsurface nutrient sources. Unweathered reduced mineral deposits allow certain bacteria to proliferate. Pieces of crust which have been heat-sterilized gradually cool, or rise, and their remaining nutrients become available to colonizing organisms. Cooked organic material (petroleum fluids) can migrate to zones where living microorganisms can utilize it. Sometimes these paths of mixing bring subsurface organisms to the surface, where they can go back on "surface time", exchange genetic material, and disperse over the surface of the globe, eventually to be swept down into the subsurface again.

From time to time, cosmic collisions flash-heat the surface of the biosphere and eliminate the fast-reacting but vulnerable interface ecosystems. Slow recolonization of the interfaces from the subsurface restarts surface biosphere processes. As the solar system ages these collisions become much less frequent, and the disruption to the surface biosphere less extreme.

Sometimes chemical conditions at the surface change drastically, and recolonizing organisms must adapt to new surface niches. The atmosphere of some planets evolves to the point where surface life is no longer possible, though life might continue in the subsurface where nutrient fluxes exist. On Earth, abundant liquid water and biological innovation leads to high-energy photosynthetic processes at the surface. Waste products from this eventually exceed the redox buffering capacity of the crust and flood the atmosphere with highly toxic oxygen. Organisms evolve ways of dealing with this substance, or become restricted to diffusion-limited habitats.

With the high-energy metabolism made possible by oxygenic processes at the surface, intracellular and multicellular symbioses eventually lead to macroorganisms, such as plants and animals (e.g., Margulis, 1993). These organisms are too large to participate directly in subsurface processes. Some of them are still trying, though.

References

Barns, S.M., R.E. Fundyga, M.W. Jeffries, and N.R. Pace. 1994. Remarkable archaeal diversity detected in a Yellowstone National Park hot spring environment. *Proc. Natl. Acad. Sci. U.S.A.,* 91:1609-1613.

Bath, A.G., N. Christofi, C. Neal, J.C. Philp, M.R. Cave, I.G. McKinley, and U. Berner. 1987. Trace element and microbiological studies of alkaline groundwaters in Oman, Arabian Gulf. A natural analogue for cement pore-waters. Report FLPU 87-2 British Geological Survey, Keyworth, England.

Boone, D.R., Y. Liu, Z.-J. Zhao, D. Balkwill, G.R. Drake, T. Stevens, and H. Aldrich. 1995. *Bacillus infernus* sp. nov., an Fe(III)- and Mn(IV)-reducing bacterium from the deep terrestrial subsurface. *Int. J. Syst. Bacteriol.,* 45:441-448.

Boston, P.J. and C.P. McKay. 1991. The deep subsurface of Mars: possible habitat for extant or recently extinct microbial life. In C.B. Fliermans and T.C. Hazen, Eds., *Proc. First Symp. Microbiology of the Deep Subsurface.* WSRC Information Services, Aiken, SC.

Boston, P.J., M.V. Ivanov, and C.P. McKay. 1992. On the possibility of chemosynthetic ecosystems in subsurface habitats on Mars. *Icarus,* 95:300-308.

Braterman, P.S., A.G. Carins-Smith, and R.W. Sloper. 1983. Photo-oxidation of hydrated Fe2+ — significance for banded iron formations. *Nature,* 303:163-164.

Brigmon, R.L., H.W. Martin, T.L. Morris, G. Bitton, and S.G. Zam. 1994. Biogeochemical ecology of *Thiothrix* spp. in underwater limestone caves. *Geomicrobiol. J.,* 12:141-159.

Brockman, F.J., T.L. Kieft, J.K. Fredrickson, B.N. Bjornstad, S.M.W. Li, W. Spangenburg, and P.E. Long. 1992. Microbiology of vadose zone paleosols in South-Central Washington State. *Microb. Ecol.,* 23:279-301.

Cairns-Smith, A.G. 1992. Mineral theories of the origin of life and an iron sulfide example. *Orig. Life Evol. Biosph.,* 22:161-180.

Carle, G.C., D.E. Schwartz, and J.L. Huntington. 1992. Exobiology in Solar System Exploration. NASA-SP-512. Natinoal Aeronautics and Space Administration, Washington, D.C.

Champ, D.R., J. Gulens, and R.E. Jackson. 1979. Oxidation-reduction sequences in groundwater flow systems. *Can. J. Earth Sci.,* 16:12-23.

Chapelle, F.H., J.L. Zelibor, Jr., D.J. Grimes, and L.L. Knobel. 1987. Bacteria in deep coastal plain sediments of Maryland: a possible source of CO2 to groundwater. *Water Resour. Res.,* 23:1625-1632.

Chapelle, F.H. and D.R. Lovley. 1990. Rates of microbial metabolism in deep coastal plain aquifers. *Appl. Environ. Microbiol.,* 56:1865-1874.

Clifford, S.M. 1991. The role of thermal vapor diffusion in the subsurface hydrologic evolution of Mars. *Geophys. Res. Lett.,* 18:2055-2058.

Deming, J.W. and J.A. Baross. 1993. Deep-sea smokers: Windows to a subsurface biosphere?, *Geochim. Cosmochim. Acta,* 57:3219-3230.

Des Marais, D.J. 1994. Tectonic control of the crustal organic carbon reservoir during the Precambrain. *Chem. Geol.,* 114:303-314.

Des Marais, D.J., H. Strauss, R.E. Summons, and J.M. Hayes. 1992. Carbon isotope evidence for the stepwise oxidation of the proterozoic environment. *Nature,* 359:605-609.

DOE, Subsurface Science Program Taylorsville Basin Working Group. 1994. D.O.E. seeks origin of deep subsurface bacteria. *EOS,* 75:390-396.

Drobner, E., H. Huber, G. Wachtershauser, D. Rose, and K.O. Stetter. 1990. Pyrite formation linked with hydrogen evolution under anaerobic conditions. *Nature,* 346:742-744.

Forterre, P., F. Confalonieri, F. Charbonnier, and M. Guguet. 1995. Speculations on the origin of life and thermophily: review of available information on reverse gyrase suggests that hyperthermophilic procaryotes are not so primitive. *Orig. Life Evol. Biosph.,* 25:235-249.

Gilbert, J., Danielopol, D., and Stanford, J., Eds. 1994. *Groundwater Ecology.* Academic Press, San Diego, CA.

Gogarten-Boekels, M., E. Hilario, and J.P. Gogarten. 1995. The effects of heavy meteorite bombardment on the early evolution. The emergence of the three domains of life. *Orig. Life Evol. Biosph.*, 25:251-264.

Gold, T. 1992. The deep, hot biosphere. *Proc. Natl. Acad. Sci. U.S.A.*, 89:6045-6049.

Gould, S.J. 1996. Microcosmos. *Natural History*, 3/96: 21-68.

Griffith, L.L. and E.L. Schock. 1995. A geochemical model for the formation of hydrothermal carbonates on Mars. *Nature*, 377:406-408.

Haldeman, D.L., B.J. Pitonzo, S.P. Story, and P.S. Amy. 1994. Comparison of the microbiota recovered from surface and deep subsurface rock, water, and soil along an elevational gradient. *Geomicrobiol. J.*, 12:99-111.

Holland, H.D. 1984. *The Chemical Evolution of the Atmosphere And Oceans*. Princeton University Press. Princeton, NJ.

Horowitz, N.H. and J.S. Hubbard. 1976. The Viking carbon assimilation experiments: interim reports. *Science*, 194:1321-1322.

Jannasch, H.W. 1995. Microbial interactions with hydrothermal fluids. in *Seafloor Hydrothermal Systems: Physical, Chemical, Biological and Geological Interactions*. Geophysical Monograph 91. American Geophysical Union, Washington, D.C.

Kennedy, M.J., S.L. Reader, and L.M. Swierczynski. 1994. Preservation records of micro-organisms. Evidence of the tenacity of life. *Microbiology (U.K.)*, 140:2513-2529.

Lauritzen, S.-E. and S. Bottrell. 1994. Microbiological activity in thermoglacial karst springs, south Spitsbergen. *Geomicrobiol. J.*, 12:161-173.

Levin, G.V. and P.A. Straat. 1976. Viking labeled release biology experiment. Interim results. *Science*, 194:1322-1329.

Maher, K.A. and D.J. Stevenson. 1988. Impact frustration of the origin of life. *Nature*, 331:612-614.

Margulis, L. 1993. Symbiosis in cell evolution. *Microbial Communities in the Archean and Proterozoic Eons*. 2nd ed. W.H. Freeman, New York.

McKay, C.P., E.I. Friedman, R.A. Wharton, and W.L. Davies. 1992. History of water on Mars: a biological perspective. *Adv. Space Res.*, 12:231-238.

McLaren, D.J. and W.D. Goodfellow. 1990. Geological and biological consequences of giant impacts. *Annu. Rev. Earth Planet Sci.*, 18:123-171.

Moore, W.S. 1996. Large groundwater inputs to coastal waters revealed by [226]Ra enrichments. *Nature*, 380:612-614.

Murphy, E.M., J.A. Schramke, J.K. Fredrickson, H.W. Bledsoe, A.J. Francis, D.S. Sklarew, and J.C. Linehan. 1992. The influence of microbial activity and sedimentary organic carbon on the isotope geochemistry of the Middendorf aquifer. *Water Resour. Res.*, 28:723-740.

Norton, C.F., T.J. McGenity, and W.D. Grant. 1993. Archaeal halophiles (halobacteria) from two British salt mines. *J. Gen. Microbiol.*, 139:1077-1081.

Olson, G.J., W.S. Dockins, G.A. McFeters, and W.P. Iverson. 1981. Sulfate-reducing and methanogenic bacteria from deep aquifers in Montana. *Geomicrobiol. J.*, 2:327-340.

Oro, J., S.L. Miller, and A. Lazcano. 1990. The origin and early evolution of life on Earth. *Annu. Rev. Earth Planet Sci.*, 18:317-356.

Parkes, R.J., B.A. Cragg, S.J. Bale, J.M. Getliff, K. Goodman, P.A. Rochelle, J.C. Fry, A.J. Weightman, and S.M. Harvey. 1994. Deep bacterial biosphere in Pacific Ocean sediments. *Nature*, 371:410-413.

Paull, C.K., A.J.T. Jull, L.J. Tolin, and T. Linick. 1985. Stable isotope evidence for chemosynthesis in an abyssal seep community. *Nature*, 317:709-711.

Pedersen, K. and S. Ekendahl. 1990. Distribution and activity of bacteria in deep granitic groundwaters of southeastern Sweden. *Microb. Ecol.*, 20:37-52.

Pedersen, K. and S. Ekendahl. 1992. Assimilation of CO2 and introduced organic compounds by bacterial communities in groundwater from southeastern Sweden deep crystalline bedrock. *Microb. Ecol.*, 23:1-14.

Sarbu, S.M., B.K. Kinkle, L. Vlasceanu, T.C. Kane, and R. Popa. 1994. Microbiological characterization of s sulfide-rich groundwater ecosystem. *Geomicrobiol. J.*, 12:175-182.

Schidlowski, M. 1988. A 3,800-million-year isotopic record of life from carbon in sedimentary rocks. *Nature*, 333:313-318.

phere: Its Origin and Evolution. Princeton University

P Init. Repts. 112,135, 138, 139.
*itribution of aerobic bacteria, protozoa, algae, and
Geomicrobiol. J.,* 7:15-31.
H.J. Morowitz. 1989. Annihilation of ecosystems by
irth. Nature, 342:139-142.
irganisms. Nature, 195:643-646.
Mars: How warm and how wet? *Science,* 265:744-749.
R.D. Eden, M. Fielder, H. Cash, and I. Vance. 1993.
ing in deep North Sea and Alaskan oil reservoirs.

*iophilic anaerobic bacteria in 2800-m-deep samples
Symp. Subsurface Microbiology. Bath, U.K. 19-24

*nd K.B. Wagnon. 1996. Rock weathering as a source
*ic microorganisms. Am. Soc. Microbiology General
*, LA.
*bility and density dependence of bacteria in terres-
for enumeration. J. Microbiol. Methods, 21:283-292.
*robial subsurface cycle: a refuge for early life. 95th
*ay 21-25, 1995. Washington, D.C.
*ckson. 1993. Bacteria associated with deep, alkaline,
Washington. *Microb. Ecol.,* 25:35-50.
*noautotrophic microbial ecosystems in deep basalt

*robial subsurface cycle: a refuge for early life. 95th
*ay 21-25, 1995. Washington, D.C.
*bility and density dependence of bacteria in terres-
for enumeration. J. Microbiol. Methods, 21:283-292.
Straub, K.L., M. Benz, B. Schink, and F. Widdel. 1996. Anaerobic, nitrate-dependent microbial
oxidation of ferrous iron. *Appl. Environ. Microbiol.,* 62:1458-1460.
Towe, K.M. 1994. Earth's early atmsophere: constraints and opportunities for early evolution.
In *Early Life on Earth.* Bengtson, S., Ed., Columbia University Press, New York.
Wächtershauser, G. 1988. Before enzymes and templates: theory of surface metabolism. *Microbiol. Rev.,* 52:452-484.
Wächtershauser, G. 1990. Evolution of the first metabolic cycles. *Proc. Natl. Acad. Sci. U.S.A.,*
87:200-204.
Wallis, M.K. and N.C. Wickramasinghe. 1995. Role of major terrestrial cratering events in
dispersing life in the solar system. *Earth Planet. Sci. Lett.,* 130:69-73.
Wetherill, G.W. 1990. Formation of the Earth. *Annu. Rev. Earth Planet Sci.,* 18:205-256.
White, D.C., D.L. Balkwill, D.R. Boone, F.S. COlwell, R.M. Lehman, J.P. McKinley, S.A. Nierzwicki-Bauer, T.C. Onstott, T.J. Phelps, S.A. Rawson, D.B. Ringelberg, B.F. Russell, and
T.O. Stevens. 1993. Evidence for the recovery of subsurface microorganisms from 2800 m
depths in the Taylorsville triassic basin sediments. Int. Symp. Subsurface Microbiology,
Bath, U.K., 19-24 September, 1993.
Woese, C.R. 1987. Bacterial evolution. *Microbiol. Rev.,* 51:221-271.

13

Movement of Bacteria in the Subsurface

Aaron L. Mills

CONTENTS

KEY WORDS: *attachment, adhesion, detachment, retention, retardation, advection, dispersion, filtration, hydrophobicity, chemotaxis, motility, hydraulic heterogeneity, porous media, models, transport.*

13.1   Introduction

Research of the past 10 years has demonstrated that there is a subsurface microbial community that is widely distributed, highly diverse, and often quite active. The discovery of the unexpectedly abundant biota at great depths has prompted the obvious question of how the microbes got there. In some cases it may be that the community has persisted through centuries or even millennia after being buried in surficial sediments. In fractured igneous or metamorphic rock it seems unlikely that a diverse community could arise from survivors of the original melt due to the high temperature of deposition. In both the sedimentary and igneous/metamorphic locations, postdepositional transport of microbes is often postulated.

0-8493-8362-5/97/$0.00+$.50
© 1997 by CRC Press LLC

For subsurface habitats close to the surface and well connected with the surface, there can be little doubt that organisms are transported from the surface or from injection sites below the surface to become widely dispersed in both saturated and unsaturated zones of the subsurface. The simple fact that sewage indicator organisms are not infrequently found downgradient from septic fields is conclusive evidence that microbes can be and are transported through subsurface environments (e.g., Gerba and Bitton, 1984; Yates and Yates, 1988).

Given the importance of microbes as mineralizers of organic carbon, and given that microbes have been demonstrated to be effective agents of biorestoration of a variety of contaminated sites, much interest is focused on the role of microorganisms as potential agents of aquifer restoration (e.g., Thomas and Ward, 1989; National Research Council, 1993). Bioremediation is already a working solution for some contaminated aquifers. Intrinsic bioremediation, in which indigenous communities degrade contaminants, and biostimulation, in which indigenous communities are stimulated to degrade a contaminant by addition of organic or inorganic nutrients, have been shown to be successful in sites including both surface and subsurface contamination of aqueous and solid phase components. Bioaugmentation, in which competent nonnative organisms are inoculated into a contaminated site, has not as yet been shown to be an effective means of reducing levels of contaminants, although competent organisms are often easily isolated and cultured, or constructed in the laboratory through genetic engineering approaches. Experience with injection of microbes in microbially enhanced oil recovery studies and studies concerning the growth and activity of microbes in porous media suggests that growth may result in clogging of the media near the injection point. Concern about adhesion of the microbial cells to the grain surfaces also suggests that transport and dispersal of bacterial cells in porous media will have a profound effect on the ability of introduced organisms to penetrate the zone of contamination to the extent that gowth of the microbes will result in remediaton of the contaminated aquifer.

Other facets of subsurface microbiology are similarly strongly influenced by bacterial transport considerations. The simple fact that abundant microorganisms have been found at great depths in both saturated and unsaturated media begs the question of the origin of the assemblage. In depositional environments, the cells could be descendents of microbes laid down during the original depositional event, or they could be of more modern origin and have been moved through the sedimentary material rapidly or slowly, to produce the observed distribution. The presence of actively metabolizing microbes at depth in fractured rock and deep unsaturated deposits suggests a transport-related origin, since depositional conditions of these materials would not seem conducive for survival of a diverse assemblage.

In near-surface environments, transport of microbes has been considered for some time as a primary means of contamination of water supplies by indicator organisms and pathogens from failed septic fields. In all of these cases, fundamental aspects of microbial transport should be similar, and application of enhanced understanding of those aspects may improve wastewater treatment and contaminant reduction in the subsurface.

13.2 Factors Affecting Transport of Microorganisms in Porous Geologic Media

The factors which control the rate and extent of bacterial transport in porous media are usually divided into two major categories: hydrogeological and biological. With a few exceptions which can be specified, transport is considered to be advective, with the major controlling factors being those related to removal of microbial cells from suspension. Some of the controls are strictly physical; clogging of narrow pore throats, for example. Others are related to the interaction of cells with the mineral grain surfaces. Properties of the cells which influence these controls will be discussed as biological phenomena, while properties of the mineral grains that influence the interactions will be included as geochemical properties.

Harvey (1991) listed the important processes and factors that form the parameters for bacterial transport through porous media. The variables included straining (pore clogging), dispersion, sorption and detachment, growth and death (grazing by bacteriovores was included in the latter), motility, and chemotaxis. Other factors which influence these particular processes, such as the mineralogy and pore water composition, pore size distribution, etc., were considered separately, as it was the modeling parameters that Harvey sought to identify. These parameters and factors have not been replaced as the primary influences on bacterial transport, rather they have been refined and examined in more detail by a number of investigators in the period since Harvey's review.

Several terms regarding the transport of microbes that are based on terms used in flow and transport of dissolved and particulate substances have come into common use. **Advection** is the movement of the bulk pore fluid and its dissolved and suspended constituents. In advective transport, nonreactive substances will be transported at a rate equal to the average linear velocity of the fluid. The term **convection** is often used in place of advection. Hydrodynamic **dispersion** is the phenomenon by which the dissolved and suspended constituents spread both longitudinally and laterally from the path expected on the basis of hydraulics alone. Dispersion is comprised of two components, molecular diffusion (or Brownian movement in the case of particles) and mechanical dispersion which results from mixing in the pore throats and from the varying distances (lengths of tortuous paths) that any individual molecule or particle may follow. **Retention** refers to the removal of cells from the pore fluid so that they do not return to suspension, i.e., in a column experiment, retained cells never emerge in the effluent. **Retardation** refers to the interaction of the cells with the porous material in such a way that the cells may be delayed in their passage past a given point in the flow path (e.g., emergence in the eluant of a column). **Adhesion** refers to the association of the cells with the surfaces of the porous medium. This term is often used synonymously with adsorption, although adsorption is really one of the mechanisms of adhesion. Adhesion can be reversible (as in equilibrium or nonequilibrium sorption) in which case it contributes to retardation, or it can be irreversible (as when a cell actually cements itself to a solid surface by secretion of a polymer) and in which case it contributes to retention. Other terms are also used frequently, and they will be defined as they are introduced.

13.2.1 Hydrogeological Controls on Bacterial Transport

13.2.1.1 *Hydraulic Heterogeneity: Transport in Preferred Flow Paths vs. Matrix Transport*

Many of the experiments associated with particle transport in porous media, including transport of bacteria, have been performed in columns containing a uniformly sized medium. Glass beads, plastic spheres, and even carefully sieved sands have been used to develop and test theory about hydraulic properties of porous media and the effect of those properties on bacterial transport. The resultant theories are simplistic in their view; even the more complicated multidimensional models usually treat the porous medium as a bed of uniformly sized spheres. Heterogeneity is recognized as being important, because experimental results in heterogeneous systems differ substantially from those obtained in hydraulically homogeneous media. (For more detail concerning heterogeneity, see Chapter 6.)

Porous media are hydraulically heterogeneous at all scales. Even if all the grains were perfectly uniform spheres of an exact diameter, the inability to pack the spheres in completely open or closed arrangement leads to heterogeneity in the distribution of pore sizes. If the pores are all substantially larger than the bacterial cells, this effect will be minimized, but the closer the size of the grains (and their interspersed pores) to the size of the bacterial cells the greater the effect on the individual cells will be. In media with a wide range of grain diameters, a high probability of large- and small-diameter flow paths exists. Bacteria will preferentially travel through the larger of the paths, and the cells in those paths will be least affected by interactions with porous medium grains. As a result, those cells will be the first to break through at some point downgradient of the inception of transport. Cells which move into the areas with the finest pores (often referred to as the matrix) will be most susceptible to straining and adhesion on grain surfaces, and thus will be delayed (retarded) in their breakthrough. Pore blockage and irreversible adhesion lead to permanent retention of the cells in the porous matrix, or at least they are retained until they are dislodged or until growth of a biofilm on the mineral grains sloughs some cells into the suspension. It can be concluded, therefore, that rapid transport of bacteria in porous media occurs primarily through preferred flow paths, but longer-term transport can occur through finer-textured matrixes, and the matrixes can also serve as sources of cells for continued inoculation of the more permeable zones. These statements should be generally true for all porous media from unconsolidated sands and gravels to fractured rock. Rapid transport of bacteria through preferred flowpaths has been demonstrated experimentally in the laboratory by Toran and Palumbo (1992), Story et al. (1996), Fontes et al. (1991), and Morley (1995). In the latter two cases, large peaks of bacterial cells eluted from columns of quartz sand containing a single vein of coarser-textured grains well in advance of the first pore volume of eluant (Figure 13.1). A second peak of bacteria containing many fewer cells than the first then followed. The first peak was thought to represent transport through the preferred flow path while the second peak represented those cells moving through the finer-textured annular matrix. Models fit to the data to test this hypothesis gave results that were consistent with the two-domain mechanism. Harvey and Garabedian (1991) attributed early initial breakthrough of bacterial cells and microspheres followed by a series of smaller peaks to the presence of multiple preferred flow paths in their field experiments in the sands and gravels on Cape Cod.

13.2.1.2 Reactive Surfaces

That bacteria attach preferentially to different mineral substrates has been known for some time. Mills and Maubrey (1981) demonstrated that bacterial attachment to limestone was faster than to sandstone in both lotic and lentic environments. Scholl et al. (1990) showed that media of different mineralogies retained different proportions of bacteria in packed column studies. In general, bacteria tended to adhere more to minerals with a greater proportion of leachable ions than to more inert substances. Scholl et al. (1990) also showed that differences in base mineralogy were less important than the nature of coatings on the mineral grains. In particular, the presence of surface coatings of iron sesquioxide dramatically increased the adhesion of bacteria to chips of limestone and sandstone suspended in bacterial suspensions over uncoated chips,

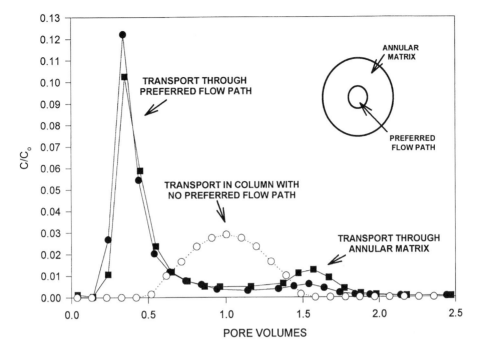

FIGURE 13.1

The effect of physical (hydraulic) heterogeneity on bacterial transport in porous media. Open circles are the results of a breakthrough experiment with bacteria in a column homogeneously packed with quartz sand sieved to a uniform size. The solid circles represent a pair of columns packed to contain a preferred flow path of coarse-textured sand surrounded by an annulus of the same sand as in the homogeneous column. For the homogeneous column, the peak breakthrough was observed at one pore volume. For the heterogeneous column, two peaks were observed. The first represents transport through the preferred flow path and the second smaller peak represents transport throughout the annulus. For both column types, the breakthrough curves are similar to those observed for a conservative tracer except for the amount of material recovered. Recovery of the conservative tracer was complete, whereas the recovery of the bacteria in the homogeneous case was very small. In both cases, bacterial breakthrough represents retention without retardation.

and also increased the retention of bacteria in packed columns over columns packed with clean quartz sand.

Mills et al. (1994) further showed that the presence of iron coatings (and by analogy, Al and Mn coatings) on clean quartz sand caused the sorption of bacteria to the sand in batch experiments to change from that demonstrating a typical equilibrium isotherm to a mode in which all added bacteria were apparently irreversibly sorbed to the coated sand grains until a saturation point was reached beyond which no additional sorption was observed. Using the observed behavior, Mills et al. (1994) offered a model which predicted the adhesion behavior of bacteria in mixtures of iron-coated and clean quartz sand, and experimental results were consistent with the predictions.

Other surface coatings can also affect the retention of bacteria on mineral surfaces. Adhesion of bacterial cells to surfaces can be inhibited by the presence of proteins on the surface or in solution (Fletcher, 1977; Feldner et al., 1983), enhanced by oxyhydroxide coatings on the surface (Scholl et al., 1990; Scholl and Harvey, 1992; Mills et al., 1994), and/or affected by nutrient conditions (McEldowney and Fletcher, 1986a; Kjellberg and Hermansson, 1984). Richardson (1994) examined bacterial attachment to iron-coated sand in the presence of humic acid and observed a high retention of cells in the presence of humic acid, but the number of cells adsorbed to the iron-coated sand in the presence of adsorbed humic-coated sand was significantly lower than adsorbed to the iron-coated sand in the absence of the humic acid. Richardson (1994) concluded that the humic acid competed with the bacteria for available sites on the iron-coated surfaces of the sand. Many of these experiments were done in batch systems, and some caution is necessary in extrapolating such results to flowing situations. For example, adhesion was found to be consistently greater in dynamic columns as compared to static batch system, even for short residence times (Rijnaarts et al., 1993). Furthermore, questions of kinetic controls on adhesion make predictions of dynamic situations (e.g., as in column experiments) with parameter estimates obtained from batch experiments unreliable, unless specifically tested for a particular situation. More information on kinetic control of sorption is offered below.

13.2.2   Biological Controls on Bacterial Transport

13.2.2.1   Cell Surface Characteristics

A substantial effort has been directed at identifying the characteristics of bacterial cells that affect their transport and quantifying their effect on the transport process. The characteristics affecting transport seem to be related either to the interaction of the bacterial cell surface with the mineral grain surface and the gas-liquid interface in the case of unsaturated media, or to the behavior of the cells as buoyant particles aside from their cell surface characteristics.

There is a fairly rich literature that discusses the roles of cell surface properties and particle surface properties on bacterial adhesion. The tendency for bacterial cells to adhere to mineral grain surfaces appears to be most strongly influenced by the surface charge on the cell, and the cell surface hydrophobicity. Hydrophobicity is quantitatively defined on the basis of water contact angle (the angle formed by a drop of water in contact with a surface). Water contacting a highly hydrophobic surface tends to "bead up", yielding a steep angle at the contact point, whereas on less hydrophobic surfaces the water tends to spread out, forming a smaller contact angle. A number of other methods for describing hydrophobicity have been used; they include partitioning of cells into nonaqueous solvents (BATH — bacterial attachment to hydrocarbons), and hydrophobic exclusion chromatography (Rosenberg and Doyle, 1990). The methods tend to be internally consistent, but do not always yield results that are consistent across methods (Figure 13.2).

Surface charges associated with colloidal particles are usually described by the zeta potential, which is derived from the migration of particles in an electric field (electrophoretic mobility). Both the sign (positive or negative) and the magnitude (distance traveled) can be expressed by this measure. Surface charges can also be determined by potentiometric titrations (Harden and Harris, 1953), but this technique is much less satisfactory for bacteria than the determination of electrophoretic mobility.

Bacterial adhesion to materials such as glass, polystyrene, sand, stainless steel, and even meat has been shown to decrease with increasing zeta potential of the bacterium or substrate (increasing zeta potential describes an increasing negative charge) (Fletcher and Loeb, 1979; Abbott et al., 1983; Van Loosdrecht et al., 1987b; Dickson and Koohmaraie, 1989). Furthermore, adhesion also tends to increase with increasing solution ionic strength, due to compression of the double layer as predicted by DLVO theory (Abbott et al., 1983; Fontes et al., 1991; Mills et al., 1994; Zita and Hermansson, 1994). Adhesion also tends to increase with increasing hydrophobicity of the surfaces (Fletcher and Loeb, 1979; Van Loosdrecht et al., 1987a; Stenstrom, 1989; Dickson and Koohmaraie, 1989; Vanhaecke et al., 1990; Sorongon et al., 1991). Bacterial adhesion to

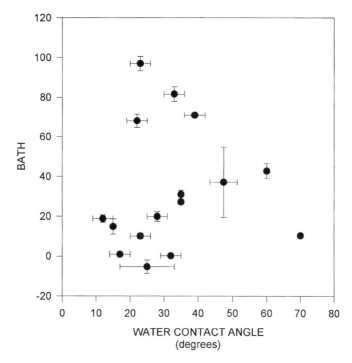

FIGURE 13.2

Comparison of cell surface hydrophobicity determined on several bacterial isolates using both the bacterial attachment to hydrocarbon (BATH) and water contact angle methods. Although there is a generally positive relationship, the weakness of the relationship ($r^2 = 0.01$) contraindicates any interchangeability of values or comparison of the values across strains. (From Richardson, R.L., M.S. Thesis, University of Virginia, Charlottesville, 1994.)

glass and Teflon® was also shown to be related to the cell surface hydrophobicity of several coryneform bacteria and that correlation extended to the chain length of mycolic acids on the bacterial surfaces (Bendinger et al., 1993).

In general, then, bacterial adhesion tends to increase with increasing hydrophobicity of the bacterial cell and substratum, and decreases as the cell or substratum becomes more negatively charged. However, hydrophobicity is the dominant factor with regard to adhesion to uncoated glass, as highly hydrophobic cells demonstrated high levels of adhesion, regardless of their surface charge (Van Loosdrecht et al., 1987b). In contrast to the majority of the literature, Mafu et al. (1991) found that bacterial hydrophobicity and zeta potential were not correlated with adhesion to polypropylene, rubber, glass, or steel, while McEldowney and Fletcher (1986b) noted that bacterial adhesion to polystyrene was not correlated with ionic strength of the bulk solution. In general, then, it would be expected that the more hydrophobic bacteria would tend to be transported less readily than hydrophilic cells (Figure 13.3). Experimental data from complex systems do not always support this hypothesis, but often, multiple confounding variables may mask the effect (e.g., Gannon et al., 1991a,b).

13.2.2.2 Other Cell Properties

Cell properties other than those measured as surface chemical characteristics have also been implicated in affecting the transport of bacteria. Some are related to cell behavior, e.g., motility and chemotaxis, whereas others are related to cell properties that might not be included in the surface chemistry category. Examples of the latter

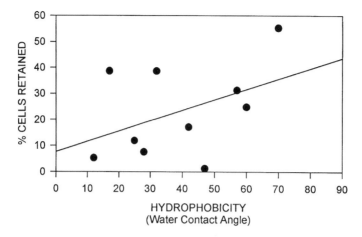

FIGURE 13.3

Relationship of cell surface hydrophobicity and the proportion of bacterial cells retained in columns of clean quartz sand. Although the relationship is generally weak ($r^2 = 0.2$) the general observation that increased hydrophobicity leads to decreased ease of transport is visible. The weakness of the relationship suggests that factors other than hydrophobicity contribute to controlling bacterial transport in porous media.

include the observation or the fact that there is a correlation between adhesion and the presence of protruding cell surface features (Uyen et al., 1988; Busscher and Weerkamp, 1987). Obviously, surface structures that can actively adhere to surfaces, or that can penetrate the electrostatic repulsive layers, will enhance attachment and thereby inhibit or at least retard transport in porous media.

13.2.2.2.1   Size and Shape

The transport of bacteria has been investigated quantitatively using colloid filtration theory (Harvey and Garabedian, 1991; Russel et al., 1989; see below). Aside from advection and dispersion in the flowing pore waters, the processes that affect the transport of bacteria are the kinetically controlled rate at which bacteria are deposited on the mineral grains of aquifer material and the kinetically controlled rate at which bacteria detach from the grain surfaces. These processes are affected by the size of cells and suggest that cells of about 1 μm in diameter may be optimal for transport through sand-sized media (Yao et al., 1971). In practice, filtration theory assumes all particles are spheres (Harvey and Garabedian, 1991; Yao et al., 1971; Hornberger et al., 1992).

While fluid dynamic forces will influence colloid particle-grain interactions, a quantitative treatment of phenomena related to particle morphology is not available; the only approach that has been used is based on empirical extrapolation from theoretical results for spherical particles (Hornberger et al., 1992). Particles with non-spherical morphologies are typically represented by an "equivalent diameter" equal to the average of the major diameters of the particle (McDowell-Boyer et al., 1986). Bacteria found in geological formations are known to have a variety of shapes, including filamentous forms, spirals, rods, and ellipsoid, ovoid, and coccoid shapes (Balkwill et al., 1989; Hirsch and Rades-Rohkohl, 1990; Ghiorse and Wilson, 1988) even though the majority of bacteria removed from sampling wells in unconsolidated aquifer materials are small cocci or coccoid rods (Balkwill et al., 1989; Ghiorse and Balkwill, 1983). To what extent different cell shapes influence bacterial transport through aquifers and how shape, size, and cell surface properties interact, is currently unknown.

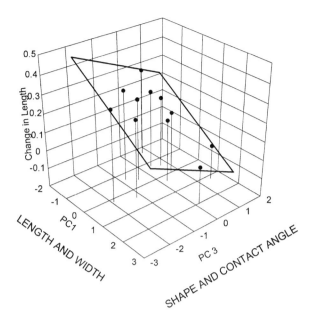

FIGURE 13.4

Effect of size, shape, and cell surface hydrophobicity on the "chromatographic" selection of shorter rounder cells as a cell suspension passes through a column of sand. Data axes were generated by a principal components analysis of the data. Each data point represents the data for a single bacterial isolate. The plane is the least squares regression surface for the data. (Reprinted from Weiss, Th.H., Mills, A.L., Herman, J.S., and Hornberger, G.M., *Environ. Sci. Technol.*, 29, 1737, 1995. With permission.)

Weiss et al. (1995) examined the effect of cell shape on the transport of bacteria in columns packed with clean quartz sand. Although a chromatographic cell-size effect on the transport of bacteria through a sand aquifer was noted by Harvey and Garabedian (1991), Weiss et al. (1995) observed a chromatographic effect for cell shape as well as size. When 14 strains of bacteria with differing size (as indicated by both cell length [L] and cell width [W]), shape (W/L), and cell surface hydrophobicity were eluted in pulses through short columns packed with clean, quartz sand of about 0.75 mm grain diameter, eluted cells for 12 of the 14 strains were more spherical (greater median cell shape indexes) than cells in the influent suspension. This observation suggests that transport models may have to be adjusted to account for cell shape.

Bacterial cells are thought to attach to mineral surfaces due to surface-surface interactions, either electrostatic or hydrophobic in nature (e.g., Harvey, 1991; Van Loosdrecht et al., 1989; McEldowney and Fletcher, 1986a,b). Although it is not clear why these attachment mechanisms would lead to preferential removal of long, rod-shaped cells, some reports suggest that cell attachment to solid surfaces indeed may be greater for elongated cells than for spherical cells (Van Loosdrecht et al., 1989; Feldner et al., 1983). Examination of a fitted-response surface for data that had been reduced by principal components analysis shows that short rods with low contact angles generally underwent the greatest decreases in cell length (Figure 13.4). There is no currently available theoretical explanation for these observed interactions, and the current theory is unlikely to advance in the absence of further experimental work.

To attain a more complete understanding of the transport behavior of bacteria in porous media, all of the important factors that can affect transport, including (but probably not limited to) size, shape, hydrophobicity, electrostatic charge, and as suggested by our results, the interactions among the factors, must be considered. The interactions, however, are not well understood. The role of particle shape has not been included in most studies thus far. Mackie and Bai (1992) recognized that particle size distributions change as the particles travel through the column and, as we do here, encouraged caution when using filtration models assuming uniform particle size. Although the major axis dimension is the most commonly selected descriptor of size properties employed in predictive transport models, it is considered in those models to be a parameter that does not evolve during the transport process. The chromatographic effect reported by Weiss et al. (1995) calls this assumption into question.

Experiments with bacterial transport in the laboratory are usually conducted at high flow rates; e.g., Fontes et al. (1991) used rates of about 14 cm h^{-1} in their column experiments. Some experiments are run at slower rates, but continuous-flow experiments in the laboratory rarely approximate the rates found in many aquifers that can be expressed in terms of a few meters per year or less. At typical laboratory flow rates, processes such as motility or settling are not likely to be demonstrated as important. In many natural situations, however, motility or settling might produce movement of cells that could approach, or even exceed, the rate of movement due to advective transport.

13.2.2.2.2   *Sedimentation*

Most considerations of bacterial transport have considered bacterial cells to be colloids and have therefore ignored sedimentation of cells as an important factor in bacterial transport. Colloid filtration theory includes a consideration of particle sedimentation as a means of moving the particles to the surface of the porous medium grains, but advective processes are usually thought to overwhelm any significant effect of sedimentation. The assumption of neutral or near-neutral buoyancy is a reasonable approximation for high flow rates and short time scales such as experienced in many column studies and probably even in forced-gradient field experiments (Hornberger et al., 1992). Wan et al. (1995) recently reported median free sedimentation rates (rates in liquid suspension) of 7 and 42 mm d^{-1} for two strains of groundwater bacteria. While they accepted the fact that actual settling velocities would be slower in the presence of particles and flowing water (measured values were 7 and 17 mm d^{-1}), short-column experiments yielded significant breakthrough of bacteria through saturated sand that was as high as 90% of the free sedimentation rate. The specific gravity measured (1.117 and 1.074) for each of the two strains used was in the middle of the range of values obtained for 25 bacterial strains (1.040 to 1.121, with a mean of 1.0882 for the 25 strains tested.). The sedimentation rates of the two strains tested were inversely related to the buoyant density, suggesting that density alone does not dictate the sedimentation rate, a conclusion consistent with the importance of the role of shape and size claimed by Weiss et al. (1995). Sedimentation of bacterial cells may indeed play a role in transport of bacteria in very slow-moving ground water over long periods of time.

13.2.2.2.3   *Motility and Chemotaxis*

Bacterial motility and chemotaxis present interesting concerns in bacterial transport. Experimental results suggest that motility may enhance the movement of bacteria in porous media, but the conclusion is by no means universal. Similarly, chemotaxis (with its implicit assumption of motility) seems to enhance movement of bacteria in porous media in some but not all cases. Motile bacteria *(Enterobacter aerogenes)* were able to penetrate saturated sandstone cores under static conditions three to eight times faster than a nonmotile strain of *Klebsiella pneumoniae* (Jenneman et al., 1985). However, because the motile strain also grew faster than the non-motile strain, the growth rate differences may have been partly responsible for the differences in penetration. In packed sand cores under static conditions, motile *Escherichia coli* penetrated the medium more quickly than the nonmotile mutants of the strain (Reynolds et al., 1989). The motile chemotactic parent strain used by Reynolds et al. (1989) moved through the cores more slowly than did the nonchemotactic mutant. It was also observed that the growth rates of the nonchemotactic mutants exceeded that of the chemotactic parent, and that the relative order of penetration was related to the growth rate of each strain. Thus, Reynolds et al. concluded that the controlling factors in the experimental system used were growth and random motility — chemotaxis was not a contributing factor. Similarly, experiments with a chemotactic strain of *Pseudomonas putida* were unable to demonstrate a significant effect of the chemotaxis in transport of the cells when compared in the presence or absence of a chemical attractant (Barton and Ford, 1995). While motility and chemotaxis must play a role in movement of microbes in some instances, those situations have not yet been defined adequately. A reasonable probability exists that these processes will be important in the movement of competent cells into contaminated areas of low permeability, where advective transport is minimal compared with adjacent preferred flow paths. Additional research is clearly required in this area.

13.2.3 Transport in Unsaturated Media

A potentially important although rarely considered aspect controlling the movement of bacteria in porous media is the role that gas-water interfaces may play in microbial transport. Wan and Wilson (1994) used glass micromodels (consisting of tiny channels between two glass plates) and flowing water to observe polystyrene beads, clay particles, and bacterial cells adhering to gas-water interfaces. The sorption to the interface appeared to increase with increasing particle hydrophobicity, solution ionic strength, and decreasing positive charge on the particles. These observations suggested an initial adsorption due to van der Waals' and electrostatic interactions, followed by an essentially irreversible adhesion due to capillary force. Wan and Wilson predicted that for a relatively hydrophobic strain of bacterium even small amounts of residual gas could dramatically reduce advective transport. An additional observation by Pitt et al. (1993) showed moving gas bubbles swept glass and polymer surfaces free of adsorbed bacteria.

Although moving gas-water interfaces could move adherent bacteria along with them, the presence of gas-water interfaces has usually been found to reduce the transport of microbes. This may be due to the tendency of bubbles to remain in place relative to water moving through nearly saturated pores, but some other explanations may also be important. Powelson et al. (1990) found that MS2 bacteriophage was not removed during passage through 1 m of saturated soil, but was 95% removed in unsaturated soil. The authors suggested that the partially hydrophobic virus adsorbed to air-water interfaces in the unsaturated soil and were degraded by physical disruption of viral structure. Poletika et al. (1995) attempted to predict retardation of MS2 bacteriophage transport in unsaturated soil from equilibrium adsorption isotherms in batch experiments with soil slurries. The adsorption experiments indicated that the virus did not attach to the soil particles. Modeling of the virus breakthrough, however, yielded a retardation factor of 254, indicating strong adsorption. These authors suggested that the difference in the slurry and unsaturated experiments may have been due to the interaction of viral particles with air-water interfaces in the unsaturated soil. Tan et al. (1992), however, were able to predict transport of bacteria through unsaturated sand columns from batch adsorption to the solids without considering additional adsorption due to air-water interfaces. Such success is not common (Mills et al., 1994), and may have been a result of the use of a weakly adsorbing bacterial strain by Tan et al. (1992). Huysman and Verstraete (1993a,b) attributed greater adsorption of bacteria in drier soil conditions to slower water flow rates, rather than interaction with air-water interfaces. Clearly, additional work on air-water interfaces is necessary before their role in bacterial transport can be defined and used in prediction.

13.3 Models of Bacterial Transport

In an effort to provide quantitative estimates of bacterial transport in porous media, scientists and engineers have turned to mathematical models to assist with prediction of transport in applied situations, and to test hypotheses of potential mechanisms of reaction and transport of bacteria in porous media. It is far beyond the scope of this chapter to review modeling of bacterial transport in detail. An excellent document specifically dealing with modeling the fate and transport of microorganisms in the environment was edited by Hurst (1991), and the reader is referred to that volume for the details of fundamental modeling of transport phenomena.

Two basic approaches have been followed to model bacterial transport in porous media. There are important interconnections between the approaches; both attempt to quantify the same processes and recognize the important factors that control transport on a microscopic scale. But there are differences that make each approach more or less useful for different applications. Filtration models are derived from theories developed for particles passing through porous media, while advection-dispersion models were developed initially to describe the transport of dissolved species.

13.3.1 Filtration Models

The fundamental concept behind filtration theory is that particles are removed from a suspension as it passes through a porous medium (Elimelech and O'Melia, 1990). The processes that govern the rate and extent of particle removal are related to the relative sizes of the suspended particles and the filter medium, and the potential for reaction of the suspended particles with the surfaces of the medium. Specifically, straining (the removal of particles too large to fit through a pore) is considered as is sorption (reversible or irreversible) of the particles to the medium surfaces. In operation, filtration models generally lump the processes together to generate a filtration coefficient. The general filtration equation has the form:

$$\frac{\partial C}{\partial x} = -\lambda C \tag{13.1}$$

where C is the concentration of particles in the suspension, and λ is the "filter coefficient". This form of the equation suggests that particle concentration should decrease in some exponential fashion as depth in the filter increases, and most data support that idea. Another term that is often used to describe filtration is the "collection efficiency". The collection efficiency explicitly incorporates the concept "sorption" or "adhesion" into its theory. In this formulation, the medium grains are considered to be "collectors" that trap particles in proportion to the collector's cross-sectional area.

A single grain is the fundamental unit, and a term "single collector efficiency" is frequently employed. The collection efficiency, η, is defined mathematically as:

$$\eta = \lambda A_c N \tag{13.2}$$

where A_c is the projected cross-sectional area of a single collector (mineral grain) and N is the total number of collectors in the filter. The term λ is referred to as the filter coefficient. This very simple model describes the filtration phenomenon reasonably well. There is difficulty, however, in relating the filter coefficient to measurable quantities associated with physical-chemical processes.

A careful analysis of the interactions between suspended particles and the collectors indicates that for viscous flow (exemplified by most ground water situations), the interactions result in trapping of particles by sedimentation, attraction, and Brownian diffusion. These processes cause the particles to cross stream lines and contact the mineral grain surface where they are retained. A number of authors have reported good results in estimating the collector efficiency by taking these processes into account. Filtration theory does not permit release of particles back to the pore fluid (entrainment), but Hornberger et al. (1992) demonstrated that use of a deposition coefficient and an entrainment coefficient accurately predicted the transport of particles in sand columns, and further suggested that the ratio of the theoretical coefficients approximated a collector efficiency. Also, simple filtration models were not formulated for situations in which the total particle mass can change, as when bacteria grow. Model terms can be placed in the equations, however, that will adjust the concentration of cells in the filter incrementally to account for changes due to growth and death.

13.3.2 Advection-Dispersion Models

Advection-dispersion (A/D) models represent transport as a combination of flow and dispersion. They are derived from fundamental equations that describe the flow of water in a porous medium. When a nonreactive tracer is added at a point in a flowing fluid, the point tends to spread as it proceeds through the porous medium. The spreading is due to molecular diffusion and to mixing processes that are caused by small-scale turbulence within pores, and by the large number of potential flow paths that exist within a porous medium. A/D-based models describe the movement of conservative (nonreactive) substances well. They were originally formulated to describe the transport of dissolved material, but can be used for colloidal or particulate material in many cases. A simple A/D equation is represented as:

$$\frac{\partial C}{\partial t} = -A\frac{\partial C}{\partial x} + D\frac{\partial^2 C}{\partial x^2} \tag{13.3}$$

where C represents the concentration of bacteria (in this case), x represents the distance through a section of porous medium, and A and D are constants that represent the advection (average linear velocity) and dispersion, respectively. This formulation is for a completely nonreactive substance. Equations for reactive materials often include a term (frequently denoted as R) that represents reactions that can remove dissolved or suspended material from the moving pore fluid:

$$\frac{\partial C}{\partial t} = -A\frac{\partial C}{\partial x} + D\frac{\partial^2 C}{\partial x^2} - R \tag{13.4}$$

where the reaction term, R, can have a variety of forms that describe the processes whereby material is removed from (or enters) the mobile fluid.

In an effort to provide a complete description of bacterial transport, Corapcioglu and Haridas (1984) presented an A/D formulation that included all of the processes thought to control bacterial transport. Effectively, the term R was expanded significantly to take all the processes into account. Terms were included to account for cell growth and cell death (both in suspension and on the particle surface), pore clogging and declogging, gravitational settling, and motility. The formulation of Hornberger et al. (1992) used the basic approach of Corapcioglu and Haridas (1984) and ignored growth, death, differential settling, and motility, but expanded the clogging-declogging concept to include all forms of deposition and entrainment.

A distinct advantage of the A/D approach as opposed to the simple filtration approach is that the former can include considerations of hydrologic and geochemical heterogeneities in expanded formulations.

In modeling the transport of dissolved substances, the concept of retardation is often used. Retardation describes the delay in breakthrough of a transported material due to interaction with the surfaces of the porous medium. Retardation theory was developed on the basis of equilibrium sorption, and a common calculation for the retardation coefficient is based on K_D, the equilibrium isotherm:

$$R_D = 1 + K_D(1-n)\frac{\rho_B}{n} \qquad (13.5)$$

where n = the porosity of the porous medium and ρ_B = the bulk density of the mineral grains. The retardation coefficient is then applied to the advection-dispersion equation as:

$$R_D\frac{\partial C}{\partial t} = -A\frac{\partial C}{\partial x} + D\frac{\partial^2 C}{\partial x^2} \qquad (13.6)$$

Breakthrough curves that display retardation appear similar to those in Figure 13.5. In laboratory column experiments on bacterial transport, breakthrough curves for bacteria often display little retardation, even though substantial retention of cells in the column is observed. Two factors may be responsible for this phenomenon. First, much of the rapid transport of microbes in porous media occurs through preferred flow paths — macropores or zones of higher permeability. Given the assumption that the pore diameters in preferred flow paths are larger than in the surrounding matrix, it could be speculated that rapid (unretarded) breakthrough of bacterial cells represents those organisms that never contact a solid surface. They are neither strained nor adsorbed from suspension. There is support for this contention in the observations of Harvey et al. (1989) and Harvey and Garabedian (1991) that bacteria tended to break through heterogeneous media before chemical tracers, suggesting that rapidly transported bacteria travel through the largest pores, while the tracers are moving through the large and fine pores. Fontes et al. (1991) also observed similar transport in preferred flow paths in laboratory simulations using constructed heterogeneities in sand columns. Thus, a large proportion of the cells that travel through these preferred flow paths are neither retained nor retarded.

Even in cases where bacterial transport is rapid, a phenomenon referred to as tailing is often observed. Tailing refers to the tendency for slow elution of bacterial cells for prolonged periods after the primary breakthrough peak has passed. Tailing is observed in all of the studies of bacterial transport in laboratory columns, but it often

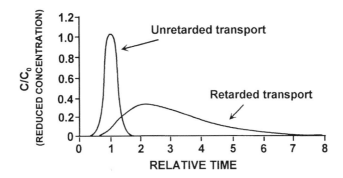

FIGURE 13.5

Retardation of transport in a porous medium. The unretarded case is expected for so-called "conservative" tracers that do not interact with the grains of the medium. The retarded case represents the effect of interaction on the breakthrough. The retarded case is often seen when equilibrium sorption processes are operative, although other mechanisms can produce similar curves. The reduced concentration (C/C_0) is the concentration at the end of the flow path divided by the initial concentration at the beginning of the flow path. The curve for retarded transport was drawn by application of Equation 6 to some idealized data.

does not show in the data because the numbers of cells eluting from the column are several orders of magnitude below that of the primary peak. When data are plotted on semilogarithmic axes, the tailing is plainly visible. In the model formulation reported by Hornberger et al. (1992), tailing could be approximated by appropriate selection of the deposition and entrainment coefficients.

An additional argument for the lack of bacterial retardation (although retention is not explained), deals with the kinetics of sorption and desorption (deposition and entrainment). Jury and Roth (1990) presented explanations for the lack of retardation of dissolved substances in laboratory studies as an artifact of the kinetics of some sorption processes. In short columns with rapid flow rates, the contact time of the sorbing material (in our case, bacterial cells) is too short to permit equilibration. While some sorption is probably occurring, it is too little to be readily observed in the breakthrough curves. Jury and Roth (1990) suggested that longer flow paths and slower flow rates should allow adequate equilibration for retardation to be observed (Figure 13.6). This argument is sometimes invoked to justify the use of the "local equilibrium" assumption for field situations, even when equilibrium is not observed in laboratory simulations. Use of deposition and entrainment coefficients as suggested by Hornberger et al. (1992) seems to provide a reasonable approach to approximation of the retardation behavior observed in laboratory simulations. Such an approach does not imply a kinetic control, and the models have not been tested for large-scale, slow-flow, field systems.

13.4 Summary

Nearly all of the data thus far collected point to the fact that microorganisms are transported through the subsurface and that dispersal of bacteria is an ongoing process. The factors which control dispersal may vary, however, depending on the spatial and temporal scales of the system being examined. For short-distance (tens to hundreds of meters) rapid transport, the hydrogeochemical and cell-associated biological factors probably dominate the dispersal process. The presence and distribution of preferred flow paths (which serve as the conduits for rapid bacterial transport) and the

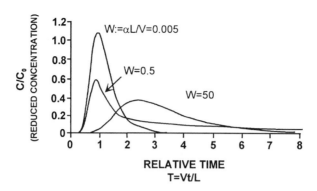

FIGURE 13.6
The effect of increasing the length of an experimental column and simultaneously decreasing the flow rate. The value W is the ratio of the column length (L) to the flow velocity (V) (α is a proportionality constant for consistency in units). Note that for large values of V and small values of L (i.e., W = 0.005) there is little retardation. Retardation is evident, however, when the ratio of L to V is increased. (Redrawn from Jury, W.A. and Roth, K., *Transfer Functions and Solute Movement Through Soil: Theory and Applications*, Birkhauser, Boston, 1990. With permission.)

distribution of geochemical phenomena (such as mineral grain coatings in combination with the cell properties of the bacterial strain of interest) will control the rate and extent of the transport of bacteria.

For regional transport that may occur over distances of kilometers and tens to hundreds or even thousands of year, survival of the microbe in the subsurface is necessarily the most important of all factors. Given adequate time, surviving microbes will permeate the subsurface environment as they have the surface; no data reported thus far contradict this hypothesis. The old adage of "everything is everywhere and the environment selects" is as true for the subsurface as it is for surface and near-surface environments.

Thus, in applied situations, for example where introduced microorganisms are considered as a part of a redemption design for a contaminated aquifer, attention should be paid to the specific cell characteristics that will enhance transport through the contaminant plume in the specific hydrogeological setting containing the contamination. For considerations of long-range, long-term transport, only persistence of viable organisms is of any real importance.

References

Abbott, A., P.R. Rutter, and R.C.W. Berkeley. 1983. The influence of ionic strength, pH and a protein layer on the interaction between *Streptococcus mutans* and glass surfaces. *J. Gen. Microbiol.*, 139:439-45.

Balkwill, D.L., J.K. Fredrickson, and J.M. Thomas. 1989. Vertical and horizontal variation in the physiological diversity of the aerobic chemoheterotrophic bacterial microflora in deep southeast coastal plain subsurface sediments. *Appl. Environ. Microbiol.*, 55:1058-1065.

Barton, J.W. and R.M. Ford. 1995. Determination of effective transport coefficients for bacterial migration in sand columns. *Appl. Environ. Microbiol.*, 61:3329-3325.

Bendinger, B., H.H.M. Rijnaarts, K. Altendorf, and A.J.B. Zehnder. 1993. Physiochemical cell surface and adhesive properties of coryneform bacteria related to the presence and chain length of mycolic acids. *Appl. Environ. Microbiol.*, 59:3973-3977.

Busscher, H.J. and A.H. Weerkamp. 1987. Specific and non-specific interactions in bacterial adhesion to solid substrata. *FEMS Microbiol. Rev.,* 46:165-173.

Corapcioglu, M.Y. and A. Haridas. 1984. Transport and fate of microorganisms in porous media: a theoretical investigation. *J. Hydrol.,* 72:149-169.

Dickson, J.S. and M. Koohmaraie. 1989. Cell surface charge characteristics and their relationship to bacterial attachment to meat surfaces. *Appl. Environ. Microbiol.,* 55:832-836.

Elimelech, M. and C.R. O'Melia. 1990. Kinetics of deposition of colloidal particles in porous media. *Environ. Sci. Technol.,* 24:1528-1536.

Feldner, J., W. Bredt, and I. Kahane. 1983. Influence of cell shape and surface charge on attachment of *Mycoplasma pneumomae* to glass surfaces. *J. Bacteriol.,* 153:1-5.

Fletcher, M. 1977. The effects of culture concentration and age, time, and temperature on bacterial attachment to polystyrene. *Can. J. Microbiol.,* 23:1-6.

Fletcher, M. and G.I. Loeb. 1979. Influence of substratum characteristics on the attachment of a marine pseudomonad to solid surfaces. *Appl. Environ. Microbiol.,* 37:67-72.

Fontes, D.E., A.L. Mills, G.M. Hornberger, and J.S. Herman. 1991. Physical and chemical factors influencing transport of microorganisms through porous media. *Appl. Environ. Microbiol.,* 57:2473-2481.

Gannon, J.T., V.B. Manlial, and M. Alexander. 1991a. Relationship between cell surface properties and transport of bacteria through soil. *Appl. Environ. Microbiol.,* 57:190-193.

Gannon, J.T., Y. Tan, P. Baveye, and M. Alexander. 1991b. Effect of sodium chloride on transport of bacteria in a saturated aquifer material. *Appl. Environ. Microbiol.,* 57:2497-2501.

Gerba, C.P. and G. Bitton. 1984. Microbial pollutants: their survival and transport pattern to groundwater. In: *Groundwater Pollution Microbiology,* G. Bitton and C.P. Gerba, Eds., John Wiley & Sons, New York, p. 65-88.

Ghiorse, W.C. and J.T. Wilson. 1988. Microbial ecology of the terrestrial subsurface. *Adv. Appl. Microbiol.,* 33:107-172.

Ghiorse, W.C. and D.L. Balkwill. 1983. Enumeration and morphological characterization of bacteria indigenous to subsurface environments. *Dev. Ind. Microbiol.,* 24:213-224.

Harden, V.P. and J.O. Harris. 1953. The isoelectric point of bacterial cells. *J. Bacteriol.,* 65:198-202.

Harvey, R.W. 1991. Parameters involved in modeling movement of bacteria in groundwater. In: *Modeling the Environmental Fate of Microorganisms,* C.J. Hurst, Ed., American Society for Microbiology, Washington, D.C., pp. 89-114.

Harvey, R.W. and S.P. Garabedian, 1991. Use of colloid filtration theory in modeling movement of bacteria through a contaminated sand aquifer. *Environ. Sci. Technol.,* 25:178-185.

Harvey, R.W., L. George, R.L. Smith, and D.L. LeBlanc. 1989. Transport of microspheres and indigenous bacteria through a sandy aquifer: results of natural- and forced-gradient tracer experiments. *Environ. Sci. Technol.,* 23:51-56.

Hirsch, P. and E. Rades-Rohkohl. 1990. Microbial colonization of aquifer sediment exposed in a ground water well in northern Germany. *Appl. Environ. Microbiol.,* 56:2963-2966.

Hornberger, G.M., A.L. Mills, and J.S. Herman. 1992. Bacterial transport in porous media: evaluation of a model using laboratory observations. *Water Resour. Res.,* 28:915-923.

Hurst, C.J. 1991. Modeling the Environmental Fate of Microorganisms. American Society for Microbiology. Washington, D.C.

Huysman, F. and W. Verstraete. 1993a. Water-facilitated transport of bacteria in unsaturated soil columns: influence of cell surface hydrophobicity and soil properties. *Soil Biol. Biochem.,* 25:83-90.

Huysman, F. and W. Verstraete. 1993b. Water-facilitated transport of bacteria in unsaturated soil columns: influence of inoculation and irrigation methods. *Soil Biol. Biochem.,* 25:91-97.

Jenneman, G.E., M.J. McInerney, and R.M. Knapp. 1985. Microbial penetration through nutrient-saturated Berea sandstone. *Appl. Environ. Microbiol.,* 50:383-391.

Jury, W.A. and K. Roth. 1990. *Transfer Functions and Solute Movement Through Soil: Theory and Applications,* Birkhauser, Boston, MA.

Kjelleberg, S. and M. Hermansson. 1984. Starvation induced effects on bacterial surface characteristics. *Appl. Environ. Microbiol.,* 48:497-503.

Mackie, R.I. and Bai, R. 1992. Suspended particle size distribution and the performance of deep bed filters. *Water Resour. Res.,* 26:1571-1575.

Mafu, A.A., D. Roy, J. Goulet, and L. Savole. 1991. Characterization of physiochemical forces involved in adhesion of *Listeria monocytogenes* to surfaces. *Appl. Environ. Microbiol.*, 57:1969-1973.

McDowell-Boyer, L.M., J.R. Hunt, and N. Sitar. 1986. Particle transport through porous media. *Water Resour. Res.*, 22:1901-1921.

McEldowney, S. and M. Fletcher. 1986a. Effect of growth conditions and surface characteristics of aquatic bacteria on their attachment to solid surfaces. *J. Gen. Microbiol.*, 132:513-523.

McEldowney, S. and M. Fletcher. 1986b. Variability of the influence of physiochemical factors affecting bacterial adhesion to polystyrene substrata. *Appl. Environ. Microbiol.*, 52:460-465.

Mills, A.L, J.S. Herman, G.M. Hornberger, and T.H. DeJesus. 1994. Effect of solution ionic strength and iron coatings on mineral grains and the sorption of bacterial cells to quartz sand. *Appl. Environ. Microbiol.*, 60:3300-3306.

Mills, A.L. and R. Maubrey. 1981. The effect of mineral composition on bacterial attachment to submerged rock surfaces. *Microb. Ecol.*, 7:315-322.

Morley, L.M. 1995. Effects of preferred flow path on transport of bacteria in laboratory columns. M.S. Thesis, University of Virginia, Charlottesville, 131 pp.

National Research Council. 1993. *In Situ Bioremediation*. National Academy Press, Washington, D.C., 207 pp.

Pitt, W.G., M.O. McBride, A.J. Barton, and R.D. Sagers. 1993. Air-water interface displaces adsorbed bacteria. *Biomaterials,* 14:605-608.

Poletika, N.N., W.A. Jury, and M.V. Yates. 1995. Transport of bromide, simazine, and MS-2 coliphage in a lysimeter containing undisturbed, unsaturated soil. *Water Resour. Res.*, 31:801-810.

Powelson, D.K., J.R. Simpson, and C.P. Gerba. 1990. Virus transport and survival in saturated and unsaturated flow through soil columns. *J. Environ. Qual.*, 19:396-401.

Reynolds, P.J., P. Sharma, G.E. Jenneman, and M.J. McInerney. 1989. Mechanisms of bacterial movement in subsurface materials. *Appl. Environ. Microbiol.*, 55:2280-2286.

Richardson, R.L. 1994. Interactive Effects of Bacterial Cell Surface Properties, Mineral Grain Coatings, and Organic Matter on Processes Affecting Transport of Bacteria in Porous Media. M.S. Thesis. University of Virginia. Charlottesville.

Rijnaarts, H.H.M., W. Norde, E.J. Bouwer, J. Lyklema, and A.J.B. Zehnder. 1993. Bacterial adhesion under static and dynamic conditions. *Appl. Environ. Microbiol.*, 59:3255-3265.

Rosenberg, M. and R.J. Doyle. Eds. 1990. Microbial cell surface hydrophobicity: history, measurement, and significance. *Microbial Cell Surface Hydrophobicity,* American Society for Microbiology, Washington, D.C., p. 1-37.

Russel, W.B., D.A. Saville, and W.R. Showalter. 1989. *Colloids Dispersions*. Cambridge University Press, Cambridge, p 525.

Scholl, M.A., A.L. Mills, J.S. Herman, and G.M. Hornberger. 1990. The influence of mineralogy and solution chemistry on the attachment of bacteria to representative aquifer material. *J. Contam. Hydrol.*, 6:321-336.

Scholl, M.A. and R.W. Harvey. 1992. Laboratory investigations on the role of sediment surface and groundwater chemistry in transport of bacteria through a contaminated sandy aquifer. *Environ. Sci. Technol.*, 26:1410-1417.

Sorongon, M.L., R.A. Bloodgood, and R.P. Burchard. 1991. Hydrophobicity, adhesion, and surface-exposed proteins of gliding bacteria. *Appl. Environ. Microbiol.*, 57:3193-3199.

Stenström, T.A. 1989. Bacterial hydrophobicity, an overall parameter for the measurement of adhesion potential to soil particles. *Appl. Environ. Microbiol.*, 55:142-147.

Story, S.P., P.S. Amy, C.W. Bishop, and F. Colwell. 1995. Bacterial transport in saturated volcanic tuff cores from Rainer Mesa, Nevada Test Site. *Geomicrobiol. J.*, 13:249-264.

Tan, Y., W.J. Bond, and D.M. Griffin. 1992. Transport of bacteria during unsteady unsaturated soil water flow. *Soil Sci. Soc. Am. J.*, 56:1331-1340.

Thomas, J.M. and C.H. Ward. 1989. *In situ* biorestoration of organic contaminants in the subsurface. *Environ. Sci. Technol.*, 23:760-766.

Toran, L. and A.V. Palumbo. 1992. Colloid transport through fractured and unfractured laboratory sand columns. *J. Contam. Hydrol.*, 9:2389-303.

Uyen, H.M., B.C. van der Mei, A.H. Weerkamp, and H.J. Busscher. 1988. Comparison between the adhesion to solid substrata of *Streptococcus mitis* and that of polystyrene particles. *Appl. Environ. Microbiol.,* 54: 837-838.

Vanhaecke, E., J.-P. Remon, M. Moors, F. Racs, D. de Rudder, and A. Van Peteghem. 1990. Kinetics of *Pseudomonas aeruginosa* adhesion to 304 and 316-L stainless steel: role of cell surface hydrophobicity. *Appl. Environ. Microbiol.,* 56:788-795.

Van Loosdrecht, M.C.M., J. Lyklema, W. Norde, G. Schraa, and A.J.B. Zehnder. 1987a. The role of bacterial cell wall hydrophobicity in adhesion. *Appl. Environ. Microbiol.,* 53:1893-1897.

Van Loosdrecht, M.C.M., J. Lyklema, W. Norde, G. Schraa, and A.J.B. Zehnder. 1987b. Electrophoretic mobility and hydrophobicity as a measure to predict the initial steps of bacterial adhesion. *Appl. Environ. Microbiol.,* 53:1898-1901.

Van Loosdrecht, M.C.M., J. Lyklema, W. Norde, and A.J.B. Zehnder. 1989. Bacterial adhesion: A physicochemical approach. *Microb. Ecol.,* 17:1-15.

Wan, J. and J.L. Wilson. 1994. Visualization of the role of the gas-water interface on the fate and transport of colloids in porous media. *Water Resour. Res.,* 30:11-23.

Wan, J., T.K. Tokunaga, and C.-F. Tsang. 1995. Bacterial sedimentation through a porous medium. *Water Resour. Res.,* 31:1627-1636.

Weiss, Th.H., A.L. Mills, J.S. Herman, and G.M. Hornberger. 1995. The effect of cell size and cell shape on bacterial retention in sand columns. *Environ. Sci. Technol.,* 29:1737-1740.

Yao, K.-M., M.T. Habibian, and C.R. O'Melia. 1971. Water and waste water filtration: concepts and applications. *Environ. Sci. Technol.,* 5:1105-1112.

Yates, M.V. and S.R. Yates. 1988. Modeling microbial fate in the subsurface environment. *CRC Crit. Rev. Environ. Control,* 17:307-344.

Zita, A. and M. Hermansson. 1994. Effects of ionic strength on bacterial adhesion and stability of flocs in a wastewater activated sludge system. *Appl. Environ. Microbiol.,* 60:3041-3048.

Section III

Applications

14

Bioremediation

Terry C. Hazen

CONTENTS

KEY WORDS: *biological treatment, biotransformation, biodegradation, intrinsic, bioremediation, engineered bioremediation, bioremoval, bioimmobilization, biomobilization, biostimulation, bioaugmentation.*

14.1   Introduction and Background

Hazardous waste is one of the most pervasive, difficult, and expensive problems confronting human society in the twentieth century. Our lack of knowledge concerning the hazards of some chemicals, lack of disposal regulations, lack of regulation enforcement, expense of treatment processes, and poor understanding of fate and effect of contaminants in the environment has led to a subsurface legacy of environmental contamination that will haunt us for hundreds of years. About 72 million U.S. citizens live within 4 mi of a USEPA Superfund National Priority List Site and 4.4 million people live within 1 mi (Bakst and Devine, 1994). The National Priority List has 1192 of the worst toxic waste sites listed and new sites are being added several times faster than sites are being removed from the list. As of December 1993, 10,624 sites were awaiting review and only 55 sites have been cleaned up and removed from the list since 1980 (Bakst and Devine, 1994). The USEPA estimated in one survey that the U.S. had more than 5 million underground storage tanks. An inspection of just 12,000 tanks revealed that 30% currently were leaking, thus as many as 300,000 to 420,000 tanks may be leaking now or will be leaking in the near future, and will require mandatory action (USEPA, 1988).

The Superfund has identified more than 36,000 toxic waste sites that will require action (Glass et al., 1995). This toxic milieu is dominated by solvents (trichloroethylene, tetrachloroethylene, vinyl chloride, carbon tetrachloride, chloroform, and chlorobenzene), petroleum products (benzene, toluene, xylene, ethyl-benzene, and total petroleum hydrocarbons), polynuclear aromatic hydrocarbons (creosote, polychlorinated biphenyls, anthracene, and benzopyrene), and metals (mercury, selenium, chromium, cadmium, lead, and radionuclides). Groundwater and soil at many waste sites, particularly landfills, contain all of these compounds above the recommended maximum contaminant levels (MCL). The health problems that these toxic waste sites represent are compounded by the increasing lack of clean surface water, thus increasing our reliance on groundwater as source water at the same time that toxic contaminant release into the terrestrial subsurface is increasing (Craun, 1986). Surveys of groundwater have revealed that 36% of more than 5000 community water sources in the U.S. had organic contaminant concentrations above the recommended MCLs for drinking water (Craun, 1986). The cost of remediating just the known contaminated sites in the U.S. is now estimated to exceed $1.7 trillion and this number is constantly being revised upwards. Bioremediation, both above and below ground, promises to be one the premier technologies for restoring the terrestrial subsurface. Indeed, bioremediation may be the only solution to clean up many of our deepest toxic plumes.

Bioremediation is the use of biological processes to return the environment to its original state. More realistically, however, the goal of bioremediation is to make the environment less toxic. In the broadest application sense, bioremediation includes use of enzymes, growth stimulants, bacteria, fungi, or plants to degrade, transform, sequester, mobilize, or contain contaminant organics, inorganics, or metals in soil, water, or air. If we accept the "Doctrine of Infallibility", i.e., there is no compound known to humans that microorganisms cannot degrade (Alexander, 1965), then bioremediation becomes one of the great solutions for our environmental problems. Unfortunately, while the Doctrine of Infallibility may be absolutely true, the rate of biodegradation or transformation of some compounds is so slow as to be negligible for bioremediation purposes. In addition, the conditions (environmental or biological) that allow certain biological reactions to take place may not be obtainable in many environments (Fewson, 1988).

The theoretical possibilities of bioremediation (at least according to the Doctrine of Infallibility), the explosive developments of biotechnology, and our great need for better, faster, cheaper, and safer new methods of remediation have resulted in an exponential proliferation of companies offering bioremediation services. At the April 1995 3rd International Symposium of *In Situ* and On Site Bioremediation in San Diego, CA there were more than 107 companies reporting results of bioremediation field studies. Indeed, bioremediation has already become standard in the tool kits of all full-service remediation companies. Unfortunately, the hope has been largely hype by many companies with "magic bugs" and "magic potions" that will "completely" biodegrade almost anything, anywhere. Because bioremediation is "biotechnology" and has received some dramatic media attention, e.g., the Exxon Valdez spill, it is generally perceived by the public to be new technology that uses special critters that can devour any toxicant that we throw at them. Indeed, the first patent for a genetically engineered organism was granted in the U.S. in 1981 for a bacterium that degrades petroleum (Atlas, 1995).

In actuality, bioremediation as a process is not particularly new or novel. The word bioremediation, however, is fairly new. A search of paper titles, abstracts, and keywords of the major abstracting services (SciSearch™, Biosis™) indicates that the word was not used in the peer-reviewed scientific literature before 1987. Composting, sewage treatment, and certain types of fermentation have been practiced by humans since recorded history began, and all of these are biological treatment methods that could be defined as bioremediation. We find evidence of kitchen middens and compost piles dating from 6000 B.C. (Senior, 1990). The Greeks used circular walled refuse bioreactors as early as 1900 B.C. The first public biological sewage treatment plant began operation more than 100 years ago (1891) in Sussex, U.K. (Senior, 1990). ZoBell (1946) showed that biotreatment of petroleum was possible more than 50 years ago. Since the early 1950s, petroleum land farming (bioremediation of petroleum sludge by tilling soil with fertilizer) has been used as a standard sludge disposal method. The first successful bioremediation using bioaugmentation (addition of biodegrading bacteria) was documented in 1968 for remediation of the oily bilge water in the Queen Mary when it was first brought to Long Beach Harbor for 'parking'. A patent for *in situ* bioremediation of groundwater contaminated with gasoline by stimulating indigenous bacteria via nutrient injection into the terrestrial subsurface was issued to Dick Raymond in 1974 (U.S. Patent 3,846,290). He successfully demonstrated this technology and began commercial applications in 1972 (Raymond et al., 1977). Clearly, bioremediation has been used successfully for more than 50 years and much is understood about where it is applicable, especially for petroleum contaminants. The really new bioremediation applications that have been done in the last 10 years are in the area of solvents, PAHs, PCBs, and metals. Bioremediation has been around for a long time; only its breadth of application in terms of types of contaminants and environments has increased in the last 10 years. This explosive proliferation of new applications and new environments in the last 10 years, especially by companies trying to establish themselves with a proprietary edge, has led to a large number of terms, many of which are highly redundant. Also, the bioremediation field applications that have been reported lack comprehensive field data, especially in the terrestrial subsurface. Though bioremediation has been used at a large number of sites, nearly all applications were completed by companies trying to do the study (1) for clients who usually wanted to remain anonymous, (2) at the least possible cost to the client and the vendor, and (3) protecting the vendor's proprietary edge for their product. This has led to a paucity of peer-reviewed data, misapplication of terminology, and confusion as to what some terms mean. The following is a series of terms and definitions important to bioremediation.

14.1.1   Terminology

Biological Treatment — Any treatment process that involves organisms or their products, e.g., bacteria or enzymes.

Biotransformation — A biological treatment process that involves changing the contaminant, e.g., valence states of metals, chemical structure, etc.

Biodegradation — A biological process of reducing a compound to simpler compounds. May be either complete, e.g., reduction of organic compounds to inorganic compounds, or incomplete, e.g., removal of a single atom from a compound.

Intrinsic Bioremediation — Unmanipulated, unstimulated, nonenhanced biological remediation of an environment; i.e., biologically natural attenuation of contaminants in the environment.

Engineered Bioremediation — Any type of manipulated or stimulated or enhanced biological remediation of an environment.

Biostimulation — The addition of organic or inorganic compounds to cause indigenous organisms to effect remediation of the environment, e.g., fertilizer.

Bioaugmentation — The addition of organisms to effect remediation of the environment, e.g., contaminant-degrading bacteria injected into an aquifer.

Bioventing — Originally defined as slow vapor extraction of unsaturated soils to increase flow of air into the subsurface via vents or directly from the surface, to increase aerobic biodegradation rates. Now defined more broadly to include the slow injection of air into unsaturated soils.

Biosparging — Injection of air or specific gases below ground, usually into saturated sediments (aquifer material) to increase biological rates of remediation.

Bioslurping — This treatment combines soil vapor extraction with removal of light nonaqueous-phase liquid contaminants from the surface of the groundwater table, thereby enhancing biological treatment of the unsaturated zone and the ground water, especially the capillary fringe zone.

Bioreactor — A contained vessel in which biological treatment takes place, e.g., fermentor.

Bioslurry Reactor — Biological treatment of soil by making a thin mixture with water and treating in a contained vessel.

Land Farming — A process of biologically treating uncontained surface soil, usually by aeration of the soil (tilling) and addition of fertilizer or organisms, hence farming.

Prepared Beds — A contained area (lined) above ground where soil can be tilled or variously manipulated to increase biological remediation, i.e., contained land farming.

Biopiles — Above-ground mounds of excavated soils that are biologically treated by addition of moisture, nutrients, air, and/or organisms.

Biofilters — Normally used to refer to treatment of gases by passing through a support material containing organisms, e.g., soil, compost, trickle filter. Sometimes used to refer to treatment of groundwater via passage through a biologically active area in the subsurface.

Composting — Treatment of waste material or contaminated soil by aerobic biodegradation of contaminants in an above-ground, contained, or uncontained environment.

Biocurtain — The process of creating a subsurface area of high biological activity to contain or remediate.

Bioremoval — A biological treatment involving uptake of the contaminant from the environment by an organism or its agent.

Bioimmobilization — A biological treatment process that involves sequestering the contaminant in the environment. There may not be biodegradation of the contaminant.

Biomobilization — A biological treatment process that involves making the contaminant more mobile in the environment. No biodegradation of the contaminant.

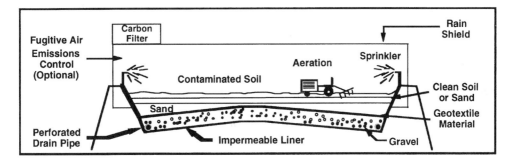

FIGURE 14.1
Prepared bed biotreatment.

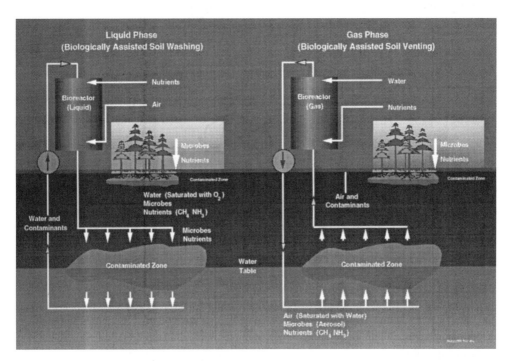

FIGURE 14.2
Bioremediation technologies.

14.1.2 Basic Schematics of Bioremediation Methods

The two schematics (Figures 14.1 and 14.2) present many of the strategies for engineered bioremediation. Figure 14.1 shows a prepared bed facility for bioremediating excavated soil. These systems consist of impermeable liner and leachate collection systems and layers of clean soil or gravel to improve drainage. A sprinkler system is often used to control moisture. Aeration, if necessary, can be as simple as rototilling or as complex as a series of vacuum blowers or compressors. Rain shields and fugitive air emissions control systems may also be necessary depending on the type of contaminants in the soil.

Figure 14.2 shows both *ex situ* and *in situ* technologies that use either liquid or gas, and treat vadose or saturated zone environments. Horizontal wells are shown, but infiltration galleries have been widely used in the past. This figure shows strategies for biofilters, bioremediation, bioventing, biosparging, bioimmobilization, bioreactors, phytoremediation, biomobilization, biocurtain, bioaugmentation, and biostimulation.

14.2 Biostimulation and Bioaugmentation

All engineered bioremediation can be characterized as either biostimulation (i.e., the addition of nutrients), bioaugmentation (i.e., the addition of organisms), or processes that use both. The problems with adding chemical nutrients to sediment and groundwater are fundamentally different from those of adding organisms. Simple infiltration of soil, and subsequently groundwater, is physically quite different in the two processes (Alfoldi, 1988; also see Chapters 13 and 16). Even the smallest bacterium has different adsorption properties than chemicals. For example, clayey soils have very low porosity and may not physically allow bacteria to penetrate. These clays may also bind the microbes that are added by, e.g., cationic bridges involving divalent metals and the net negative charge on the surface of the bacteria and the surface of the clay. In some soils, inorganic chemicals that are injected may precipitate metals, swell clays, change redox potentials and conductivity, thereby having a profound effect on groundwater flow and biogeochemistry of the environment.

Biostimulation is dependent on the indigenous organisms and thus requires that they be present and that the environment be capable of being altered in a way that will have the desired bioremediation effect. In most terrestrial subsurface environments, the indigenous organisms have been exposed to the contaminant for extended periods of time and have adapted, or degradative organisms have even become enriched in relative community abundance through selection. This is not surprising, as many contaminants, especially organic compounds, are naturally occurring or have natural analogs that occur in the environment. Rarely can a terrestrial subsurface environment be found that does not have a number of organisms already present that can degrade or transform most contaminants that might be present. Indeed, even pristine environments have bacteria with an increasing number of plasmids with sediment depth in response to increasing recalcitrance of the organics present (Fredrickson et al., 1988).

Our ability to enhance bioremediation of any environment is directly proportional to knowledge of the biogeochemistry of the site. Finding the limiting conditions for the indigenous organisms to carry out the desired remediation is the most critical step. As with surface environments, the parameters that are usually limiting organisms are required nutrients, inorganic and organic. Of these, the most common are phosphorus, oxygen, nitrogen, and water. In the terrestrial subsurface, water can be limiting but usually is not, even in vadose zone environments. Oxygen is quite often limiting since the contaminant can be used as a carbon and energy source by the organisms and the contaminant concentration greatly exceeds the oxygen input needed by the organisms. Introduction of air, oxygen, or hydrogen peroxide via infiltration galleries, tilling, sparging, or venting have proven to be extremely effective in bioremediating petroleum contaminants and a variety of other organic compounds that are not particularly recalcitrant (Thomas and Ward, 1992). However, if the environment has been anaerobic for extended periods of time, and the contaminant has a high carbon content, it is likely that denitrification has reduced the overall nitrogen content of the environment, making this nutrient limiting. Nitrogen has been successfully introduced into the terrestrial subsurface for biostimulation using ammonia, nitrate, urea, and nitrous oxide (USEPA, 1989). Phosphorus is naturally quite low in concentration in most environments, and in terrestrial subsurface environments even if phosphorus concentrations are high it may be in a mineral form that is biologically unavailable, e.g., apatite. Several inorganic and organic forms of phosphate have been successfully used to biostimulate contaminated environments (USEPA, 1989). In environments where the contaminant is not a good carbon or energy source and other sources of carbon or energy are absent or unavailable, it may be necessary to add an additional source of carbon (Horvath, 1972). An additional source of organic carbon will also be required if the total organic carbon concentration in the environment falls below 1 ppm and the contaminant cleanup levels have still not been met. Methane, methanol, acetate, molasses, sugars, agricultural compost, phenol, and toluene have all been added as secondary carbon supplements to the terrestrial subsurface to stimulate bioremediation (National Research Council, 1993). Even plants, e.g., poplar trees, have been used to biostimulate remediation of subsurface environments (Schnoor et al., 1995). In this latter case the plants act as solar-powered nutrient pumps stimulating rhizosphere microbes to degrade contaminants (Anderson et al., 1993).

Biostimulation strategies will be limited most by our ability to deliver the stimulus to the environment. The permeability of the formation must be sufficient to allow perfusion of the nutrients and oxygen through the formation. The minimum average hydraulic conductivity for a formation is generally considered to be 10^{-4} cm/s (Thomas and Ward, 1989). The stimulants required must be compatible with the environment. For example, hydrogen peroxide is an excellent source of oxygen, but it can cause precipitation of metals in soils and such dense microbial growth around the injection site that all soil pores are plugged. It is also toxic to bacteria at high concentrations >100 ppm (Thomas and Ward, 1989). Ammonia can also be problematic in that it adsorbs rapidly to clays, causes pH changes in poorly buffered environments, and can cause clays to swell, decreasing permeability around the injection point. Many of these problems can be handled at some sites by excavating the soil or pumping the groundwater to the surface and treating it in a bioreactor, prepared bed, land farming, bioslurry reactor, biopile, or composting. In these cases, the permeability can be controlled or manipulated to allow better stimulation of the biotreatment process. It is generally accepted that soil bacteria need a C:N:P ratio or 30:5:1 for unrestricted growth (Paul and Clark, 1989). Stimulation of soil bacteria can generally be achieved when this nutrient ratio is achieved following amendment addition. The actual injection ratio used is usually slightly higher, 100:10:2 (Litchfield, 1993), because these nutrients must be bioavailable, a condition that is much more difficult to measure and control in the terrestrial subsurface. It may also be necessary to remove light nonaqueous-phase liquid (LNAPL) contaminants that are floating on the water table or smearing the capillary fringe zone. This process has been named bioslurping (Keet, 1995). This strategy greatly increases the biostimulation response time by lowering the highest concentration of contaminant the organisms are forced to transform.

Bioaugmentation may provide significant advantages over biostimulation for (1) environments where the indigenous bacteria have not had time to adapt to the contaminant; (2) particularly recalcitrant contaminants that only a very limited number of organisms are capable of transforming or degrading; (3) environments that do not allow a critical biomass to establish and maintain itself; (4) applications where the desired goal is to plug the formation for contaminant containment, e.g., biocurtain; and (5) controlled environments where specific inocula of bacteria with high rates of degradation will greatly enhance the process, e.g., bioreactors, prepared beds, composting, bioslurry reactors, and land farming. Like biostimulation, a major factor affecting the use of bioaugmentation in the terrestrial subsurface is hydraulic conductivity, the 10^{-4} cm/s limit for biostimulation will need to be an order of magnitude higher for bioaugmentation and may need to be higher yet, depending on the size and adherence properties of the organism being applied (Baker and Herson, 1990; also see Chapter 13). Recent studies have shown that less adherent strains of some contaminant-degraders can be produced, allowing better formation penetration (DeFlaun et al., 1994). However, the ability to rapidly clog a formation is a significant advantage of bioaugmentation in applications where containment is a primary goal. The oil industry has been using this strategy to plug fluid loss zones and enhance oil recovery for a number of years (Cusack et al., 1992). Above-ground applications allow manipulation of the permeability in order to overcome most of these problems.

A number of novel organisms have been successfully injected into the subsurface for *in situ* bioremediation of PCBs, chlorinated solvents, PAHs, and creosote (National Research Council, 1993). Surface applications of bioaugmentation for petroleum contaminants in prepared beds and land farming are routine since they help jump-start the bioremediation process.

For controlled and carefully optimized environments, e.g., bioreactors, biofilters, biopiles, and bioslurry reactors, bioaugmentation is preferred since it is easier to control and achieves higher rates of transformation or degradation. Bioaugmentation suffers the dilemma of being indistinguishable from biostimulation in many environments, since nutrients are often injected with the organisms and since dead cells are an excellent source of nutrients for most indigenous organisms. For many applications it is difficult, if not impossible, to determine if the added organisms provided a significant advantage over nutrient stimulation alone. Even some of the best-controlled bioaugmentation field studies, e.g., caisson studies of PCB biodegradation in Hudson River sediment, could not show a significant advantage for bioaugmentation over biostimulation alone (Harkness et al., 1993). Given the high cost of producing the organisms for inoculation, and the delivery problems, bioaugmentation applications will probably remain limited. However, bioaugmentation may have a very significant advantage when genetically engineered microorganisms (GEMs) are used. It is possible that a GEM could be constructed with unique combinations of enzymes to facilitate a sequential biotransformation or biodegradation of a contaminant. This would be particularly helpful for contaminants that are extremely recalcitrant (e.g., PCBs), or under limited conditions (where the contaminant can only be degraded anaerobically, e.g., tetrachloroethylene and carbon tetrachloride). In addition, GEMs could be modified with unique survival or adherence properties that would make them better suited to the environment where they are to be applied.

14.3 Treatability and Modeling

Determining and demonstrating the ability of an environment to be biostimulated or bioaugmented to effect the remediation of a contaminant is critical to the successful application of a bioremediation technique. Treatability studies need to be done to determine the (1) biodegradability or biotransformability of the contaminants, both anaerobically and aerobically; (2) effectiveness of the proposed amendments; (3) compatibility of the proposed amendment additions with the soil and groundwater matrix at the site; (4) abiotic losses, e.g., volatilization, sorption, leaching; and (5) final toxicity of the environmental material (USEPA, 1989). All of these determinations will require collection of a large number of representative field samples (see Chapter 6). Field sample collection is critical to validate the process chosen and may require aseptic and anoxic sampling techniques for terrestrial subsurface sediments (see Chapter 3) and immediate access to laboratory facilities (see Chapter 5). A number of treatability study protocols have been published by USEPA for different types of contaminants and environmental conditions (USEPA, 1989). The reactor, flask, pan, or soil column chosen for the treatability study must mimic, in so far as possible, the environmental conditions of the site. They must also be of sufficient size to minimize "bottle effect" and subsampling errors. The systems used must also provide simultaneous testing of positive (inoculated), negative (sterile), and no-treatment (native microbiota) controls of the treatability protocol (Nelson et al., 1994). The types of systems chosen will be largely dependent on the historical data and process knowledge from the site, as well as the initial characterization. For example, when particularly volatile contaminants (e.g., VC, TCE, PCE) are considered, it may be necessary to spike samples with concentrations high enough to measure during the treatability study. This leads to inherent problems since the original contaminant source may no longer be available, forcing the use of reagent-grade chemicals which may have quite different compositions. In addition, the contaminants at the site may have been "weathered", i.e., exposed to leaching and low-level biodegradation or biotransformation and soil chemical reactions for extended periods of time, resulting in a contaminant chemical composition that is quite unique. These problems have contributed greatly to the unreliability of treatability studies to predict bioremediation in the field (Nelson et al., 1994).

Mass balance is one of the more challenging measurements that needs to be made in a treatability study. The total mass of the contaminant in the soil/ground water must first be determined. For each lab experiment, controls and amendments, one must know where all the contaminant went. This means all daughter products must be measured, as well as adsorption, volatilization, leaching, etc. Measuring only the contaminant of concern can give erroneous results. For example, tetrachloroethylene (PCE) can be reductively dechlorinated to TCE, then DCE, then vinyl chloride (VC); however, there are few situations where VC can be further reduced. If we measured only for PCE in our anaerobic treatability study, we might see its complete disappearance suggesting complete bioremediation of the sample, yet an equal quantity of VC was produced. Since VC is more toxic than PCE this would clearly be an undesirable outcome. This also illustrates the need for toxicity tests at the end of the experiment to determine if the amendments (biological or chemical) produced changes in the environment that really decreased its toxicity.

Measurements of degradation kinetics are also critical to accurately predict the amendments needed, their concentrations, rate of application, and time necessary to reach the cleanup goals for the site. These measurements can be made indirectly by modeling the mass balance through time, by direct measurement of enzyme activities or organism density changes in the reaction vessel, or by real-time measurements of terminal electron acceptor concentrations. This last technique has been increasing in popularity in recent years, especially when less volatile contaminants are considered and the desired effect is complete mineralization. These measurements are made with microrespirometers which can measure very small real-time changes in carbon dioxide, oxygen, and methane in reaction vessels (Pietro et al., 1992). This allows calculations of respiration rates for controls and amendments and, if appropriate controls can be done, a calculation of the rate of carbon dioxide or methane production from the contaminant. Two other techniques can provide excellent kinetics and mass balance information in treatability studies. Unfortunately they are so expensive as to be prohibitive by normal bioremediation practitioners. These are radiolabeled contaminants (pulse and chase) and stable isotopic ratios of carbon. The later measurement has proven useful in both treatablity studies and in field monitoring to determine mineralization rates of petroleum contaminants (Hinchee et al., 1991). Radiolabeled contaminants allow measurement of all daughter products, end products, adsorption, and even incorporation into cell components; unfortunately, these measurements are quite expensive and introduce some error from spiking of the sample with reagent-grade, unweathered contaminants. Caution must also be exercised in choosing the isotope and how the contaminant is labeled. For example, a dodecane (10 carbons) that has only 1 carbon labeled (^{14}C) will only produce 10% radiolabeled carbon dioxide molecules. Radiolabeled mineralization measurements have been used successfully both in treatability studies and in field monitoring for petroleum contaminants and chlorinated solvents (Palumbo et al., 1995).

Modeling of the bioremediation process has become increasingly important in determining the fate and effect of contaminants and predicting the outcome of different amendment scenarios. The models will only be as good as the data they receive from the characterization studies and the treatability studies. However, models can also be used to suggest treatability studies that should be performed from a minimum of characterization data. The simple kinetic models using Monod or Michaelis-Menten functions of 15 years ago are completely inadequate for current bioremediation applications in the terrestrial subsurface. One- and two-dimensional models of aerobic biodegradation of organic contaminants in ground water did not appear until quite recently (Molz et al., 1986; Widdowson et al., 1987). These models incorporated advective and dispersive transport factors coupled with an assumption of microcolonies. Widdowson et al. (1988) later added nitrate respiration as an option to their model. Perhaps the best documented and most widely used model for bioremediation has been the BIOPLUME model (Borden and Bedient, 1986). This model, now in its third version, uses a series of simultaneous equations to simulate growth, decay, and transport of microorganisms, oxygen, and hydrocarbons. The original model was used to simulate PAH biodegradation at a Texas Superfund site (Borden and Bedient, 1986). Rifai et al. (1987) later modified this model (BIOPLUME II) to incorporate the USGS two-dimensional model (Konikow and Bredehoeft, 1978). BIOPLUME II has been used to model biodegradation of aviation fuel at the U.S. Coast Guard Station at Traverse City, Michigan (Rifai et al., 1988) and to characterize benzene biodegradation over three years in another shallow aquifer (Chiang et al., 1989). Most recently, Travis and Rosenberg (1994) used a numerical simulation model to successfully predict aerobic bioremediation of chlorinated solvents in the ground water and vadose zone using methane biostimulation at the U.S. Department of Energy Savannah River Site near Aiken, South Carolina. Their model also used a series of simultaneous equations for microbial growth, nutrient limitations, and contaminant, microbe, and nutrient transport. The model predicted the amount of TCE that was biodegraded during a 14-month, full-scale demonstration, and was validated by 5 other methods (Hazen et al., 1994). As there is an increased emphasis on intrinsic bioremediation as a solution, models like these are becoming increasingly important to understand the terrestrial subsurface "black box" of bioremediation. These types of models, along with rigorous treatability studies, are required for intrinsic bioremediation to be acceptable, particularly as a solution for bioremediation of terrestrial subsurface environments.

TABLE 14.1

Bioremediation Characterization and Monitoring Methods

Measurements	Methods
Biomass	
Viable counts	Plate counts, MPN, enrichments, BIOLOG™
Direct counts	AODC, FITC, DFA
Signature compounds	PLFA, DNA, RNA
Bioactivity and bioremediation	
Daughter products	Cl, CO_2, CH_4, stable isotopic C, reduced contaminants
Intermediary metabolites	Epoxides, reduced contaminants
Signature compounds	PLFA, ribosome probes, BIOLOG™, phosphatase, dehydrogenase, INT, acetylene reductions, recalcitrant contaminants
Electron acceptors	O_2, NO_3, SO_4, (microrespirometer)
Conservative tracers	He, CH_4, Cl, Br
Radiolabeled mineralization	^{14}C, ^3H-labeled contaminants, acetate, thymidine
Sediment	
Nutrients	PO_4, NO_3, NH_4, O_2, total organics, SO_4
Physical/chemical	Porosity, lithology, cationic exchange, redox potential, pH, temperature, moisture, heavy metals
Groundwater	
Nutrients	PO_4, NO_3, NH_4, O_2, SO_4
Physical/chemical	pH, redox potential, conductivity, temperature, heavy metals, flow (vertical and horizontal vectors)
Toxicity	Microtox™, Mutatox™

14.4 Characterization and Monitoring

Characterization and monitoring of bioremediation can be as simple as maintaining a fermentor for above-ground processes like prepared beds, land farming, bioslurry reactors, composting, and bioreactors. Monitoring the terrestrial subsurface is much more difficult, however, due to its sampling problems, poorly defined interfaces, and spatial heterogeneity. For any type of bioremediation, careful consideration and planning must be given to the remediation objectives, sampling, the types of samples, frequency, cost, priority, and background literature for method verification. The microbiology and chemistry may be of less overall importance to the remediation of the site than the hydrology, geology, meteorology, toxicology, and engineering requirements. All of these things must be integrated into the plan for characterization and monitoring of any site. Some methods that have been used for the measurement of various parameters are listed in Table 14.1. For examples of bioremediation test plans see Hazen (1991), Lombard and Hazen (1994), and Nelson et al. (1994).

The type of sample used for monitoring and characterization of sediment or groundwater can have a significant impact on a bioremediation project. Hazen et al. (1991) demonstrated that deep oligotrophic aquifers have dense, attached communities of bacteria that are not reflected in the groundwater from that aquifer. This has serious implications for the *in situ* bioremediation of deep contaminated aquifers, since monitoring of groundwater is the principal method used to characterize and control biodegradation by indigenous bacteria stimulated by nutrient infiltration. Groundwater monitoring may not indicate community or population numbers, or physiological activity of the microbes attached to sediment, and these microbiota may be the most biologically active component of these aquifers. Harvey et al. (1984) and Harvey and George (1987) have shown that shallow, eutrophic, rapidly moving aquifers behave quite differently in that there are no significant differences between groundwater and attached sediment communities. This is reasonable because attachment in such an environment would have no significant advantage, unlike the oligotrophic deep aquifers. Fortunately, most bioremediation applications are shallow and eutrophic due to the nature of the waste mix usually deposited. Enzien et al. (1994) further underscored the need for careful sampling when they showed significant anaerobic reductive dechlorination processes occurring in an aquifer whose bulk groundwater was aerobic (>2 mg/l O_2).

Determining the rate and amount of contaminant that is bioremediated in any environment is one of the most difficult measurements. Many of the problems and measurements discussed above (for mass balance in treatability studies) also apply here. In recent years, bioremediation studies have focused on measurement of biodegradation products, rather than on the organisms, because of the difficulty in measuring organisms. Soil and ground water measurements of microorganisms often require long incubations or long preparation times and the measurements are usually not specific to contaminant-degrading bacteria. Several methods have been used to determine the rate and amount of biodegradation: monitoring of conservative tracers, measurement of by-products of anaerobic activity, intermediary metabolite formation, electron acceptor concentration, stable isotopic ratios of carbon, and ratio of nondegradable to degradable substances. Helium has been used at a number of sites as a conservative tracer since it is nonreactive, nonbiodegradable, and moves like oxygen (National Research Council, 1993). By simultaneously injecting He with O_2 at known concentrations and comparing the subsurface ratios over time, the rates of respiration can be calculated. This technique has also been used to measure rates of injected methane consumption (Hazen, 1991). Bromide has been successfully used as a conservative tracer for liquid injection comparisons with nitrate, sulfate, and dissolved oxygen (National Research Council, 1993). By-products of anaerobic biotransformation in the environment have been used to estimate the amount of biodegradation that has occurred in anaerobic environments (e.g., PCB-containing sediments). These by-products include methane, sulfides, nitrogen gas, and reduced forms of iron and manganese (Harkness et al., 1993).

Measurements of chloride changes have also proven useful in indicating the amount of chlorinated solvents that have been oxidized or reduced (Hazen et al., 1994). Consumption of electron acceptors (O_2, NO_3^-, or SO_4^-) has been used for measuring rates of biodegradation and bioactivity at some bioremediation sites (National Research Council, 1993; Smith et al., 1991). Bioventing remediations of petroleum-contaminated sites rely on stable isotopic ratios of carbon, carbon dioxide production, and oxygen consumption to quantify biodegradation rates in the field (Hinchee et al., 1991; Hoeppel et al., 1991). Mixtures of contaminants, e.g., petroleum hydrocarbons, can have their own internal standard for biodegradation. By comparing concentrations of nonbiodegradable components of the contaminant source with concentrations of degradable components from both virgin and weathered sources, the amount of contaminant degraded can be calculated. These measurements have been used on the Exxon Valdez spill cleanup (Glasser, 1994) and at a number of other petroleum-contaminated sites (Breedveld et al., 1995).

Microbial ecologists have continually struggled with methods for measuring the diversity, number, and activity of organisms in the environment. For bioremediation, this information is needed regarding the contaminant-degrading microbiota. It is also critical to know if there are other organisms present that are important in terms of biogeochemistry and what proportion of the total community the degrading bacteria or other important groups represent.

Plate counts can only provide a measurement of the microbiota that will grow on the media used and under the specific conditions of incubation. Given the large number of possible media and the large number of possible incubations, this leads to an infinite number of possible interpretations. Generally, heterotrophic plate counts have been used to show that bacterial densities in the sediment or groundwater increase in response to biostimulation (Litchfield, 1993). Using contaminant enrichment media and either plates or most probable number (MPN) extinction dilution techniques, the number of contaminant-degrading bacteria can be estimated (National Research Council, 1993). However, there are serious fallacies in the underlying assumptions of many of these assays, e.g., diesel-degrading bacteria may be determined using a minimal medium with a diesel-soaked piece of cotton taped to the top of the petri dish. Are the colonies that are observed utilizing the diesel, or are they merely tolerant to the volatile components of the diesel fuel? In contrast, MPN assays have been used to conservatively measure methanotroph densities in soil and groundwater at chlorinated solvent-contaminated sites by sealing each tube under an air/methane headspace and then scoring as positive only those tubes that are turbid and have produced carbon dioxide and used methane (Fogel et al., 1986). The main drawback with these techniques is that the incubation time for plate count and MPN contaminant-degrading assays is 1 to 8 weeks, thus negating their use for real-time monitoring and control.

A number of direct count assays have been tried on contaminant-degrading bacteria, including direct fluorescent antibody staining (DFA), acridine orange direct counts (AODC), and fluorescein isothiocyanate (FITC) direct counts. The fluorochrome stains only indicate the total numbers of organisms present in the sample, they do not indicate the type of organism or its activity. However, these techniques have been used in bioremediation studies to determine changes in the total numbers of organisms (Litchfield, 1993). Increases in total counts have been found when contaminated environments are biostimulated. DFA shows promise but requires an antibody that is specific to the contaminant-degrading bacteria that are in that environment. The environment must be checked for organisms that may cross-react with the antibody and for contaminant-degrading bacteria that do not react with the antibody. DFA will be most useful in monitoring specific organisms added for bioaugmentation, though it has been used in biostimulation applications (Fliermans et al., 1994). Since the assay time is reduced to hours with these direct techniques, they have significant advantages for real-time monitoring and rapid characterization.

Biological activity at bioremediation sites has been determined in a number of ways: INT activity/dehydrogenase, fatty acid analyses, acetate incorporation into lipids, ^{3}H-thymidine incorporation into DNA, BIOLOG™, phosphatase, and acetylene reduction. The INT test has been used in combination with direct counts because INT-formazan crystals can be detected in the cells. (Cells with INT-formazan crystals are assumed to be actively respiring; the reaction occurs within the electron transport system and is associated with activity.) The assay requires only a 30-min incubation; however, it can only be used in groundwater samples since particles in sediment samples cause too much interference with interpretation of the intracellular crystals. Barbaro et al. (1994) used this technique to measure microbial biostimulation of the Borden Aquifer in Canada.

Phospholipid fatty acid analyses (PLFA) have been used for characterization and monitoring at a number of bioremediation sites (for more information on this topic see Chapter 8). The PLFAs (signature compounds) that an organism has may be unique to that species or even to that strain, or they may be conserved across physiological groups, families, or even a kingdom. Certain groups of fatty acids (cis and trans isomers) may also change in response to the physiological status of the organism. PLFAs have been used at bioremediation sites to provide direct assays for physiological status (cis/trans ratio), total biomass estimates, presence and abundance of particular contaminant-degraders and groups of organisms, e.g., methanotrophs, actinomycetes, and anaerobes (Phelps et al., 1989; Heipieper et al., 1995; Ringelberg et al., 1994). PLFAs would seem to be a panacea for characterization and monitoring of bioremediation; unfortunately, the assays require $-70\,°C$ sample storage, long extraction times, have a fairly high detection limit (10,000 cells), and require expensive instrumentation. This technique merits careful consideration since it is so versatile and is a direct assay technique.

Radiolabeled acetate and thymidine incorporation into lipids and DNA, respectively, have been used at bioremediation sites to provide measurements of total community metabolic and growth responsiveness (Fliermans et al., 1988; Palumbo et al., 1995). These techniques require incubation, extraction, purification, and radiolabeled substrates, making interpretation of results difficult.

The BIOLOG™ assay has also been adapted to determine the activity of bacteria in groundwater and soil samples to contaminants. The assay consists of a 96-well microtiter plate with carbon sources and an electron transport system (ETS) indicator. It can be used to identify isolates and to examine the overall activity of a soil or water sample to a particular substrate. Gorden et al. (1993) adapted the assay to determine activity to different contaminants by using both contaminants and ETS indicator alone and adding contaminants to the plates with substrates to determine co-metabolic activity. The assay provides more rapid screening than other viable count techniques but it suffers from some of the same problems, e.g., incubation conditions and repeatability. It is also difficult to determine if the contaminants are being transformed or tolerated.

Phosphatase and dehydrogenase enzyme assays have also been used to access bioactivity in soil and groundwater during bioremediation of terrestrial subsurface sites. Acid and alkaline phosphatase have been linked to changes in ambient phosphate concentrations and bioactivity at contaminated sites caused by biostimulation (Lanza and Dougherty, 1991). The incubation, extraction, and interference caused by pH differences in samples make results difficult to interpret. Acetylene reduction has been used to indicate nitrogenase activity in a few bioremediation studies; however, the importance of nitrogen fixation at most bioremediations is probably insignificant, unless the site is oligotrophic (Hazen et al., 1994).

Nucleic acid probes provide, at least theoretically, one of the best ways to characterize and monitor organisms in the environment (Amy et al., 1990; Brockman, 1995; Hazen and Jiménez, 1988). Since many contaminants, especially the more recalcitrant ones, are degraded by only a few enzymes, it is possible to produce DNA or even RNA probes that will indicate the amount of that gene in the environment. This tells us if the functional group that can degrade or transform the contaminant is present, and its relative abundance. Since probes have also been developed for species, families, and even kingdoms, this allows soil and groundwater communities to be monitored. Recently, conserved regions in ribosomes have also been found, allowing samples to be probed for the relative abundance of ribosomes and, hence, the bioactivity of the total community (Ruminy et al., 1994). Bowman et al. (1993) demonstrated that probes for methanotrophs indicated their presence in soil at TCE-contaminated sites in South Carolina and Tennessee. Brockman et al. (1995) also showed that methane/air injection at the South Carolina site increased the methanotroph probe signal in sediment near the injection point in the aquifer. The probe signal increase for methanotrophs coincided with increases in the MPN counts for methanotrophs. Thus, sediment can be directly extracted and probed with DNA and RNA for bioremediation characterization and monitoring. As more nucleic acid sequences are described and mapped, it will be possible for us to construct complementary sequences that can be used as probes. These probes can be used to determine the abundance of organisms in an environment that carry a degradative gene of interest, or the number genes themselves. This clearly will allow bioremediation injection strategies to have better control of the process, in terms of effecting the desired changes in the functional group responsible for the bioremediation process.

Unfortunately, nucleic acid probe technology has some serious obstacles to overcome before it becomes practical: (1) for many applications, the direct detection of nucleic acids in soil and ground water requires lysis, extraction, and purification; (2) these processes may not be equally efficient for all microbial and sediment types; (3) soil humics and groundwater pH may interfere with nucleic acid signals; and (4) the number of cells needed for these techniques may exceed those found in natural environments. The extraction and purification steps also add greatly to the cost and analysis time. These problems are not insurmountable but will impede realistic use of nucleic acid probes for bioremediation. Certainly, research in this area needs to be encouraged, especially given the sound theoretical advantages that these techniques provide for bioremediation.

14.5 Intrinsic Bioremediation

Intrinsic bioremediation is developing rapidly as an important alternative for many contaminated environments. This strategy of natural attenuation by thorough characterization, treatability studies, risk assessment, modeling, and verification monitoring of contaminated environments was first proposed by John Wilson of EPA's Kerr Lab in the early 1990s. Wilson organized the first "Symposium on Intrinsic Bioremediation" in August 1994; development and regulatory acceptance has been exponential ever since. Certainly, much of the rapid deployment of intrinsic bioremediation has been due to the crushing financial burden that environmental cleanup represents and our need to use more risk-based cleanup goals for the thousands of new contaminated sites identified every year. Intrinsic bioremediation as a strategy carries with it a burden of proof of: (1) risk to health and the environment and (2) a model that will accurately predict the unengineered bioremediation of the environment. Thus, applications of intrinsic bioremediation have been confined to environments with few risk receptors and containing contaminants with relatively low toxicity, such as petroleum in fairly homogeneous and confined, i.e., predictable, subsurface environments. The EPA reported for 1995 that intrinsic bioremediation was already in use at 29,038 leaking underground petroleum storage tank (LUST) sites in 33 states (Tremblay et al., 1995). This represents 28% of the 103,479 LUST being remediated in 1995 and an increase of more than 100% since 1993. Intrinsic bioremediation has also been implemented at a creosote-contaminated methanogenic aquifer in Florida (Bekins et al., 1993) and in three TCE-contaminated, reducing aquifers (Martin and Imbrigotta, 1994; Wilson et al., 1994; Major et al., 1994).

The coupling of intrinsic bioremediation to engineered bioremediation has been proposed but not yet tried. Nearly all engineered bioremediation projects could substantially reduce costs by stopping the biostimulation or bioaugmentation process early and allowing intrinsic bioremediation to finish the cleanup process. The only projects that would not benefit from such a strategy would be those where immediate risk to health and the environment demanded an emergency response. Intrinsic bioremediation has the same requirements for treatability, modeling, and characterization as engineered bioremediation discussed above. The only difference is that a greater emphasis is put on risk assessment, predictive modeling, and verification monitoring. Once an intrinsic bioremediation project has been started, verification monitoring of the predictive model is initially quite rigorous. Afterwards, if the model holds true, monitoring frequency and numbers of parameters gradually decline until the site is cleaned up.

14.6   Conclusions: Bioremediation — The Hope and the Hoax

The public perception that bioremediation uses the natural cleansing capacity of the environment and can clean up any toxicant, anywhere, is such a gross oversimplification that it really represents: a *Hoax*. The popular press (newspapers and magazines) have added to this hoax in recent years by portraying smiling, hungry little blobs happily eating piles of garbage and drinking toxic chemical cocktails (see Snyder, 1993). This trivializes bioremediation in an effort to provide the public with the "cutest" understanding. The "black box" philosophy of some of the first bioremediation demonstrations may also have helped to manifest this perception, since basically all that was known was that if fertilizer or air was injected the contaminants in the aquifer and soil would go away. This has led to a proliferation of companies selling "magic bugs" or "magic potions". These enterprises are usually based on a patent for a single organism they have discovered that will degrade some toxicant, or some nutrient mix that will stimulate indigenous organisms to degrade toxicants in a jar of dirt or water. These companies are doomed to failure because bioremediation is a strategy that involves a thorough understanding of the entire environment and all the synergistic interactions therein, including a thorough study of risk. There have been and will continue to be bioremediation failures; unfortunately, those are almost never documented.

Bioremediation, both engineered and intrinsic, offers reduced cost, reduced time, improved safety, reduced risk, and better, or at least more environmentally friendly, cleanup of many of our contaminated environments, and herein lies the *Hope*. The advances in the field of bioremediation in the last 10 years have been startling. Contaminants, that 10 years ago were considered to be completely recalcitrant, can now be remediated by proven bioremediation techniques. Indeed, new methods of gaseous nutrient injection, fatty acid analyses, nucleic acid probes, 3D subsurface models, and genetically engineered microbes promise even broader and more efficient applications of bioremediation, especially in terrestrial subsurface environments (Miller and Poindexter, 1994). Regulatory acceptance and rapid implementation of intrinsic bioremediation has made it a primary driver for research and study of contaminated and pristine environments. Only through a better understanding of these environments will we be able to build successful predictive models of the environment for both engineered and intrinsic bioremediation.

References

Alexander, M. 1965. Biodegradation: problems of molecular recalcitrance and microbial fallibility. *Adv. Appl. Microbiol.*, 7:35-80.

Alfoldi, L. 1988. Groundwater microbiology: problems and biological treatment-state-of-the-art report. *Water. Sci. Technol.*, 20:1-31.

Amy, P. S., M. V. Staudaher, and R. J. Seidler. 1990. Comparison of a gene probe with classical methods for detecting 2,4-dichlorophenoxyacetic acid (2,4-D)-biodegrading bacteria in natural waters. *Appl. Environ. Microbiol.*, 95-101.

Anderson, T. A., E. A. Guthrie, and B. T. Walton. Phytoremediation. 1993. *Environ. Sci. Technol.*, 27:2630-2636.

Atlas, R. M. 1995. Bioremediation. *Chem. Eng. News*, 73:32-42.

Baker, K. H. and D. S. Herson. 1990. In situ bioremediation of contaminated aquifers and subsurface soils. *Geomicrobiol. J.*, 8:133-146.

Bakst, J. S. and K. Devine. 1994. Bioremediation: Environmental Regulations and Resulting Market Opportunities, P. E. Flathman., D. E. Jerger, and J. H. Exner, Eds., Lewis Publishers, Boca Raton, FL, pp. 11-48.

Barbaro, S. E., H. J. Albrechtsen, B. K. Jensen, C. I. Mayfield, and J. F. Barker. 1994. Relationships between aquifer properties and microbial populations in the Borden Aquifer. *Geomicrobiol. J.*, 12:203-219.

Bekins, B. A., E. M. Godsy, and D. F. Goerlitz. 1993. Modeling steady-state methanogenic degradation of phenols in groundwater. *J. Contam. Hydrol.*, 14:279-294.

Borden, R. C. and P. B. Bedient. 1986. Transport of dissolved hydrocarbons influenced by reaeration and oxygen limited biodegradation. I. Theoretical development. *Water Resour. Res.*, 22:1973-1982.

Bowman, J. P., L. Jiménez, I. Rosario, T. C. Hazen, and G. S. Sayler. 1993. Characterization of the methanotrophic bacterial community present in a trichloroethylene-contaminated subsurface groundwater site. *Appl. Environ. Microbiol.*, 59:2380-2387.

Breedveld, G. D., G. Olstad, T. Briseid, and A. Hauge. 1995. Nutrient Demand in Bioventing of Fuel Oil Pollution, R. E. Hinchee, R. N. Miller, and P. C. Johnson, Eds., Battelle Press, Columbus, Ohio, pp. 391-399.

Brockman, F. J. 1995. Nucleic-acid-based methods for monitoring the performance of in-situ bioremediation. *Mol. Ecol.*, 4:567-578.

Brockman, F. J., W. Payne, D. J. Workman, A. Soong, S. Manley, and T. C. Hazen. 1995. Effect of gaseous nitrogen and phosphorus injection on *in situ* bioremediation of a trichloroethylene-contaminated site. *J. Hazard. Mater.*, 41:287-298.

Chiang, C. Y., J. P. Salanitro, E. Y. Chai, J. D. Colthart, and C. L. Klein. 1989. Aerobic biodegradation o-benzene, toluene, and xylene in sandy aquifer: data analysis and computer modeling. *Ground Water*, 27:823-834.

Cusack, F., S. Singh, C. McCarthy, J. Grieco, M. Derocco, D. Nguyen, H. Lappinscott, and J. W. Costerton. 1992. Enhanced oil-recovery — 3-dimensional sandpack simulation of ultramicrobacteria resuscitation in reservoir formation. *J. Gen. Microbiol.*, 138:647-655.

Craun, G. F. 1986. *Waterborne Diseases in the United States*, CRC Press, Boca Raton, FL, 295 pp.

DeFlaun, M. F., B. M. Marshall, E. P. Kulle, and S. B. Levy. 1994. Tn5 insertion mutants of *Pseudomonas fluorescens* defective in adhesion to soil and seeds. *Appl. Environ. Microbiol.*, 60:2637-2642.

Enzien, M. V., F. Picardal, T. C. Hazen, R. G. Arnold, and C. B. Fliermans. 1994. Reductive dechlorination of trichloroethylene and tetrachloroethylene under aerobic conditions in a sediment column. *Appl. Environ. Microbiol.*, 60:2200-2205.

Fewson, C. A. 1988. Biodegradation of xenobiotic and other persistent compounds: the causes of recalcitrance. *Trends BIOTECH*, 6:148-153.

Fliermans, C. B., J. M. Dougherty, M. M. Franck, P. C. McKinsey, and T. C. Hazen. 1994. *Immunological Techniques as Tools to Characterize the Subsurface Microbial Community at a Trichloroethylene Contaminated Site*. R. E. Hinchee, D. B. Anderson, F. B. Metting, Jr., and G. D. Sayles, Eds., CRC Press, Boca Raton, FL, pp. 186-203.

Fliermans, C. B., T. J. Phelps, D. Ringelberg, A. T. Mikell, and D. C. White. 1988. Mineralization of trichloroethylene by heterotrophic enrichment cultures. *Appl. Environ. Microbiol.*, 54:1709-1714.

Fogel, M. M., A. R. Taddeo, and S. Fogel. 1986. Biodegradation of chlorinated ethenes by a methane-utilizing mixed culture. *Appl. Environ. Microbiol.*, 51:720-724.

Fredrickson, J. K., R. J. Hicks, S. W. Li, and F. J. Brockman. 1988. Plasmid incidence in bacteria from deep subsurface sediments. *Appl. Environ. Microbiol.*, 54:2916-2923.

Glass, D. J., T. Raphael, R. Valo, and J. Van Eyk. 1995. *International Activities in Bioremediation: Growing Markets and Opportunities*, R. E. Hinchee, J. A. Kittel, and H. J. Reisinger, Eds., Battelle Press, Columbus, Ohio. pp. 11-33.

Glasser, J. A. 1994. *Engineering Approaches Using Bioremediation to Treat Crude Oil-Contaminated Shoreline Following the Exxon Valdez Accident in Alaska*, P. E. Flathman, D. E. Jerger, and J. H. Exner, Eds., Lewis Publishers, Boca Raton, FL, pp. 81-106.

Gorden, R. W., T. C. Hazen, and C. B. Fliermans. 1993. Rapid screening for bacteria capable of degrading toxic organic compounds. *J. Microbiol. Methods*, 18:339-347.

Harkness, M. R., J. B. McDermott, D. A. Abramowicz, J. J. Salvo, W. P. Flanagan, M. L. Stephens, F. J. Mondello, R. J. May, J. H. Lobos, K. M. Carroll, M. J. Brennan, A. A. Bracco, K. M. Fish, G. L. Warner, P. R. Wilson, D. K. Dietrich, D. T. Lin, C. B. Margan, and W. L. Gately. 1993. In situ stimulation of aerobic PCB biodegradation in Hudson River sediments. *Science*, 259:503-507.

Harvey, R. W. and L. H. George. 1987. Growth determination for unattached bacteria in a contaminated aquifer. *Appl. Environ. Microbiol.*, 53:2992-2996.

Harvey, R. W., R. L. Smith, and L. H. George. 1984. Effect of organic contamination upon microbial distributions and heterotrophic uptake in a Cape Cod, Mass. aquifer. *Appl. Environ. Microbiol.*, 48:1197-1202.

Hazen, T. C. 1991. Test Plan for *In Situ* Bioremediation Demonstration of the Savannah River Integrated Demonstration Project, DOE/OTD TTP No.: SR 0566-01. WSRC-RD-91-23. 82 pp. WSRC Information Services, Aiken, SC.

Hazen, T. C., K. H. Lombard, B. B. Looney, M. V. Enzien, J. M. Dougherty, C. B. Fliermans, J. Wear, and C. A. Eddy-Dilek. 1994. Summary of *In Situ* Bioremediation Demonstration (Methane Biostimulation) Via Horizontal Wells at the Savannah River Site Integrated Demonstration Project, *Proc. 33rd Hanford Symp. Health and the Environment: In-Situ Remediation: Scientific Basis for Current and Future Technologies*, G. W. Gee and N. R. Wing, Eds., Battelle Press, Columbus, OH. p 135-150.

Hazen, T. C. and L. Jimenez. 1988. Enumeration and identification of bacteria from environmental samples using nucleic acid probes. *Microbiol. Sci.*, 5:340-343.

Hazen, T. C., L. Jimenez, G. López De Victoria, and C. B. Fliermans. 1991. Comparison of bacteria from deep subsurface sediment and adjacent groundwater. *Microb. Ecol.*, 22:293-304.

Heipieper, H. J., B. Loffeld, H. Keweloh, and J. A. M. Debont. 1995. The cis/trans isomerization of unsaturated fatty-acids in *Pseudomonas putida* s12 — an indicator for environmental stress due to organic compounds. *Chemosphere*, 30:1041-1051.

Hinchee, R. E., D. C. Downey, R. R. Dupont, P. K. Aggarwal, and R. N. Miller. 1991. Enhancing biodegradation of petroleum-hydrocarbons through soil venting. *J. Hazard. Mater.*, 27:315-325.

Hoeppel, R. E., R. E. Hinchee, and M. F. Arthur. 1991. Bioventing soils contaminated with petroleum hydrocarbons. *J. Ind. Microbiol.*, 8:141-146.

Horvath, R. S. 1972. Microbial co-metabolism and the degradation of organic compounds in nature. *Bacteriol. Rev.*, 36:146-155.

Keet, B. A. 1995. *Bioslurping State of the Art*, R. E. Hinchee, J. A. Kittel, and H. J. Reisinger, Eds., Battelle Press, Columbus, Ohio, pp. 329-334.

Konikow, L. F. and J. D. Bredeheoft. 1978. Computer Model of Two Dimensional Solute Transport and Dispersion in Ground Water: Automated Data Processing and Computations. Techniques of Water Resources Investigations, U.S. Geological Survey, Washington, D.C.

Lanza, G. R. and J. M. Dougherty. 1991. Microbial enzyme-activity and biomass relationships in soil ecotoxicology. *Environ. Toxicol. Water Qual.*, 6:165-176.

Litchfield, C. D. 1993. *In Situ Bioremediation: Basis and Practices*, M. A. Levin and M. A. Gealt, Eds., McGraw-Hill, New York, pp. 167-196.

Lombard, K. H. and T. C. Hazen. 1994. Test Plan for the sOILS Facility Demonstration — Petroleum Contaminated Soil Bioremediation Facility (U). WSRC-TR-94-0179. Westinghouse Savannah River Company, Aiken, SC.

Major, D., E. Cox, E. Edwards, and P. W. Hare. 1994. The complete dechlorination of trichloroethene to ethene under natural conditions in a shallow bedrock aquifer located in New York state. In: Proc. EPA Symp. Intrinsic Bioremediation of Ground Water, August 30-September 1, Denver, CO. EPA/540/R-94/515. U.S. Environmental Protection Agency, Washington, D.C.

Martin, M. and T. E. Imbrigiotta. 1994. Contamination of ground water with trichloroethylene at the building 24 site at Picatinny Arsenal, New Jersey. In: Proc. EPA Symp. Intrinsic Bioremediation of Ground Water, August 30-September 1, Denver, CO. EPA/540/R-94/515. U.S. Environmental Protection Agency, Washington, D.C.

Miller, R. V. and J. S. Poindexter. 1994. *Strategies and Mechanisms for Field Research in Environmental Bioremediation*, American Society for Microbiology, Washington, D.C.

Molz, F. J., M. A. Widdowson, and L. D. Benefield. 1986. Simulation of microbial growth dynamics coupled to nutrient and oxygen transport in porous media. *Water Resour. Res.*, 22:1207-1216.

National Research Council. 1993. *In Situ Bioremediation: When Does it Work?* National Academy Press, Washington, D.C.

Nelson, M. J. K., G. Compeau, T. Maziarz, and W. R. Mahaffey. 1994. *Laboratory Treatability Testing for Assessment of Field Applicability*, P. E. Flathman, D. E. Jerger, and J. H. Exner, Eds., Lewis Publishers, Boca Raton, FL, pp. 59-80.

Palumbo, A. V., S. P. Scarborough, S. M. Pfiffner, and T. J. Phelps. 1995. Influence of nitrogen and phosphorus on the in-situ bioremediation of trichloroethylene. *Appl. Biochem. Biotechnol.*, 51:635-647.

Paul, E. A. and F. G. Clark. 1989. *Soil Microbiology and Biochemistry*, Academic Press, San Diego.

Phelps, T. J., D. Ringelberg, D. Hedrick, J. Davis, C. B. Fliermans, and D. C. White. 1989. Microbial biomass and activities associated with subsurface environments contaminated with chlorinated hydrocarbons. *Geomicrobiol. J.*, 6:157-170.

Pietro, M. G., S. Claudia, D. Daniele, B. Grazia, and R. Lorenzo. 1992. Activity and Evolution of mixed microbial culture degrading molinate. *Sci. Total Environ.*, 123:309-323.

Raymond, R. L., V. W. Jamison, and J. O. Hudson. 1977. Beneficial stimulation of bacterial activity in groundwater containing petroleum hydrocarbons. *Am. Inst. Chem. Eng. Symp. Ser.*, 73:390-404.

Rifai, H. S., P. B. Bedient, R. C. Borden, and J. F. Haasbeek. 1987. BIOPLUME II Computer Model of Two-Dimensional Contaminant Transport Under the Influence of Oxygen Limited Biodegradation in Groundwater User's Manual Version 1.0. National Center for Ground Water Research, Rice University, Houston, TX.

Rifai, H. S., P. B. Bedient, J. T. Wilson, K. M. Miller, and J. M. Armstrong. 1988. Biodegradation modeling at an aviation fuel spill. *ASCE J. Environ. Eng.*, 114:1007-1029.

Ringelberg, D. B., G. T. Townsend, K. A. Deweerd, J. M. Suflita, and D. C. White. 1994. Detection of the anaerobic dechlorinating microorganism *Desulfomonile tiedjei* in environmental matrices by its signature lipopolysaccharide branched-long-chain hydroxy fatty acids. *FEMS Microbiol. Ecol.*, 14:9-18.

Ruimy, R., V. Breittmayer, V. Boivin, and R. Christen. 1994. Assessment of the state of activity of individual bacterial-cells by hybridization with a ribosomal-RNA-targeted fluorescently labeled oligonucleotidic probe. *FEMS Microbiol. Ecol.*, 15:207-213.

Senior, E. 1990. *Microbiology of Landfill Sites*, CRC Press, Boca Raton, FL.

Schnoor, J. L., L. A. Licht, S. C. McCutcheon, N. L. Wolfe, and L. H. Carreira. 1995. Phytoremediation of organic and nutrient contaminants. *Environ. Sci. Technol.*, 29:318A-323A.

Smith, R. L., B. L. Howes, and S. P. Garabedian. 1991. In situ measurement of methane oxidation in groundwater by using natural-gradient tracer tests. *Appl. Environ. Microbiol.*, 57:1997-2004.

Snyder, J. D. 1993. Off-the-shelf bugs hungrily gobble our nastiest pollutants. *Smithsonian*, 24:66-76.

Thomas, J. M. and C. H. Ward. 1989. In situ biorestoration of organic contaminants in the subsurface. *Environ. Sci. Technol.*, 23:760-766.

Thomas, J. M. and C. H. Ward. 1992. Subsurface microbial ecology and bioremediation. *J. Hazard. Mater.*, 32:179-194.

Travis, B. J. and N. D. Rosenberg. 1994. Numerical simulations in support of the *in situ* biore-mediation demonstration at Savannah River. Los Alamos National Laboratory Technical Report: LA-UR94-716, Los Alamos, NM, 43 p.

Tremblay, D., D. Tulis, P. Kostecki, and K. Ewald. 1995. Innovation skyrockets at 50,000 LUST sites, EPA study reveal technology use at LUST sites. *Soil Groundwater Cleanup,* December 1995:6-13.

USEPA. 1988. Clean-up of releases from petroleum USTs: selected technologies. Report-530-UST-88-001. U.S. Environmental Protection Agency, Washington, D.C.

USEPA. 1989. Bioremediation of hazardous waste sites workshop. CERI-89-11. U.S. Environ-mental Protection Agency, Washington, D.C.

Widdowson, M. A., F. A. Molz, and L. D. Benefield. 1987. Development and application of a model for simulating microbial growth dynamics coupled to nutrient and oxygen trans-port in porous media. in *Proc. Assoc. Groundwater Scientists and Engineers/Int. Ground Water Model Center,* National Ground Water Association, Dublin, OH, p. 28-51.

Widdowson, M. A., F. J. Molz, and L. D. Benefield. 1988. A numerical transport model for oxygen- and nitrate-based respiration linked to substrate and nutrient availability in porous media. *Water Resour. Res.,* 24:1553-1565.

Wilson, J. T., J. W. Weaver, and D. H. Kampbel. 1994. Intrinsic bioremediation of TCE in ground water a an NPL site in St. Joseph, Michigan. In: *Proc. EPA Symp. Intrinsic Bioremediation of Ground Water,* August 30-September 1, Denver, CO. EPA/540/R-94/515. U.S. Environ-mental Protection Agency, Washington, D.C.

ZoBell, C. E. 1946. Action of microorganisms on hydrocarbons. *Bacteriol. Rev.,* 10:1-49.

15

European Microbiology Related to the Subsurface Disposal of Nuclear Waste

Nick Christofi and Jim C. Philp

CONTENTS

KEY WORDS: *radioactive waste, radionuclides, nuclear waste management, subsurface waste, disposal, biodegradation, biodeterioration, corrosion, gas production, biofilm, chelating/complexing agents, repository.*

Abbreviations

AECL	Atomic Energy of Canada Limited.
AERE	Atomic Energy Research Establishment, Harwell Laboratory, U.K.
BGS	British Geological Survey, Keyworth, U.K.
BNL	Brookhaven National Laboratory.
CEC	Commission of the European Communities.
DOE	Department of the Environment (U.K.).
FLPU	Fluid Processes Research Unit.
FOA	Swedish Defense Research Institute.
GTS	Grimsel Test Site.
IGS	Institute of Geologic Sciences.
IUR	International Union of Radioecologists.
MIND	Microorganisms in Nuclear Waste Disposal.
MRS	Materials Research Society, Pittsburgh, PA.
NAGRA	Nationale Genossenschaft fur die Lagerung Radioaktiver Abfalle (National Co-operative Association for the Storage of Radioactive Waste), Switzerland.
NEA	Nuclear Energy Agency, OECD.
NERC	Natural Environment Research Council (U.K.).
NIREX	Nuclear Industries Radioactive Waste Executive.
NNC	National Nuclear Corporation.
NSARP	Nirex Safety Assessment Program, Harwell, U.K.
OECD	Organisation for Economic Co-operation and Development.
SKB	Swedish Nuclear Fuel and Waste Company.

15.1 Introduction

The processing and safe disposal of radioactive waste material from nuclear power plants and other sources such as research labs are a matter of urgency in both Europe and the rest of the world. Nuclear waste must be confined from human communities because of the detrimental effects of short- and long-lived radioisotopes present.

Radioactive waste can be classified into one of three categories; high-level (HLW), intermediate-level (ILW), and low-level (LLW) (Day et al., 1985; HMSO, 1990; NAGRA, 1985). The disposal of the three categories of waste in subsurface geological environments is now the preferred option, with depths ranging from tens of meters to over one kilometer. Details on the storage of nuclear waste in deep repositories utilizing the multibarrier approach are given by, for example, Atkinson and Marsh (1988).

Safety assessment research programs in Europe and in countries in other continents are considering the possible effects of microorganisms on waste and waste isolation materials disposed of in deep geological environments. Possible microbiological effects include the degradation of organic wastes producing chelating/complexing agents capable of solubilizing radionuclides making them more mobile; corrosion of waste isolation materials such as steels and concretes; alteration of the groundwater chemistry; and microbial gas generation and mobilization of radionuclides after sorption onto motile microorganisms (Christofi, 1991).

This review examines the microbiological studies in Europe since the early 1980s when work began. Initial work concentrated on delineating the potential role of microorganisms in a waste disposal vault based on the known versatility of these ubiquitous organisms. This was followed by an examination of relevant groundwater and sediment samples for the presence of microorganisms and microbial activity. A range of laboratory experiments have been devised to determine the biodegradation of organic wastes and the biodeterioration of waste isolation materials, particularly steel. These laboratory studies have been supplemented with an examination of relevant natural analogue sites.

Deep microbiological studies (West, 1994; West et al., 1992a) did not really take off until 1985 when the U.S. Department of Energy started the research initiative entitled "The Subsurface Science Program" with a specific subprogram entitled "Deep Subsurface Microbiology". The aim was to obtain information on the presence, types, and activity of microorganisms in a range of substrata, and examine their potential use in the detoxification of contaminated subterranean environments used in the disposal of energy and defense waste (Ghiorse and Wobber, 1989; U.S. DOE, 1987). Although not directly linked to nuclear waste disposal, such microbiological studies will provide invaluable information on factors affecting microbial activity in deep geological environments. A number of publications are available documenting preliminary results on the types of microorganisms and activity delineated (Balkwill, 1989; Balkwill et al., 1988; Beloin et al., 1988; Bone and Balkwill, 1988; Francis et al., 1989; Frederickson et al., 1989; Hicks and Fredrickson, 1989; Jones et al., 1989; Kaiser and Bollag, 1990; Madsen and Bollag, 1989; Phelps et al., 1989; Sargent and Fliermans, 1989; Sinclair and Ghiorse, 1989). For deep subsurface microbiology studies, there has been emphasis on the recovery of uncontaminated samples (see, e.g., Russell et al., 1989) although in terms of nuclear waste disposal it is not the origin of the organisms which is important but whether they will be active within the repository environment. Results from the deep sites (up to depths of more than 500 m) reveal a diverse microbial community which is both abundant and metabolically active.

15.2 The Possible Role of Microorganisms in Nuclear Waste Disposal

Disposal options exist for a number of countries worldwide, whether or not they have a nuclear program. In order to assess the safety of disposal options, both individual and joint projects are carried out by the various countries. The aims of the research programs around the world are to determine the long-term safety of the deep disposal of nuclear waste. National research programs are available in Australia, Belgium, Canada, Finland, France, Germany, Italy, Japan, Spain, Sweden, Switzerland, U.S. and Russia (including East European "partners" and Cuba). The list of countries involved is not exhaustive but represents the major players involved. A review of the organizations and programs in each country responsible for waste management is given in Christofi (1991), NNC (1987), and Rosevear (1991).

In this review, the role of microorganisms on the integrity of ILW and LLW disposal options (Bachofen, 1990; McCabe, 1987) will be primarily examined although some of the work done with respect to HLW is relevant to the disposal of other waste forms. Subsurface microbiology in Europe in relation to groundwater pollution by agricultural and other pollutants is important (Boesten et al., 1991; Chilton and West, 1992; Helweg, 1990, 1992; Kinniburgh et al., 1994; Leistra and Boesten, 1989; Lewis et al., 1992) but will not be considered in this review.

15.2.1   Waste Management and Microorganisms

Low-level waste constitutes by far the greatest bulk of radioactive waste. It consists of low concentrations of radionuclides produced by various industries, hospitals, universities, and other research organizations. A multitude of radionuclides, heavy metals, contaminated equipment, and organic material are present. The bulk is composed of discarded contaminated clothing, paper, scintillation fluids, animal carcasses, and other substances. This material has been considered suitable for shallow land disposal or direct discharge to the environment in liquid or gaseous forms. There are containment and contamination problems with such disposal options. Current containment plans include its packaging and emplacement in engineered repositories (HMSO, 1990; NAGRA, 1985).

The disposal option in many countries for ILW and LLW is in deep underground sites, although shallow, engineered trenches have been used in the past and were envisaged for future disposal of some types of waste. United Kingdom Nirex Ltd (Nirex) is responsible for providing and operating a repository for the disposal deep underground of ILW and LLW in accordance with government policy. Nirex aims to open the first permanent repository for ILW by the year 2010. This may be within formations more than 600 m below Sellafield in Cumbria, England where most of the waste is produced. The safety of this disposal site will be assessed by constructing (subject to planning permission) an adjacent deep underground research laboratory (Rock Characterization Facility) to study in detail the geology and groundwater movement (NIREX, 1993). The construction of underground laboratories for safety assessment research purposes is (or will be) taking place in Belgium (Mol), France, Germany, Sweden (Aspo), and Switzerland (Grimsel) (Christofi, 1991; Pedersen, 1993). The overall aim is to isolate radionuclides from the human environment (biosphere) until such time that they have decayed sufficiently to be "nontoxic". To fulfil this aim research has been initiated to assess the suitability of various deep geological environments to contain these radioactive wastes using a multibarrier approach which includes the encapsulated waste form, backfill/buffer materials, and the geological formation.

The integrity of the barriers used can be compromised by a multitude of physical and chemical factors. It was not until around 1982 that serious consideration was given to the potential role of biological factors on radioactive waste isolation in deep environments (West et al., 1982a). An important biological factor is the natural presence or introduction of microorganisms to the repository environment and the possible role of such organisms in enhancing the release and migration of radionuclides to the biosphere.

The microbial effect, however, can be considered both detrimental and beneficial. Detrimental effects include the corrosion/degradation of the metal containers or immobilizing matrix (e.g., cement, bitumen) thus allowing for the release of radionuclides. In terms of metal corrosion, attack can be from within a canister containing a microbially contaminated waste, or attack can be external (e.g., originating from a microbial population present in groundwater and/or the backfill/buffer material). Microbial attack of materials can be exacerbated by nonbiological (physical and/or chemical) attack. The extent of microbial attack will depend on the activity of the given microbial contaminants in the repository environment (McGahan, 1987; Rushbrook, 1985).

Backfill materials are chosen to prevent the ingress of groundwater to the waste and for their ability to retard the migration of any radionuclides released (NIREX, 1993). Any microbial degradation of backfill materials such as clays (e.g., bentonite), bitumen, and concrete may enhance radionuclide migration. In addition, degradation of organic waste or backfill material may lead to the production of complexing/chelating agents, enhancing the movement of radionuclides away from the near-field. Particles such as microorganisms can be transported through porous media (McDowell-Boyer et al., 1986) (see Chapter 13). A number of microorganisms are motile, and given sufficiently sized pore flow paths within the near- and far-fields of a repository, microbes can transport radionuclides sorbed onto their cell surfaces or taken up internally (Goyette, 1987). Microbial activity can alter the groundwater physicochemical environment, affecting the speciation of released radionuclides through pH and Eh changes. Microbial gas production through metabolic processes can also have detrimental effects.

Beneficial effects of microbial activity include the generation of anaerobiosis, which may enhance waste stability, retarding radionuclide movement. Microbial growth and extracellular polysaccharide production can lead to blocking of pores (Christofi, 1991; Lappin-Scott and Costerton, 1990), retarding radionuclide movement (see Chapter 17). The growth of microbes on surfaces along groundwater flow paths (biofilm formation) can also retard radionuclide movement via sorption processes. In addition, small concentrations of organic substances present in groundwaters can be assimilated by microorganisms, making them unavailable in complexing reactions. This is particularly important in deep groundwaters which are normally microbe free but which may become contaminated during excavation processes.

15.2.2 Factors Affecting Microbial Activity in a Waste Repository

Microbial activity in any environment is governed by a range of physical and chemical environmental conditions. These include the presence and availability of water, the quality and quantity of nutrient and energy sources, pH, redox conditions, and temperature. In relation to radioactive waste disposal in deep environments, factors such as hydrostatic/lithostatic pressure and radiation will have an effect. Radioactivity associated with ILW and LLW should not restrict microbial growth if other growth factors are adequate (West, 1995).

Nutritional diversity in the microbial world is virtually limitless but chemoorganotrophic bacteria may not be able to proliferate without adequate organic carbon. This point is highlighted because of the fact that deep geological formations (particularly crystalline rock formations) usually contain low concentrations of, or recalcitrant forms of, organic carbon (Christofi and Philp, 1991). In addition, most of the organic material is of high molecular weight (humics and fulvics) and difficult to assimilate. It is theoretically possible, however, for some microorganisms to proliferate in an organic-poor habitat utilizing inorganic compounds for energy and building up their organic cellular carbon (growth) from the assimilation (fixation) of carbon dioxide. Effective growth, however, is also a function of the availability of nutrients such as nitrogen, phosphorus, and sulfur and smaller quantities of other elements and electron acceptors.

Apart from the appropriate concentration of nutrient and energy substrates, other factors inherent to a radioactive waste vault will affect the survival and activity of microorganisms. Microorganisms use a variety of electron acceptors including O_2, NO_3, Fe, Mn, SO_4, and CO_2. The use of alternative electron acceptors is controlled by the prevailing redox (Eh) conditions of a repository. As the repository environment changes from aerobic to anaerobic after waste emplacement and backfilling, there may be a successional shift in microorganisms from aerobic to anaerobic species affecting the chemistry of the pore water. The rapidity of this scenario will depend on the rate of resaturation of the repository, and Sharland and Tasker (1986) have made calculations on the duration of the aeration period for a hypothetical nuclear waste repository. A similar but more rapid change is observed in landfills. The production of low molecular weight organic acids, principally acetate, under anaerobic conditions may solubilize radionuclides, possibly affecting their mobility. Although ground waters exist which have a high dissolved oxygen concentration (maintained over long periods of time) (Winograd and Robertson, 1982), there is no doubt that microbial activity such that would have an effect on radionuclide containment would deplete any O_2 trapped during waste emplacement, or naturally present. It must also be considered that the amount of O_2 present in groundwaters would depend on the type of geological formation to be used for waste disposal. It is unlikely that clay-type formations would have very much O_2 associated with groundwater.

Oxidizing conditions may evolve in repository environments due to radiolytic processes caused by radionuclide release. Free radicals such as superoxides and other oxidizing species such as hydrogen peroxide can have a killing effect on anaerobic microorganisms which do not have the protective superoxide dismutase and catalase enzymes of the aerobic bacteria. Oxidizing species produced by radiolysis can also enhance the breakdown (oxidation) of organic compounds.

In terms of ILW and LLW, it appears that the major constraint to microbial growth in the near-field will be the high pH of cement or concrete pore water, either when waste is immobilized in cement-based matrices (Glasser et al., 1988, 1989), or when used as a backfill or engineered barrier (Atkinson et al., 1988; Atkinson and Guppy, 1988; Atkinson and Marsh, 1988). Initially, the pH of the cement or concrete will be above pH 13. In the long term, the pH may often be maintained around pH 10.5, but possibly over pH 12 (Atkinson and Guppy, 1988). It is also appreciated that voids exist in cement matrices where the pH will be lower than expected. Immobilizing waste such as cellulose in cement may prolong the equilibration of waste pH with the cement pH. This will have important implications to waste degradation either through alkaline hydrolysis or through microorganisms. Microorganisms may survive above pH 13, but are incapable of growth (Krulwich and Guffanti, 1983) in such alkaline conditions. However, they are capable of growth at a pH value of 10.5, expected during the life of a waste repository. There are microbial species reported capable of growth at values between 10.5 and 13, with most favoring values lower than pH 12.

15.2.3   Origin of Energy and Nutrients

A natural source of growth substances for microorganisms will be the groundwater. The quality and quantity of nutrients present will rely on the geological formation. It is generally accepted that lower concentrations of organic compounds are present in igneous and metamorphic rock formations than in sedimentary rock formations (Christofi and Philp, 1991). Some data suggest, however, that this may not be the case (Mayfield and Barker, 1982).

Groundwater may be supplemented by nutrients introduced during excavation processes such as blasting (nitrates), influx of soil surface nutrients, and organics from drilling/excavation fluids. The quantities will unlikely be sufficient to have a significant effect on microbial growth in a repository. Most nutrients (organic) will be introduced to a repository as the waste itself (particularly for LLW) or be present in backfill/buffer materials such as clay and bitumen. It seems that natural clays and shales contain sufficient organic and inorganic (N, P, and S) nutrients to support significant microbial activity (Mayfield and Barker, 1982). Bitumen is primarily organic material of high molecular weight which is difficult to degrade. Microbial proliferation would rely on significant bitumen mineralization under potentially anaerobic conditions. Although bitumen has been shown to be degraded, it is a refractive material which is difficult to break down anaerobically.

Some inorganic nutrients and energy sources could arise from metal corrosion products. These include Fe(II) and Fe(III) and H_2. Cement and concrete immobilization matrices will also contain some microbial growth substances.

ILW and LLW will contain much organic material that can be utilized by microorganisms. The organics are both labile and refractive. Waste embedded in concrete/cement providing a high pH environment can be hydrolyzed (alkaline hydrolysis) (Rosevear, 1986). Cellulosic material is prone to such attack, leading to assimilable products.

15.2.4   Types of Microorganisms Likely to Be Active in a Deep Repository

Until the early 1980s, there was generally a paucity of information concerning the presence of microorganisms in deep geological environments. Most of the work carried out dealt with the enumeration and identification of specific groups of microorganisms in groundwaters used for potable water supply and those associated with oil deposits (West et al., 1982b; and Chapter 4). One of the problems with deep subsurface microbiology work was the inability, until recently, to obtain uncontaminated material from depth. It is now accepted that microorganisms are naturally present in deep geological formations and much work has been done through the Subsurface Science Program, initiated by the U.S. Department of Energy in 1985.

The types of microorganisms found at depth include both prokaryotic and eukaryotic types. In one study of a deep aquifer at a site in South Carolina, a range of bacteria, fungi, protozoa, and algae (and possibly viruses associated with these) were isolated (Balkwill, 1989). Bacteria were the most numerous, representing around 95% of the microorganisms present. In this study, highest populations were associated with transmissive aquifer sediments with lower numbers in nontransmissive confining layers.

The general consensus is that bacteria will have the most prominent effect on waste repositories. Other organisms such as algae require light for growth and reproduction, a factor which will be absent underground. Other eukaryotic microorganisms will be restricted by lack of light, organic matter, and aerobic conditions.

15.3   European Nuclear Waste Microbiology Research

15.3.1   Microbiology and Microbial Activity in Subsurface Environments

Deep subsurface microbiology studies in Europe have been funded by national agencies such as Nirex and the Department of the Environment (DOE) in the U.K.,

NAGRA (National Co-operative for the Storage of Radioactive Waste) in Switzerland, and SKB (Swedish Nuclear Fuel and Waste Co.) in Sweden, and by the European Union through, for example, programs such as MIRAGE (Migration of Radionuclides in the Geosphere) (Come, 1985). Studies on relevant sites in Europe have been carried out using material such as groundwater and sediment from boreholes and existing or specially excavated mines (Christofi, 1991; Christofi et al., 1984a, 1985, 1984b, 1983; Ekendahl et al., 1994; Pedersen, 1987, 1989; Pedersen and Ekendahl, 1990, 1992a, 1992b; Pedersen et al., 1991; Rushbrook, 1987; West et al., 1985b, 1985c, 1989). Although quite a large body of knowledge has accumulated with regard to microbial impact on radioactive waste repositories, much of the research carried out for waste management authorities is commercial-in-confidence and not available in the open literature.

Belgium is primarily involved in the study of clay geological formations for the disposal of HLW. An excavated gallery in Boom clay at the underground research laboratory (Hades) near Mol was sampled for microorganisms (Christofi et al., 1985). Microorganisms could not be isolated from recovered clay cores, although there was evidence of sulfate-reducing bacteria in enrichment cultures. It was not known whether these were contaminants from water percolating down the shaft from the water-bearing sand layers above the clay formation, but the sand layers contained an estimated aerobic heterotrophic bacterial population of 1.2×10^3 CFU (Colony Forming Units) ml^{-1} (Christofi et al., 1985; West et al., 1985c).

A number of researchers in Germany are examining the microbiology of underground environments, and inherent problems in doing this, but not necessarily with respect to radioactive waste disposal (Hirsch and Rades-Rohkohl, 1988; Kolbel-Boelke et al., 1988a, 1988b; Marxsen, 1988). Recently, Kutzner and Herrman (1990) carried out a review of the possible effects of microorganisms on radionuclide transport. The microbiology of relevant sites in Germany was studied as part of the MIRAGE program. Ground water from the iron-ore mine at Konrad and samples from a salt mine at Asse were tested for the presence of microorganisms and microbial activity (Christofi et al., 1985; West et al., 1985c). The numbers in Konrad groundwaters were low, ranging from undetectable to 7.5×10^2 CFU ml^{-1}. Microorganisms could not be isolated from the Asse salt mine samples. The ground waters could not support a significant microbial activity.

In Italy, epifluorescence direct counting techniques have been used on uncontaminated Plio-Pleistocene clay samples (Brondi, 1987). Microorganisms were isolated from all samples examined, with populations ranging between 2.05 to 6.26×10^7 organisms per gram of sample. Sulfate-reducing bacteria were also recovered from these samples. Future studies were aimed at assessing *in situ* activity (Come and Chapman, 1986).

Sweden has been involved in microbiological studies relating to HLW, ILW, and LLW (Roffey, 1990). Studies of relevant subsurface sites were carried out by Christofi et al. (1985) with respect to HLW disposal. During these studies, the microbiology and microbial activity in groundwaters recovered from the Stripa iron-ore mine (granite) were investigated. The Stripa mine, located around 250 km west of Stockholm, is being used as an international test facility for deep *in situ* experiments related to the disposal of radioactive waste. The mine is not intended to be used as a final disposal site. The groundwaters of the Stripa granite contained a significant number of microorganisms, up to 1.3×10^5 CFU ml^{-1}, including corrosion-important sulfate-reducing bacteria. The groundwaters were shown to be nutrient limited and sustained microbial activity required external nutrient inputs (Christofi et al., 1985). More recently, studies using granitic groundwaters have been undertaken (Pedersen, 1987, 1989, 1990) revealing similar populations. The mean populations estimated by epifluorescence counting techniques were 3.0×10^5 CFU ml^{-1} and recovered from borehole depths ranging from 129 to 860 m. Enrichment culture experiments suggested the presence of methanogenic and sulfate-reducing bacteria (Pedersen, 1990). The study also provided evidence for the presence of microbial biofilms on rock surfaces and radiographic techniques showed that the bulk water-phase population was inactive and possibly originated from rock biofilms (through detachment). An examination of autotrophic and heterotrophic activity by attached and unattached bacterial populations in deep fractured crystalline bedrock was studied (Ekendahl and Pedersen, 1994; Pedersen and Ekendahl, 1992b). Bacteria were able to grow slowly under anoxic, high pH (9–10), and low redox conditions utilizing CO_2 and lactate. The calculated quotient for CO_2/organic carbon utilization was between 0.07 and 0.25, indicating that autotrophy could not support the observed growth rates and that heterotrophy dominated (Ekendahl and Pedersen, 1994).

The study of microbiology is an important part of the Swiss radioactive waste program (Knecht et al., 1990; NAGRA, 1985). The microbiology program in Switzerland was started after that in the U.K., with general literature reviews on the possible effects of microorganisms (Bachofen and Luscher, 1984; McKinley et al., 1985; West and McKinley, 1984) and their specific role on the degradation of backfill/buffer materials such as bitumen (Bachofen et al., 1984). Reviews have also been commissioned by NAGRA on the role of microbes in complexant formation and radionuclide solubilization (Birch and Bachofen, 1990a,b). Initial research involved the examination of mines and materials of relevance to waste disposal (Christofi and Milner, 1990; Christofi and Philp, 1991; Christofi et al., 1985; West and Grogan, 1986). All sites examined contained microorganisms and the prominent types varied with site. Groundwaters were generally organic-carbon limited and there was also low concentration of nitrogen and phosphorus. Hence microbial activity was significantly affected by external nutrient input. Autotrophic metabolism was shown using groundwater samples supplemented with $NaH^{14}CO_3$, and it was suggested that organic carbon compounds could be increased through CO_2 fixation.

The microbial examination of the groundwaters of the granite experimental mine at Grimsel (Grimsel Test Site — GTS) has revealed interesting information on the role of microorganisms in radionuclide migration experiments. Microorganisms were detected in groundwater within isolated fractures in granite used to follow the migration of radionuclides (Christofi and Milner, 1990). It was not known whether these microbes were naturally present within fractures or whether they were introduced when the migration experiment was set up. If the latter was true, then it casts doubt on the validity of migration models based on these results. It is likely that natural deep granitic environments may be microbe free (Christofi et al., 1983) and microbial effects on radionuclide transport processes may be nonexistent or negligible. Even if microbial contamination does occur there is no evidence to suggest that microbial populations would be as high as those found during the GTS microbiology study. There was indeed large numbers of bacteria within the flow path isolated by the borehole sampling equipment (Packers). In addition a significant microbial biofilm could have developed along the fracture surfaces leading to radionuclide sorption. This would have underestimated rates of radionuclide movement.

Information on the microbiology of relevant geological environments by other European countries was not revealed during literature surveys, although in studies on the chemical environment of a waste repository by Finnish workers, the role of microorganisms was considered (Snellman, 1984; Snellman and Uotila, 1984).

Tables 15.1 and 15.2 present data on the groundwater chemistry and populations of microorganisms determined for some sites in Europe. It is seen that significant populations are present in some waters, with undetectable levels in others, particularly the high-salinity environment of Asse, Germany. These populations may be contaminant surface species introduced through excavation and drilling. However, whether autochthonous or allochthonous, such populations need to be sustained by ground-water nutrients or those present in wastes or waste isolation materials to be of concern for repository stability. Techniques used to detect bacterial presence are cultural in the main, however attempts to detect specific groups of microorganisms using molecular characterization and 16S rRNA PCR methodology have been carried out (Ekendahl et al., 1994). In our laboratory we have developed rapid PCR techniques to detect bacteria of the genus *Rhodococcus*. PCR methodology fails to ascertain whether microorganisms detected are viable or indeed native to the subsurface environment and work currently in progress in our laboratory is attempting to solve the problem utilizing fluorescent *in situ* hybridization methodology (FISH). Research on microbial activity by resident populations has shown that ground waters are nutrient limited. Organic carbon, although present, would not support a significant microbial activity. The TOC presented in Table 15.1 does not differentiate between labile and refractive forms, but in experiments microbial activity could be stimulated by an exogenous input of organic carbon (West et al., 1986a). This organic carbon could originate from waste material or waste isolation matrices but no adequate experimentation has been carried out to verify or discount this. Philp et al. (1984) and Rushbrook (1987) examined the microbiology of Oxford clay formations. Significant populations and activity were obtained in laboratory studies using the clay material.

A range of microorganisms have been isolated from various sites. These have been utilized in experiments to determine the effects of extreme conditions of pH, temperature, pressure, and radiation. West and Arme (1985) studied the effect of temperature, pressure, and radiation on isolated sulfate-reducing and sulfur-oxidizing bacteria. They used static laboratory systems under an ambient temperature of up to 80°C and pressure of 250 bar. Radiation doses offered to the bacteria varied up to single doses of 105 rad gamma irradiation over 4 h. They found that sulfate-reducing bacteria were more tolerant than sulfur-oxidizers to pressure, temperature, and radiation, and concluded that gradual acclimatization to extreme conditions may lead to adaptation and success of the microbes to the repository conditions. Similar work is described in other publications (West et al., 1987a, 1985a).

Activity experiments utilizing single isolated species are considered inappropriate in generating data for mathematical modeling purposes because any microbial effect *in situ* will inevitably be impacted by interactions. Complete mineralization of some organic compounds results from the co-metabolic activities of microbial consortia. Experiments to date, utilizing microbial systems, have been unrealistic in that microbial inocula have been utilized to assess degradation of primarily cellulosic materials in neutral and high pH systems (Christofi, 1991; Colasanti et al., 1989; Ewart et al., 1989). Indeed, without microbial additions the production of CH_4 is difficult to show (Biddle and Moreton, 1992; Rees, 1989).

15.3.2   Biodegradation of Waste

Nuclear waste, particularly ILW and LLW, contains a variety of organic materials including cellulose, polyvinylchloride, and neoprene and hypalon rubbers which will be subject to radiolytic, chemical, and possibly microbial attack. Greenfield et al. (1991) reviewed the possibility of such attack at high pH and the generation of soluble organic complexing agents affecting radionuclide migration. It is accepted that alkaline hydrolysis will generate a range of low molecular weight compounds, but experiments have not been attempted which show that microbes are involved in their

TABLE 15.1

Chemical Data for European Groundwaters Examined for Microorganisms and Microbial Activity

Groundwater	Depth (m)	pH	HCO_3^- (mg/l)	Cl^- (mg/l)	SO_4^{2-} (mg/l)	SiO_2 (mg/l)	Na^+ (mg/l)	K^+ (mg/l)	Ca^{2+} (mg/l)	Mg^{2+} (mg/l)	TIN[a] (mg/l)	PO_4^{3-} (mg/l)	Total Fe (mg/l)	TOC[b] (mg/l)
Harwell, U.K.	165	8.75	31	30	61	0.8	18	3.5	35	3.1	22.4	0.05	0.05	ND[c]
Altnabreac, U.K.	259–281	7.8	124	20	4	28	31	0.7	26	1.81	0.4	0.03	0.18	1.0
Mol, Belgium	190	12.7	7	2,800	18	2	350	140	1700	0.05	8	0.5	0.1	29
Konrad, Germany	1100	6	5	132,000	820	6	64,400	300	12,600	2640	71	1.2	47	50
Asse, Germany	750	ND[c]	203	307,700	25,980	2.8	84,450	140	31.4	95,060	69	0.1	4.84	1115
Stripa, Sweden	340	8.84	86	40	4.9	11.8	46	0.59	14	0.23	0.23	0.03	0.006	1.1

[a] TIN = total inorganic nitrogen (NH_4^+, NO_2^-, NO_3^-).

[b] TOC = total organic carbon.

[c] ND = not determined/not detected.

TABLE 15.2

Location, Geology, and Numbers of Bacteria of European Sites Examined for Microorganisms

Type	Location	Geology	Depth[a]	Bacterial Count[b]
Boreholes	Harwell, U.K.	Oxford clay	165–331	8.6×10^3 to 3.5×10^5
(groundwater)	Altnabreac, U.K.	Granite	10–281	9.4×10^5
	Aspo, Sweden	Granite	129–1078	20 to 1.5×10^4 [d]
	Stripa, Sweden	Granite	340	3×10^1 to 1.2×10^5
			799–1240	20 to 90^e
Used/disused mines	Cumbria, U.K.	Anhydrite	ND	ND
	Derbyshire, U.K.	Limestone	61	ND
	Cornwall, U.K.	Granite	80–800	2.5×10^2 to 2.5×10^4
	Moll, Belgium	Boom clay	190–223	0 to 1.2×10^3
	Asse, Germany	Salt	750	Not detected
	Konrad, Germany	Iron ore	1100	0 to 7.5×10^2
	Grimsel, Switzerland	Granite	ND[c]	9.5×10^1 to 9×10^4
	Felsenau, Switzerland	Gypsum/anhydrite	ND[c]	1×10^1 to 1.6×10^4

Note: ND = not determined/unknown.

[a] meters below ground level.
[b] beneath mountains.
[c] aerobic heterotrophs (CFU ml^{-1}) — viable counts.
[d] total counts = 1.5×10^4 to 1.8×10^6 (Pedersen, 1993).
[e] total counts = 1.8×10^3 to 1.2×10^5 (Pedersen, 1993).

production or utilization under realistic conditions. West (1985) examined the literature for information on acetate and EDTA degradation. This work was related to the shallow glacial sand horizons at the Drigg radioactive waste storage in Cumbria (West et al., 1989a).

The major work on waste degradation has been done in the U.K. and is related to the Nirex Safety Assessment Research Programme (NSARP) examining the radiological impact of a LLW and ILW repository on humans and the environment. The CASCADE exercise (Comparative Assessment of Concepts and Areas for Deep Emplacement) requires data for modeling every area of repository safety addressed by Nirex. The microbiological work carried out, primarily at the Harwell Laboratory but with input from higher-education establishments, deals with two of the NSARP areas. These are the "Evolution of the Near-field Environment" and "Near-field Radionuclide Chemistry" programs (Colasanti et al., 1989; Ewart et al., 1988, 1989). Microbiology within CASCADE has looked at the survival and overall metabolism of microorganisms under representative repository conditions. It is true to say that as yet realistic experiments have not been attempted. This statement is endorsed by Greenfield et al. (1991).

15.3.3   Biodeterioration of Waste Isolation Materials

15.3.3.1   *Corrosion of Metal Containers*

It is axiomatic that corrosion of metal containers will take place (Atkinson and Marsh, 1988), and this will be facilitated primarily by nonbiological electrochemical means, and potentially by microorganisms or their metabolic by-products (Duddridge, 1986a; Newman, 1987). U.K. research on corrosion of metal containers has concentrated on various grades of mild carbon and stainless steel. In Sweden, copper may be utilized for HLW containment and work is considering the corrosion of this metal. The only microbiological research on metal corrosion was that carried out by Napier University in collaboration with the British Geological Survey and Materials

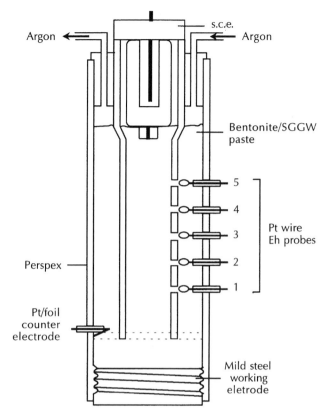

FIGURE 15.1

Diagrammatic representation of the corrosion cell used to assess the long-term effects of sulfate-reducing bacteria on the biodeterioration of mild carbon steel; s.c.e. — standard calomel electrode; SGGW — synthetic granitic groundwater; all Eh probes were platinum wire.

Development Division, Harwell (Philp et al., 1986, 1991; Philp and Taylor, 1987). The approach in this work was to model the potential repository environment of HLW by examining the possibility of enhancement of corrosion of mild carbon steel embedded in a clay-granitic groundwater matrix with or without SRB, the physiological group most likely to be implicated, given the physicochemical constraints of the repository. A corrosion cell was designed as shown in Figure 15.1. Enhancement of corrosion (up to three times the rate of nonmicrobial systems) was seen (Philp et al., 1991) but it must be remembered that these experiments were carried out without heat-generating radioisotopes, at a temperature of 50°C, and atmospheric pressure. In addition, the study could not detect SRB growth above pH 9.5. Intuitively, corrosion would be less than that obtained during these studies, however, even at the high rates obtained, there would not be sufficient corrosion to breach the thickness of the steel within the expected lifetime of the container.

Although considered, microbial corrosion of steel drums or other metal containers for ILW and LLW has not been examined in the U.K. Because of the high alkaline environment expected in LLW and ILW, Nirex funded a microbiological study of natural ultramafic spring waters in the Sultanate of Oman. The aim was to delineate the types of microorganisms present in permanently high-pH waters (pH 11.5) and hence "likely" to develop in a waste repository. A range of alkalitolerant and some alkaliphilic bacteria were found in these waters including sulfate-reducing bacteria important in corrosion, sulfate reducers could not be induced to grow significantly above pH 9.5 (Bath et al., 1987).

The use of copper for the containment of HLW has been adopted in Sweden. Copper is subject to corrosion and research is being done to examine processes affecting the deterioration of this metal (KBS, 1978). Many of the corrosion studies relate to the examination of analogue systems. Hallberg et al. (1984) examined the corrosion of copper plates embedded in soil, acting as lightning conductors. Of the three plates examined, two exhibited pitting corrosion while the third was unaffected after 50 years in the ground. A bronze cannon embedded in clay sediments was also found to exhibit little corrosion after more than 300 years. The maximum corrosion for the bronze matrix was estimated at less than 10 mm in 100,000 years (Hallberg et al., 1988).

15.3.3.2   Concrete/Cement Biodegradation

Cements and concretes can play an important role in the chemical and physical containment of radionuclides in ILW and LLW (Glasser et al., 1988, 1989; Sharland, 1986). The high pH (Loadsman et al., 1988) affects the solubility such that release is retarded. Low solubility of radionuclides may not be guaranteed throughout a cementitious matrix where voids and compacted waste may exist at values less than the minimum expected of pH 10.5 (Atkinson and Marsh, 1988; Sharland, 1986). Within these areas, microbial activity may occur. Although experiments are being undertaken to examine microbial presence and activity in cement-waste systems as part of the NSARP at Harwell, no data are available on concrete/cement degradation due to microbial action.

Microbial growth and gas production can theoretically disrupt concretes, particularly under fully saturated conditions. It may be prudent to calculate the total volume of gas produced based on stoichiometric conversion of organic compounds in waste. If one assumes that all carbon in the organics is converted to CO_2 and CH_4 over the lifetime of a repository, then volumes of gases given in Table 15.3 are possible. The inventory does not take account of various material referred to as 'other organics' (McGinnes, 1992). The volumes of gas produced, assuming complete degradation of cellulosic materials, are 3.5×10^7 m^{-3} and 5.3×10^6 m^{-3} for LLW and ILW, respectively. Approximately 9×10^7 m^{-3} and 3.5×10^7 m^{-3} of H_2 will be produced from LLW and ILW. This compares with an approximate total volume of 10^9 m^{-3} which could be produced from the steel inventory corrosion assuming a corrosion rate of 1 μm y^{-1} under anaerobic conditions (Agg et al., 1993; Biddle and Moreton, 1992; Table 15.4; and see Section 15.3.5).

In cement/concrete biodegradation studies in France, fungal isolates have been utilized to examine the effect of carboxylic acid production (e.g., citrate, oxalate, and lactate) on decalcification processes. Inorganic acid-producing bacteria (nitrifiers and sulfur-oxidizers) have also been utilized (Perfettini, 1989; Perfettini et al., 1991). It has been shown that direct contact of microorganisms with cement surfaces is not necessary for dissolution to occur.

15.3.3.3   Bitumen Biodegradation

Bitumen is potentially an important source of microbial nutrients if used in LLW and ILW conditioning. It is a complex organic material which has been shown to be biodegradable under both aerobic and anaerobic conditions (Ait-Langomazino et al., 1991; Bachofen et al., 1984; Brunner et al., 1987; Buckley et al., 1985; Clegg, 1982; Roffey and Norquist, 1991; Wolf and Bachofen, 1991). Bitumen biodegradation studies

TABLE 15.3

Gas Production Based on the Inventory of Organics

Organic Waste[a]	Representative Formulas[b]	LLW (High Arisings)		ILW	
		Waste Quantity (T)	Gas Production[c] (m^3)	Waste Quantity (t)	Gas Production[c] (m^3)
Plastics (PVC etc.)	$_n[CH_2 CH Cl]_n$	41,900	3.0×10^7	12,700	9.1×10^6
Ion-exchange resin	$[(C_6 H_5 CH CH_2)_2$ $(N(CH_3)_3 OH \cdot SO_3 \cdot H)]_n$	3,600	4.6×10^6	13,000	1.6×10^7
Cellulose	$[C_6H_{10}O_5]_n$	42,300	3.5×10^7	6,400	5.3×10^6
Rubber	$[C_5H_8]_n$	12,300	2.0×10^7	3,300	5.4×10^6

[a] Inventory from McGinnes (1992).
[b] Formulas from Grogan and McKinley (1990), except rubber (Merck Index, 1989).
[c] $CO_2 + CH_4$.

TABLE 15.4

Best Estimates of Total Gas Volumes Generated in the Degradation of Wastes Under Laboratory Conditions[a]

Waste	Gas(es)	Total Volume (m^3 at STP)
LLW		
Paper and wood	CH_4, CO_2	10^7
Metals	H_2	10^9
ILW		
Magnox	H_2	10^9
Zircaloy	H_2	10^9
Steels	H_2	10^7–10^8

Note: Metal corrosion will add to the volumes of H_2 generated, but is not expected to influence the overall rate of gas generation.
[a] Data from Rees and Rodwell (1988).

have been carried out at the University of Zurich (Brunner and Bachofen, 1987; Wolf, 1989; Wyndham and Costerton, 1981). These experiments used a percolation system containing a soil inoculum and bitumen, and found significant biodegradation under aerobic conditions. It is unlikely that the rates under repository conditions would be as high as those obtained in the continuous-flow system employed (Buckley et al., 1985). Roffey (1990) carried out a literature survey on asphalt and bitumen biodegradation and Roffey and Hjalmarsson (1983, 1984) utilized a theoretical approach to assess microbial activity in ILW and LLW repositories. Roffey (1990) describes experiments used to determine microbial bitumen degradation. In some countries the use of bitumen and cements as immobilization and isolation matrices is considered, and so is the possible effects of microorganisms on such materials. These countries include Denmark (Brodersen et al., 1983), Finland (Hellmuth, 1989a,b), Czechoslovakia (Kristofikova et al., 1989), and Russia (Zhukova et al., 1982).

15.3.3.4 Biodegradation of Other Materials

Other waste isolation materials include clays and various polymer solidification matrices (Duddridge, 1986a). Work on clays relates to the use of clay formations for disposal and clay as a backfill in repositories (Gilling et al., 1987a,b; Higgo, 1986). Research in the U.K. on clays has concentrated on the detection of microorganisms in

Oxford clay and demonstration of activity (Philp et al., 1984; Rushbrook, 1987). Although microorganisms and microbial activity are evident in natural clay samples, experimentation to show their role in alteration of clay structure and chemistry is not documented. Christofi and Philp (1991) failed to isolate microorganisms from a highly compacted (consolidated) Boom clay sample from Mol, Belgium. It is still not clear whether significant microbial activity will occur in clay backfill materials or clay formations in the deep subsurface. Philp and Taylor (1987), however, demonstrated enhanced corrosion of steel embedded in a laboratory clay matrix experiment.

Little information is available on other backfill/buffer materials and microbiological effects in the U.K. (Duddridge, 1986b). Again, this is due to the fact that it has been decided to use cements as isolation materials and also it avoids duplication of research effort.

15.3.4 Microbial Production of Chelating/Complexing Agents

Metabolic by-products of microorganisms may lead to the solubilization and increased mobility of radionuclides (Birch and Bachofen, 1990a,b; Francis, 1980, 1990a,b; Francis et al., 1980a; Francis and Dodge, 1990; Francis et al., 1980b; West, 1985; West and Rowe, 1986). These are in addition to naturally present forms including fulvic and humic acids which may have an effect on radionuclide movement (Ewart and Williams, 1986; Means, 1978; Oscarson et al., 1986; Wilkins, 1988).

In the NSARP at Harwell, work has concentrated on the degradation of organic waste material such as cellulose and the production of water-soluble organic species able to complex with long-lived radionuclides. Decomposition of organics is likely to be a function of the high pH, radiation, and microbial activity (Greenfield et al., 1991).

In chemical and radiation degradation studies, production of water-soluble complexing agents is delineated by increased solubility of plutonium in the water in equilibrium with the leachate (Ewart et al., 1989). Tests have shown enhancement of plutonium solubility by such degradation pathways. Microbiological degradation studies using a range of experimental models have generated potential microbial products including gases (mainly CO_2 and CH_4) and volatile fatty acids (acetate, n-butyrate, and propionate). Not all microbial products were identified (Ewart et al., 1989). It is questionable whether experimental techniques utilized to date unequivocally show that microbes are capable of attacking and degrading organic waste leading to formation of complexing agents under repository conditions. The experiments conducted suggest potential production, and further work is needed to demonstrate a microbial involvement (Greenfield et al., 1991) and the likelihood that chelating agents produced by microorganisms would have a significant effect on radionuclide solubility.

15.3.5 Microbial Gas Production

Gas production can affect the integrity of a repository. Nonbiological gas production and that due to microorganisms is postulated for waste repositories (Biddle et al., 1987; Norris, 1987; Rees and Rodwell, 1988). Nonbiological processes include the anaerobic corrosion of metals and alloys, producing H_2, and radiolysis. The latter is not expected to produce significant volumes of gas. Microbial gas production from organic degradation is possible and the NSARP aims to quantify this. Experiments at Harwell have been commissioned to examine gas evolution from LLW under varying moisture and pH regimes. The major gaseous products expected within a closed repository are CO_2 and CH_4 as anaerobic conditions are likely to develop quickly.

Estimated volumes of the major gases formed by microbial degradation of organics and metal corrosion are shown in Tables 15.3 and 15.4. The major gas produced will be H_2 from the corrosion of the metal canisters.

A series of static (batch) and flow-through column experiments have been carried out at Harwell where microbial activity has been assessed by gas production. Growth under defined conditions using small batch systems examined variation in water activity (A_w), organic and inorganic nutrients, and pH on microbial growth (Colasanti et al., 1989). Such "screening" experiments showed that environmental conditions are adequate for slow microbial growth and that this may be accelerated by "common" supplements. Microbial action was found on "damp" waste and it was suggested that microbial degradation may occur on stored material as well as on that which is buried.

Large-scale batch tests were initiated using 100 g simulated LLW incubated for 2.5 y. Optimum conditions were chosen where the waste mix contained supplements of SO_4, PO_4, and NO_3. The study suggests that at neutral pH there would be significant gas and soluble organic production (similar rates to a domestic landfill). Realistic repository conditions were not used in these and other experiments carried out (Colasanti et al., 1989) but it was concluded that CO_2 and CH_4 will be produced by microbial action with CH_4 predominating, as CO_2 is expected to dissolve. It was also suggested that H_2 (e.g., produced through metal corrosion) would be depressed by microbial action (methanogenesis and sulfate reduction) leading to more CH_4 production. Data were obtained using sealed bottles containing tap water and a representative LLW mix (3.5 g of soil-waste and 25 ml water). The experiment was carried out over 200 days and samples were analyzed for headspace gases, leachate pH, soluble carbohydrates, and organic metabolites. Leachate was used to examine the solubility of plutonium. CO_2 and CH_4 were shown to be produced and plutonium solubility increased by the leachate.

15.3.6   Microbial Uptake of Radionuclides

Released radionuclides can interact with the geology and waste isolation materials (McKinley and West, 1983). Radionuclides may also interact with microorganisms in two ways. Firstly, they may become adsorbed externally onto the microbial cell wall (Beveridge, 1989; Mayers and Beveridge, 1989; Walker et al., 1989) and appendages (van Loosdrecht and Zehnder, 1990) such as fimbriae (pili) and flagella, and secondly there may be active uptake into the microbial cell (Gadd, 1990). Either way, if the pores within the near-field and far-field are large enough to allow microbial passage, radionuclide migration can be enhanced due to motile bacteria or their products (Champ, 1985, 1984). Bacteria of 1 μm or less in size can act as colloidal material (McCabe, 1991). A beneficial effect may be due to adhesion of microbes onto geological surfaces, sorption/uptake of radionuclides, and retardation. Such biofilm formation can block flow paths (see Chapter 17). It is accepted that microbial growth on a surface imparts an advantage to them under conditions of nutrient limitation (Bar-Or, 1990; Berkeley et al., 1980; Delaquis et al., 1989).

Experiments have been carried out at BGS, Keyworth on the effect of microorganisms on sorption processes in the presence of host rock material (sand). Two organisms were used, namely a sulfate-reducing bacterium (*Desulfovibrio* sp.) and a thermophilic sulfur-oxidizer (*Sulfolobus solfataricus*). The method used was a modified standard batch reaction technique developed to study radionuclide sorption on naturally disaggregated rock material (McKinley and West, 1983; West et al., 1986a, 1987b, 1991). In this standard technique, rock material is equilibrated with groundwater spiked with gamma-emitting radionuclides under controlled environmental conditions. Radionuclide partitioning between solid and liquid phases is determined after centrifugation, using gamma spectrometry. In modified systems, microorganisms were mixed with radionuclides in the presence or absence of rock (West et al., 1987b, 1991). It was found that sorption of radionuclides such as ^{137}Cs onto rock is influenced by the presence of bacteria and there is an increase in the partition coefficient. It was not possible to assess the importance of radionuclide retardation in these experiments. Further experiments have been undertaken examining the influence of sulfate-reducing bacteria on ^{137}Cs sorption in Fuller's Earth (calcium montmorillonite) groundwaters. The preliminary results (West et al., 1987a) showed that the presence of bacteria lowered the distribution coefficients but that the extent depended on the number and activity of bacteria in the test batch sorption systems. West et al. (1987a,b) further showed that microbes naturally present in Fuller's Earth did not influence sorption data, but that increases in microbes in the aqueous phase altered sorption properties in a complex manner.

15.3.7   Microbial Movement Through Isolation Materials and the Geosphere

Under nutrient limitation, microorganisms may become dormant with concomitant reduction in cell size and volume (Lappin-Scott and Costerton, 1990; Morita, 1990) (see Chapter 11). Such cells are better suited to survival and may be more easily dispersed through low-porosity matrices (see Chapter 13). There appears to be no data on the migration of microorganisms through concretes, clays, or other isolation materials. It may not be surprising that research on this has not been initiated given the likely low porosity of engineered barriers — considerably less than the size of microbes. Fractured engineered barriers and geological formations may present a problem in terms of microbial migration.

Pedersen and Albinsson (1990) examined the role of bacteria in the transport of radionuclides in groundwater. The approach was to study elements with a high sorption capacity (e.g., Cs, lanthanides, and tri- and tetravalent actinides). They utilized promethium, a trivalent lanthanide with similar chemistry to trivalent actinides, especially americium. The overall aim of the work was to examine sorption of promethium by bacteria and determine the Kd (distribution ratio) of the lanthanide between the aqueous and bacterial phases under differing conditions of pH, bacterial numbers, and lanthanide concentrations. The bacterial species *Shewanella putrefaciens*, isolated from groundwater at a depth of 463 m, was used for these studies (Pedersen and Albinsson, 1990, 1991).

The results showed that sorption of lanthanides onto bacteria is a reversible cell surface phenomenon affected by pH. There was a negative linear relationship between promethium sorption and cell numbers, and between Kd and cell numbers. Sorption tended to decrease at higher pH values, which may be important in higher-pH repository environments.

In clay backfilled environments it is concluded that microorganisms and microbial activity will have little or no influence on radionuclide mobility when bentonite-sand or compacted pure bentonite are used as buffer materials (Jacobsen and Pusch, 1978).

15.3.8   Biofilm Formation

The attachment to and growth of microorganisms on surfaces has important implications to their survival and success in natural environments. It is beneficial for organisms growing under nutrient limitations. The attachment of bacteria and biofilm formation is important in degradation processes. The subject of attachment and its importance has been reviewed (Bar-Or, 1990; van Loosdrecht and Zehnder, 1990). Biofilm formation in the near- and far-field environments of waste repositories can also be important for radionuclide migration. Biofilms can block pores and reduce radionuclide movement (see Chapter 17). Surface growth can also lead to corrosion of metal structures, enhancing the release of contained waste, and in high-pH environments biofilms can retard hydrolytic reactions due to acidic exopolymer barriers produced (Greenfield et al., 1991). Colasanti et al. (1989) reported surface growth on concrete (slime formation) used in column perfusion experiments.

Work in Sweden examined biofilm development of natural populations present deep granitic groundwaters to a maximum depth of 860 m (Pedersen and Ekendahl, 1990, 1992b). With a groundwater flow of 10^{-3} m s^{-1}, a biofilm with 1.2×10^6 to 7.1×10^6 cells cm^{-2} developed (Pedersen and Ekendahl, 1922b).

15.3.9   Experimental Model Systems

The various research groups in the U.K. have utilized a range of laboratory experimental systems to delineate the extent of microbial activity in a real repository. A discussion paper was presented by Christofi at the 2nd MIND Meeting in Stockholm, Sweden (McCabe, 1989) examining the methods used and their problems in generating realistic data for the mathematical modelers. It was suggested that most of the experimental systems used up to 1989 generated data for "worst case" scenarios and there was probably an argument for dispensing with such experiments. Further, it may be more useful to use published data on microbial processes and, given a potential waste inventory, calculate optimum gas production, complexant formation, etc. In experiments to assess microbial growth and activity, systems have been set up containing representative waste mixes and waste isolation materials spiked with pure or mixed cultures of microorganisms. Continuous, semicontinuous, and static-flow experiments have been done. In a deep repository, static systems best represent the environment where little or no replenishment of nutrients occurs. Any movement of nutrients would be via advection, diffusion, and pressure effects (due to gas production) (Colasanti et al., 1989, 1991).

Experimental studies at Harwell have used heterogeneous and homogeneous systems. Heterogeneous systems included sealed-bottle experiments, recycle and single-pass columns, and the use of agar plates. Agar plate experiments are basically useful for screening microbial isolates and examining their growth requirements. They do not provide information on the ability of isolates to carry out particular reactions in a real repository. The other three heterogeneous systems used are considered suitable for general screening of effects of environmental and repository conditions. Homogeneous systems include the use of covered flasks, fermenters, multiwell plates, and the Gradostat. Again, such systems do not represent repository conditions adequately. Further information on the Harwell experimental studies are available (Colasanti et al., 1989; Ewart et al., 1988, 1989).

The model system (Figure 15.1) used to examine corrosion of mild carbon steel (Philp et al., 1991) utilizes an expected static environment. It suffers, however, by the use of a single isolated microbial species and not a mixed microbial population. In addition, the environmental conditions were not realistic in terms of radiation flux and pressure. To incorporate these factors into experimental design would require a considerable increase in sophistication of equipment.

Microbial isolates have been tested for survival under high temperature, pressure, and radiation using batch culture systems (West et al., 1985a, 1986a; West and Arme, 1985) and their influence on radionuclide sorption onto geological material from the deep subsurface (West et al., 1987a,b, 1991). Such studies did not realistically reproduce waste repository conditions.

15.3.10   Mathematical Modeling

A number of mathematical models are available describing microbial activity in various natural environments (Lensing et al., 1994). Their effect on geochemical processes in the near- and far-fields needs to be included in models describing repository performance. Simple modeling exercises have been carried out at the British Geological Survey at Keyworth and in collaboration with NAGRA to predict performance over time (Agg, 1993; Arter et al., 1991; Capon and Grogan, 1991; Colasanti et al., 1989, 1991; Ewart et al., 1988, 1989; Grogan and McKinley, 1990; McKinley and Grogan, 1991; McKinley et al., 1985a,b,c; 1984; West et al., 1986b).

The general approach to modeling at BGS considers the near-field and a unit repository volume containing the waste, backfill/buffer materials, steel containers, and in the case of ILW/LLW, cement and concrete. It is assumed that all these constituents are available as nutrient and energy sources. Such an approach represents a worst-case modeling approach and has been used for HLW, ILW, and LLW (McKinley and Grogan, 1991; McKinley et al., 1985b). It is concluded that microbial effects would be negligible in a HLW repository. For ILW and LLW, it has been predicted that phosphorus would limit microbial growth. Computer codes are now available based on an energy and nutrient inventory calculation approach in Switzerland called EMMA (Capon and Grogan, 1991) and in the U.K. called MGSE — Microbial Growth in Subsurface Environments (Rochell et al., 1993).

The modeling approach at Harwell was based on the availability of carbon in the waste form (Colasanti et al., 1989). The carbon is considered as labile or refractive. Soluble waste is considered to be in the form of glucose; insoluble waste is represented by glucose polymers such as cellulose, and recalcitrant waste includes lignin and other polymers. The model describes the degradation of the waste by physical, chemical, and biological means leading to stable products such as recalcitrant material or gas. Microbial biomass and product formation utilizes Monod kinetics. Variations in the model describe aerobic and anaerobic conditions and changes from aerobic to anaerobic metabolism. Details of the Harwell model and attempts to validate it are available (Colasanti et al., 1989; Ewart et al., 1989). Generally, such a simple modeling technique can be used to estimate the maximum potential for substrate disappearance and microbial biomass and by-product formation.

Recently, a gas generation model (still under development) encoded in the GAMMON computer program has been proposed (Agg, 1993). The gas generation model includes a cellulose degradation model and cellulose degradation is considered to be the principal source of microbiological gas production. The rate at which the hydrolytic breakdown of cellulose and intermediates is achieved is assumed to be governed by eight microbial processes (Table 15.5). The activity of the microbial populations is considered a function of various physicochemical environmental factors including pH, total nitrogen, water, other growth-requiring substances, and inhibitors. A potential growth rate has been proposed for each process and generation times for organisms involved are calculated (Table 15.5).

15.3.11   Analogue Studies

A number of natural analogue sites have been studied in the performance assessment of the disposal of radioactive waste (Bath et al., 1987; Chapman, 1988; Chapman et al.,

TABLE 15.5

Microbial Reactions Used in the Gas Generation Model[a]

Microbe	Process	Reaction	Growth Rate (h^{-1})	Generation Time (h)
1	Aerobic decomposition	$C_6H_{12}O_6 + 6\ O_2 \rightarrow 6CO_2 + 6H_2O$	5×10^{-1}	1.39
2	Nitrate decomposition	$5C_6H_{12}O_6 + 24HNO_3 \rightarrow 30CO_2 + 12N_2 + 42H_2O$	4×10^{-1}	1.73
3	Acidogenesis	$C_6H_{12}O_6 + H_2O \rightarrow 2CO_2 + 2H_2 + C_2H_5OH + CH_3COOH$	3×10^{-1}	2.31
4	Acetogenesis	$C_2H_5OH + H_2O \rightarrow CH_3COOH + 2H_2$	3.6×10^{-2}	19.25
5	Methanogenesis 1	$CH_3COOH \rightarrow CO_2 + CH_4$	8.33×10^{-3}	83.19
6	Methanogenesis 2	$CO_2 + 4H_2 \rightarrow CH_4 + 2H_2O$	1.67×10^{-1}	4.15
7	Sulfate reduction 1	$CH_3COOH + H_2SO_4 \rightarrow H_2S + 2CO_2 + 2H_2O$	3.5×10^{-2}	19.8
8	Sulfate reduction 2	$4H_2 + H_2SO_4 \rightarrow H_2S + 4H_2O$	2.3×10^{-1}	3.01

Note: Starting populations in wastes and groundwaters unknown.

[a] Based on Agg et al. (1993)

1992; Come and Chapman, 1986; Cramer, 1986; Hellmuth, 1989a,b; Hooker and Chapman, 1989; Milodowski et al., 1990). Natural analogue studies in the U.K. have been conducted in collaboration with other national programs.

Bath et al. (1987) report on the cement analogue site studies in the Sultanate of Oman. Spring water emanating from ultramafic rock deposits with pH around 11.5 have been examined for the presence of alkalitolerant and alkaliphilic bacteria. A range of such bacteria were isolated. Activity studies showed that these organisms, although capable of growth at high pH, were limited by availability of nutrients in the springwaters. Maintenance of the population present was considered a function of the relatively rapid throughput of water, and hence nutrients, in these oligotrophic (nutrient-poor) ecosystems. In a real repository, microbial activity would rely on indigenous nutrient sources in a more or less static system.

Switzerland is involved in a number of international programs examining analogue systems, including the Oman (Bath et al., 1987) and the Pocos de Caldas projects (Chapman, 1988; West et al., 1995, 1992, 1990). The Pocos de Caldas project involving NAGRA, SKB, the U.K. Department of the Environment, the U.S. Department of Energy, and supported by Nuclebras in Brazil, involved a microbiological element. The study included an examination of two sites, the Osamu Utsumi open-cast uranium mine, and the nearby thorium and rare-earth deposit of Morro do Ferro (Chapman, 1988). West et al. (1992) report on the microbiological methods used during the project. The aims were to determine the presence and activity of microbes known to be important in the uranium and other biogeochemical cycles and how microorganisms can affect the movement of radionuclides across redox fronts present in these sites, which are analogous to those found in the near- and far-field and deep repositories for HLW or spent fuel. Results show microorganisms in all samples examined, including sulfur-cycling bacteria involved in uranium cycling. Low microbial activity is postulated based on nutrient levels and it is concluded that the main role of the microorganisms is in catalysis of specific redox reactions and not in element mobilization through organic by-product formation (West et al., 1988, 1990).

Uranium geochemistry studies are being undertaken in southwest Scotland by BGS to obtain information to validate the thermodynamic databases and equilibrium speciation codes used in the mathematical modeling of radionuclide transport. During these studies, biological structures similar to fungal hyphae have been observed in geological material and associated with these are uranium-mineralized material.

Such structures have been attributed to the activity of filamentous microorganisms (Milodowski et al., 1990). Microbial uptake of radionuclides, whether internal or external, or microbially induced precipitation reactions, are likely to retard radionuclide migration and such observations are important.

BGS has also carried out microbiological studies in natural high-pH groundwaters in the Middle-East-Magarin, North Jordan (West et al., 1995).

15.4 Conclusions

There have been numerous reviews on the potential role of microorganisms in nuclear waste disposal in Europe. Based on available data on microbial growth and survival under various physicochemical environmental conditions, it is conceivable that microorganisms can survive a deep subsurface waste disposal environment. Even the high-pH near-field environment generated through the use of cements and concretes as barriers may contain microsites (voids) of sufficiently low pH to allow microbial activity. It has yet to be shown, however, whether such environments will support a significant microbial population and activity to compromise waste disposal.

There is a need for good experimental model systems and microbial activity measurements to delineate a microbial role in biodegradation, biodeterioration, and gas production above nonbiological processes. Also the biological production of complexing/chelating organic compounds has yet to be shown under the realistic conditions of a waste repository. Mathematical modeling techniques describing microbial activity in relation to nuclear waste disposal are at the infancy stage. Any model, however complex, would need to be verified with good experimental data which at present are not available.

It is difficult to envisage microbial movement through the expected near-field containing barriers of low porosity. Even though microorganisms can uptake radionuclides it is unlikely that they will transport them to the biosphere. It is more likely that any uptake of radionuclides will lead to beneficial immobilization.

Any microbial growth within the near-field is likely to lead to biofilm formation, leading to a favorable reduction in the porosity of the waste isolation environment. The role of biofilms in the far-field may be important after radionuclides have been released from the near-field through biological or nonbiological means. Experiments need to be done to show whether biofilms will retard or enhance radionuclide transport in the far-field.

Analogue studies can be important in assessing a waste disposal option. Some of the analogue sites examined have not been relevant to "real" repositories and others should be identified and considered for study.

Acknowledgments

Most of the work published by the authors and reported in this review was supported by research grants received from the United Kingdom DOE and NIREX, and, NAGRA, Switzerland. We are thankful for their support.

References

Agg, P. J. 1993. Application of the gas generation program GAMMON in the preliminary assessment of Sellafield. U.K. NIREX Report NSS/B004.

Agg, P. J., A. D. Moreton, J. H. Rees, W. R. Rodwell, and P. J. Sumner. 1993. NSARP Reference Document: Gas Generation and Migration, January 1992, U.K. NIREX Report NSS/B142.

Ait-Langomazino, N., R. Sellier, G. Jouquet, and M. Trescinski. 1991. Microbial degradation of bitumen, *Experientia*, 47: 533.

Arter, H. E., K. W. Hanselmann, and R. Bachofen. 1991. Modeling of microbial degradation processes. The behaviour of microorganisms in a waste repository, *Experientia*, 47: 578.

Atkinson, A., N. M. Everitt, and R. M. Guppy. 1988. Evolution of pH in a radwaste repository. Internal reactions between concrete constituents, U.K. DOE Report, DOE/RW/89.025 Part 2.

Atkinson, A. and R. M. Guppy. 1988. Evolution of pH in a radwaste repository. Leaching of modified cements and reactions with groundwater, U.K. DOE Report, DOE/RW/89.025 Part 3.

Atkinson, A. and G. P. Marsh. 1988. Engineered barriers: current status, U.K. NIREX Report NSS/G102.

Bachofen, R. 1990. Microorganisms in nuclear waste disposal. Introduction, *Experientia*, 46: 777.

Bachofen, R., A. C. Dubach, A. W. Tesch, and D. Luscher. 1984. Literaturstudie uber den Abbau von Bitumen durch Mikroorganismen, NAGRA Technical Report — NTB 83-18.

Bachofen, R. and D. Luscher. 1984. Mogliche mikrobiologische Vorgange in unterirdischen Kavernen im Hinblick auf die Endlagerung radio aktiver Abfalle (Literaturstudie), NAGRA Technical Report — NTB 84-07.

Balkwill, D. L. 1989. Numbers, diversity and morphological characteriztics of aerobic, chemoheterotrophic bacteria in deep subsurface sediments from a site in South Carolina, *Geomicrobiol. J.*, 7: 33.

Balkwill, D. L., F. R. Leach, J. T. Wilson, J. F. McNabb, and D. C. White. 1988. Equivalence of microbial biomass measures based on membrane lipid and cell wall components, adenosine tri-phosphate and direct counts in subsurface aquifer sediments, *Microb. Ecol.*, 16: 73.

Bar-Or, Y. 1990. The effects of adhesion on survival and growth of microorganisms, *Experientia*, 46: 823.

Bath, A. H., N. Christofi, C. Neal, J. C. Philp, M. R. Cave, I. G. McKinley, and U. Berner. 1987. Trace element and microbiological studies of alkaline groundwater in Oman, Arabian Gulf. A natural analogue for cement porewater, NERC, British Geological Survey Report FLPU 87-2; also NAGRA Technical Report NTB 88-02.

Beloin, R. M., J. L. Sinclair, and W. C. Ghiorse. 1988. Distribution and activity of microorganisms in subsurface sediments of a pristine site in Oklahoma, *Microb. Ecol.*, 16: 85.

Berkeley, R. C., J. Lynch, L. Melling, P. Rutter, and B. Vincent, Eds., 1980. *Microbial Adhesion to Surfaces*, Ellis Horwood, Chichester.

Beveridge, T. J. 1989. Role of cellular design in bacterial metal accumulation and mineralisation, *Annu. Rev. Microbiol.*, 43: 147.

Biddle, P. and A. D. Moreton. 1992. A review of the results and understanding achieved by the Nirex Research Programme to September 1991 on gas generation in a repository for radioactive wastes, NIREX Report NSS/R210.

Biddle, P., D. McGahan, J. H. Rees, and P. E. Rushbrook. 1987. Gas generation in repositories, U.K. AERE, R 12291.

Birch, L. D. and R. Bachofen. 1990a. Complexing agents from microorganisms, *Experientia*, 46: 827.

Birch, L. D. and R. Bachofen. 1990b. Effects of microorganisms on the environmental mobility of radionuclides, in *Soil Biochemistry*, Vol. 6, Bollag, J.-M. and Stotzky, G., Eds., Marcel Dekker, New York.

Boesten, J. J. T. I., L. S. T. Van der Pas, J. Smelt, and M. Liestra. 1991. Transformation rate of methyl isothiocyanate and 1,3-dichloropropane in water-saturated subsoil, *Neth. J. Agric. Sci.*, 39:179.

Bone, T. L. and D. L. Balkwill. 1988. Morphological and cultural comparison of microorganisms in surface soil and subsurface sediments at a pristine site in Oklahoma, *Microb. Ecol.*, 16: 49.

Brodersen, K., B. M. Pedersen, and A. Vinther. 1983. Comparative study of test methods for bituminised and other low- and medium-level solidified waste materials, RISO National Laboratory, Report RISO-M-2415.

Brondi, A. 1987. Geochemical and microbiological studies on the sand-clay series of Orte. Final Report to the Commission of the European Community, Contract 375-83-7 WASI.

Brunner, C., M. Wolf, and R. Bachofen. 1987. Enrichment of bitumen degrading microorganisms, *FEMS Microbiol. Lett.*, 43: 337.

Buckley, L. P., B. C. Clegg, and W. K. Oldham. 1985. Microbial activity on bituminised radioactive waste, *Radioact. Waste Manage. Nucl. Fuel Cycle*, 6: 19.

Capon, P. and H. A. Grogan. 1991. EMMA: A user guide and description of the program. NAGRA Internal Report NIB 91-87.

Champ, D. R. 1984. Microbial mediation of radionuclide transport — significance for the nuclear fuel waste management program, AECL Report, AECL-8566.

Champ, D. R. 1985. Effect of microbial activity on vault chemistry, in Proc. 19th Info. Meet. Nuclear Fuel Waste Management Program, AECL Report, TR-350.

Chapman, N. A., Ed. 1988. Pocos de Caldas Project; Second Year Final Report (June 1987-May 1988), U.K. DOE Report, DOE/RW/89.019.

Chapman, N. A., I. G. McKinley, M. E. Shea, and J. A. T. Smellie. 1992. The Pocos de Caldas Project. Natural analogues of processes in a radioactive waste repository, *J. Geochem. Explor.*, 45 (Special Vol.).

Chilton, P. J. and J. M. West. 1992. Aquifers as environments for microbial activity, in *Proc. Int. Symp. Environ. Aspects of Pesticide Microbiology*, Swedish University of Agricultural Sciences, Uppsala.

Christofi, N. 1991. A review of microbial studies, U.K. DOE Report DOE/HMIP/RR/92/008, Parts 1-3.

Christofi, N. and C. R. Milner. 1990. Microbiological examination of groundwater from the Grimsel test site: the results of two studies carried out by Napier Polytechnic of Edinburgh, NAGRA Internal Report NIB 89-23.

Christofi, N. and J. C. Philp. 1991. Microbiology of subterranean waste sites, *Experientia*, 47: 524.

Christofi, N., J. C. Philp, and J. M. West. 1984a. Microorganisms and microbial activity in groundwater from geological environments of relevance to nuclear waste disposal, in *Role of Microorganisms on the Behaviour of Radionuclides in Aquatic and Terrestrial Systems and their Transfer to Man*, Proceedings of a Workshop held in Brussels, Belgium, 25-27 April, 1984, Bonnyns van Gelder, E. and Kirchmann, R., Eds., IUR/CEC, 239.

Christofi, N., J. M. West, and J. C. Philp. 1985. The geomicrobiology of European mines relevant to nuclear waste disposal, U.K. NERC, British Geological Survey Report FLPU 85-1.

Christofi, N., J. M. West, J. C. Philp, and J. E. Robbins. 1984b. The geomicrobiology of used and disused mines in Britain, U.K. NERC, British Geological Survey Report FLPU 84-5.

Christofi, N., J. M. West, J. E. Robbins, and I. G. McKinley. 1983. The geomicrobiology of the Harwell and Altnabreac boreholes, U.K. NERC, Institute of Geological Sciences Report FLPU 83-4.

Clegg, B. C. 1982. The Effect of Microbial Action on Nuclear Waste Management: Is There Enhanced Leaching From Bitumen and Increased Radionuclide Movement Through Geologic Media, Masters Thesis, Department of Civil Engineering, University of British Columbia, Vancouver, BC.

Colasanti, R., D. Coutts, S. Y. R. Pugh, and A. Rosevear. 1989. Microbiology and Radioactive Waste Disposal — Review of the Nirex Research Programme — January 1989, U.K. NIREX Report NSS/R131.

Colasanti, R., D. Coutts, S. Y. R. Pugh, and A. Rosevear. 1991. The microbiology programme for U.K. Nirex, *Experientia*, 47: 560.

Come, B., Ed. 1985. Mirage project: Second summary progress report. CEC Report, EUR 100023 EN. This report is the third of a series which already includes: the description of the MIRAGE Project (EUR 9304); the first summary report for 1983 (EUR 9543).

Come, B. and N. A. Chapman. 1986. CEC Natural analogue working group, CEC Report, EUR 10671.

Cramer, J. J. 1986. A natural analogue for a fuel waste disposal vault, Proc. 2nd Int. Conf. Radioactive Waste Management, Winnipeg, Canada, September 7-11, 697.

Day, D. H., A. E. Hughes, J. W. Leake, J. A. C. Marples, G. P. Marsh, J. Rae, and B. O. Wade. 1985. The management of radioactive wastes, *Rep. Prog. Phys.,* 48: 101.

Delaquis, P. J., D. E. Caldwell, J. R. Lawrence, and A. R. McCurdy. 1989. Detachment of *Pseudomonas fluorescens* from biofilms on glass surfaces in response to nutrient stress, *Microb. Ecol.,* 18: 199.

Duddridge, J. 1986a. A theoretical evaluation of the consequences of microbiological activity on the corrosion of carbon steel overpacks, U.K. AERE Report 12379.

Duddridge, J. 1986b. Cement and polymer matrices for the immobilisation of intermediate level radioactive waste. A preliminary assessment of their potential biodeterioration during storage and disposal, U.K. DOE Report DOE/RW/86.065.

Ekendahl, S., J. Arlinger, F. Stahl, and K. Pedersen. 1994. Characterisation of attached bacterial populations in deep granitic groundwater from the Stripa research mine by 16S rRNA gene sequencing and scanning electron microscopy, *Microbiology,* 140: 1575.

Ekendahl, S. and K. Pedersen. 1994. Carbon transformations by attached bacterial populations in granitic groundwater from deep crystalline bed-rock of the Stripa research mine, *Microbiology,* 140: 1565.

Ewart, F. T., S. Y. R. Pugh, S. J. Wisbey, and D. R. Woodwark. 1988. Chemical and microbiological effects in the nearfield: current status, U.K. NIREX Report NSS/G103.

Ewart, F. T., S. Y. R. Pugh, S. J. Wisbey, and D. R. Woodwark. 1989. Chemical and microbiological effects in the near field: current status, U.K. NIREX Report NSS/G111.

Ewart, F. T. and S. J. Williams. 1986. A literature survey of the possible effects of humic and fulvic acids on the disposal of LLW and ILW, U.K. AERE R12023.

Francis, A. J. 1980. Microbial transformation of low-level radioactive waste, Brookhaven National Laboratory, Report BNL-29762.

Francis, A. J. 1990a. Characteristics of nuclear and fossil energy wastes, *Experientia,* 46: 794.

Francis, A. J. 1990b. Microbial dissolution and stabilization of toxic metals and radionuclides in mixed wastes, *Experientia,* 46: 840.

Francis, A. J., S. Dobbs, and R. F. Doring. 1980a. Biogenesis of tritiated and carbon-14 methane from low-level radioactive waste, *Nucl. Chem. Waste Manage.,* 1: 153.

Francis, A. J. and C. J. Dodge. 1990. Anaerobic microbial remobilization of toxic metals co-precipitated with iron oxide, *Environ. Sci. Technol.,* 24: 373.

Francis, A. J., C. R. Iden, B. J. Nine, and C. K. Chang. 1980b. Characterisation of organics in leachates from low-level radioactive waste disposal sites, *Nucl. Technol.,* 50: 158.

Francis, A. J., J. M. Slater, and C. J. Dodge. 1989. Denitrification in deep subsurface sediments, *Geomicrobiol. J.,* 7: 103.

Frederickson, J. K., T. R. Garland, R. J. Hicks, J. M. Thomas, S. W. Lie, and K. M. McFadden. 1989. Lithotrophic and heterotrophic bacteria in deep subsurface sediments and their relation to sediment properties, *Geomicrobiol. J.,* 7: 53.

Gadd, G. M. 1990. Heavy metal accumulation by bacteria and other microorganisms, *Experientia,* 46: 834.

Ghiorse, W. C. and F. J. Wobber. 1989. Introductory comments, *Geomicrobiol. J.,* 7: 1.

Gilling, D., N. L. Jefferies, and T. R. Lineham. 1987a. An experimental study of solute transport in mudstones, U.K. NIREX Report NSS/R109.

Gilling, D., N. L. Jefferies, and T. R. Lineham. 1987b. Laboratory measurements of the solute transport properties of samples from the Bradwell, Elstow, Fulbeck and Killingholme site investigations, U.K. AERE Report 12884.

Glasser, F. P., D. MacPhee, M. Atkins, N. Beckley, S. Carson, G. McHugh, N. J. Mattingley, C. C. Naish, and C. R. Wilding. 1988. Immobilization of radwaste in cement based matrices (Progress Report October 1986-April 1987), U.K. DOE Report, DOE/RW/88.020.

Glasser, F. P., D. MacPhee, M. Atkins, C. Pointer, J. Cowie, C. R. Wilding, N. J. Mattingley, and P. A. Evans. 1989. Immobilization of radwaste in cement based matrices (Progress Report October 1988-March, 1989), U.K. DOE Report, DOE/RW/89.058.

Goyette, D. 1987. The Effect of Microorganisms on the Transport Of Iodine And Technetium in a Biogeochemical Simulation of a Radioactive Waste Disposal Vault, Masters Thesis, Department of Chemical Engineering and Applied Chemistry, University of Toronto.

Greenfield, B. F., A. Rosevear, and S. J. Williams. 1991. Review of the microbiological, chemical and radiolytic degradation of organic material likely to be present in inter-mediate level and low level radioactive wastes, U.K. DOE Report No: DOE/HMIP/RR/91/002.

Grogan, H. A. and I. G. McKinley. 1990. An approach to microbiological modeling: application to the near-field of a Swiss low/intermediate level waste depository, NAGRA Technical Report, NTB 89-06.

Hallberg, R., A.-G. Engvall, and T. Wadsten. 1984. Corrosion of copper lightning conductor plates, *Br. Corros. J.*, 19: 85.

Hallberg, R. O., P. Ostlund, and T. Wadsten. 1988. Inferences from a corrosion study of a bronze cannon applied to high level nuclear waste. *Appl. Geochem.*, 3:273-280.

Hellmuth, K.-H. 1989a. Natural analogues of bitumen and bitumenised radioactive waste, Finnish Centre for Radiation and Nuclear Safety, Report STUK-B-VALO 58.

Hellmuth, K.-H. 1989b. The long-term stability of natural bitumen — a case study at the bitumen-impregnated limestone deposit near Holzen/Lower Saxony, Germany, Finnish Centre for Radiation and Nuclear Safety, Report STUK-B-VALO 59.

Helweg, A. 1990. Degradation of ^{14}C-labeled atrazine in subsurface soil, 7th Congr. Pesticide Chemistry, IUPAC, Hamburg, Germany, August 5-10.

Helweg, A. 1992. Degradation of pesticides in subsurface soil, in *Proc. Int. Symp. Environ. Aspects of Pesticide Microbiology*, Swedish University of Agricultural Sciences, Uppsala.

Hicks, R. J. and J. K. Fredrickson. 1989. Aerobic metabolic potential of microbial populations indigenous to deep subsurface environments, *Geomicrobiol. J.*, 7: 67.

Higgo, J. J. W. 1986. Clay as a barrier to radionuclide review, U.K. DOE Report, DOE/RW/86.082.

Hirsch, P. and E. Rades-Rohkohl. 1988. Some special problems in the determination of viable counts of groundwater microorganisms, *Microb. Ecol.*, 16: 99.

HMSO. 1990. Radioactive Waste Management Advisory Committee, 11th Annual Report, Her Majesty's Stationery Office, London.

Hooker, P. J. and N. A. Chapman, Eds. 1989. Fault detection, organics and radionuclide mobility, and natural analogue studies, U.K. DOE Report, DOE/RW/89.021.

Jacobsen, A. and R. Pusch 1978. Long term mineralogical properties of bentonite/quartz buffer substances, SKB Technical Report, KBS-TR-32.

Jones, R. E., R. E. Beeman, and J. M. Suflita. 1989. Anaerobic metabolic processes in the deep terrestrial subsurface, *Geomicrobiol. J.*, 7: 117.

Kaiser, J.-P. and J.-M. Bollag. 1990. Microbial activity in the terrestrial subsurface, *Experientia*, 46: 797.

Kinniburgh, D. G., I. N. Gale, P. L. Smedley, W. G. Darling, J. M. West, R. T. Kimblin, A. Parker, J. E. Rae, P. J. Aldous, and M. J. O'Shea. 1994. The effects of historic abstraction of ground-water from the London Basin aquifers on groundwater quality, *Appl. Geochem.*, 9:175.

Knecht, B., I. G. McKinley, and P. Zuidema. 1990. Disposal of low- and intermediate-level waste in Switzerland: Basic aspects of potential relevance to microbial effects, *Experientia*, 46: 787.

Kolbel-Boelke, J., E-M. Anders, and A. Nehrkorn. 1988a. Microbial communities in the satu-rated groundwater environment. II. Diversity of bacterial communities in a Pleistocene sand aquifer and their *in vitro* activities, *Microb. Ecol.*, 16: 31.

Kolbel-Boelke, J., B. Tienken, and A. Nehrkorn. 1988b. Microbial communities in the saturated groundwater environment. I. Methods of isolation and characterization of heterotrophic bacteria, *Microb. Ecol.*, 16: 17.

Kristofikova, L., L. Tibensky, V. Vollek, and F. Krejci. 1989. Influence of soil microorganisms on radioactive waste solidification, *Biologia (Bratislava)*, 44: 1069.

Krulwich, T. A. and A. A. Guffanti. 1983. Physiology of acidophilic and alkalophilic bacteria, *Adv. Microb. Physiol.*, 24: 173.

Kutzner, H. J. and D. Herrman. 1990. Untersuchungen zum moeglichen Einfluss von Mikro-organismen auf die Ausbreitung von Radionukliden aus Endlagen (Literaturstudie). Schlussbericht, (Investigations of the possible influence of microorganisms on the migration of radionuclides from repositories (a literature study), Final Report, Institute of Microbiology, Technische Hochschule, Darmstadt (Germany).

Lappin-Scott, H. M. and J. W. Costerton. 1990. Starvation and penetration of bacteria in soils and rocks, *Experientia*, 46: 807.

Leistra, M. and J. J. T. I. Boesten. 1989. Pesticide contamination of groundwater in Western Europe, *Agric. Ecosyst. Environ.*, 26: 369.

Lensing, H. J., M. Vogt, and B. Herrling. 1994. Modeling of biologically mediated redox processes in the subsurface. *J. Hydrol.*, 159: 125.

Lewis, K. J., F. J. Lewis, K. Tagaki, F. P. E. Anderson, M.-C. Dictor, and G. Soulas. 1992. Microbial diversity and activity of subsoil: characterization of microflora from the unsaturated zone, in *Proc. Int. Symp. Environ. Aspects of Pesticide Microbiology*, Swedish University of Agricultural Sciences, Uppsala.

Loadsman, R. V. C., D. H. Acres, C. J. Stokes, and L. Wadeson. 1988. A study on the water permeability of concrete structures, U.K. DOE Report, DOE/RW/88.052.

Madsen, E. L. and J.-M. Bollag. 1989. Aerobic and anaerobic microbial activity in deep subsurface sediments from the Savannah River Plant, *Geomicrobiol. J.*, 7: 93.

Marxsen, J. 1988. Investigations into the number of respiring bacteria in groundwater from sandy and gravely deposits, *Microb. Ecol.*, 16: 65.

Mayers, I. T. and T. J. Beveridge. 1989. The sorption of metals to *Bacillus subtilis* walls from dilute solutions and simulated Hamilton Harbour (Lake Ontario) water, *Can. J. Microbiol.*, 35: 764.

Mayfield, C. I. and J. F. Barker. 1982. An evaluation of the microbiological activities and possible consequences in a fuel waste disposal vault. A literature review, AECL Technical record TR-139.

McCabe, A.M., Ed. 1989. MIND-Group formulation of a collaborative technical program. Proceedings of the 2nd meeting of the MIND (Microorganisms in Nuclear Waste Disposal) Working Group, Stockholm, Sweden, May 1988, U.K. Central Electricity Generating Board Research Report, TPRD/B/6181/R89.

McCabe, A. M. 1987. The potential significance of microbial activity in radioactive waste disposal, U.K. Central Electricity Generating Board Report, TPRD/B/0951/R87.

McCabe, A. M. 1991. The potential significance of microbial activity in radioactive waste disposal, *Experientia*, 46: 779.

McDowell-Boyer, L. M., J. R. Hunt, and N. Sitar. 1986. Particle transport through porous media, *Water Resour. Res.*, 22: 1901.

McGahan, D. J. 1987. Survey of microbiological effects in low-level radioactive waste disposed of to land, U.K. AERE, R12477.

McGinnes, D. F. 1992. Waste volumes, radionuclide content, heat output and material composition for RSD. U.K. NIREX Report 277.

McKinley, I. G. and H. A. Grogan. 1991. Consideration of microbiology modeling the near field of a L/ILW repository, *Experientia*, 47: 573.

McKinley, I. G., H. A. Grogan, and J. M. West. 1985a. Quantitative modeling of the effects of microorganisms on radionuclide transport from a HLW repository, INEA workshop on the Effects of Natural Organic Compounds and of Microorganisms on Radionuclide Transport, Paris, June 1985, 50.

McKinley, I. G., H. A. Grogan, and J. M. West. 1985b. Quantitative modeling of the effects of microorganisms on radionuclide transport from a HLW repository, OMEA workshop on the Effects of Natural Organic Compounds and of Microorganisms on Radionuclide Transport, Paris, June 1985, 50-56.

McKinley, I. G., F. Van Dorp, and J. M. West. 1984. Modeling microbial contamination of a deep geological repository for HLW. In *Role of Micro-organisms on the Behaviour of Radionuclides in Aquatic and Terrestrial Systems and Their Transfer to Man*, Proceedings of a workshop held in Brussels 25-27 April, 1984, Bonnyns van Gelder, E. and Kirchmann, R., Eds., IUR/CEC, 245.

McKinley, I. G. and J. M. West. 1983. Radionuclide sorption/desorption processes occurring during groundwater transport, U.K. NERC, Institute of Geological Sciences, Report FLPU 83-2.

McKinley, I. G., J. M. West, and H. A. Grogan. 1985c. An analytical overview of the consequences of microbial activity in a Swiss HLW Repository, NAGRA Technical report (NTB) 85-43.

Means, J. L. 1978. Migration of radioactive wastes: Radionuclide mobilization by complexing agents, *Science*, 200: 1477.

Merck Index. 1989. *An Encyclopedia of Chemicals, Drugs, and Biochemicals*, 11th ed., Merck and Co., Rahway, NJ.

Milodowski, A. E., J. M. West, J. M. Pearce, E. K. Hyslop, I. R. Basham, and P. J. Hooker. 1990. Unusual uraniferous hydrocarbons associated with uranium mineralising microorganisms found in south west Scotland, *Nature*, 347: 465.

Morita, R. Y. 1990. The starvation-survival state of microorganisms in nature and its relationship to the bioavailable energy, *Experientia*, 46: 813.

NAGRA. 1985. Project Guarantee 1985, Final repository for low- and intermediate-level radioactive wastes. The system of safety barriers. NAGRA Report, NGB-85-07.

Newman, R. C. 1987. The influence of thiosulfate ions on the localised corrosion of stainless steel, U.K. AERE, R12664.

NIREX. 1993. Nirex deep waste repository project: scientific update, U.K. NIREX Report 525.

NNC (National Nuclear Corporation). 1987. Direct Disposal of Spent Nuclear Fuel; the Current Status of Technology, U.K. DOE Report, DOE/RW/87.021.

Norris, G. H. 1987. A review of literature relevant to gas production in radioactive waste. U.K. DOE Report DOE/RW/87.109.

Oscarson, D. W., S. Stroes-Gascoyne, and S. C. H. Cheung. 1986. Effects of organic matter in clay sealing materials on the performance of a nuclear fuel waste disposal vault, AECL Report, AECL-9078.

Pedersen, K. 1987. Preliminary investigations of deep ground water microbiology in Swedish granitic rock, SKB Technical Report, TR 88-01.

Pedersen, K. 1989. Deep groundwater microbiology in Swedish granitic rock and its relevance for radionuclide migration from a Swedish high level nuclear waste repository, SKB Technical Report, TR 89-23.

Pedersen, K. 1990. Potential effects of bacteria on radionuclide transport from a Swedish high level nuclear waste repository, SKB Technical Report TR 90-05.

Pedersen, K. 1993. Bacterial processes in nuclear waste disposal, *Microbiol. Eur.*, September/October, 18.

Pedersen, K. 1993. The deep subterranean biosphere, *Earth-Sci. Rev.*, 34: 243.

Pedersen, K. and Y. Albinsson. 1990. Effect of cell number, pH and lanthanide concentration on the sorption of promethium by *Shewanella putrefaciens*, SKB Technical Report, TR 90-24.

Pedersen, K. and Y. Albinsson. 1991. Effect of cell number, pH and lanthanide concentration on the sorption of promethium by *Shewanella putrefaciens*, *Radiochim. Acta*, 54: 91.

Pedersen, K. and S. Ekendahl. 1990. Distribution and activity of bacteria in deep granitic groundwaters of southeastern Sweden, *Microb. Ecol.*, 20: 37.

Pedersen, K. and S. Ekendahl. 1992a. Assimilation of CO_2 and introduced organic compounds by bacterial communities in groundwater from southeastern Sweden deep crystalline bedrock, *Microb. Ecol.*, 23: 1.

Pedersen, K. and S. Ekendahl. 1992b. Incorporation of CO_2 and introduced organic compounds by bacterial populations in groundwater from the deep crystalline bedrock of the Stripa mine, *J. Gen. Microbiol.*, 138: 369.

Pedersen, K., S. Ekendahl, and J. Arlinger. 1991. Microbes in crystalline bedrock. Assimilation of CO_2 and introduced organic compounds by bacterial populations in groundwater from deep crystalline bedrock at Laxemar and Stripa, SKB Technical Report, 91-56.

Perfettini, J. V. 1989. Etude de l'Alteration d'un Materiau d'Enrobage des Dechets Radioactifs, le Ciment CPA, par des Microorganismes Heterotrophes Isoles de Milieux Naturels, PhD Thesis, University of Marseille, France.

Perfettini, J. V., E. Revertegat, and N. Langomazino. 1991. Evaluation of the cement degradation induced by the metabolic products of two fungal strains, *Experientia*, 47: 527.

Phelps, T. J., E. G. Raione, D. C. White, and C. B. Fliermans. 1989. Microbial activities in deep subsurface environments, *Geomicrobiol. J.*, 7: 79.

Philp, J. C., N. Christofi, K. J. Taylor, and J. M. West. 1986. Studies on the effects of sulfate-reducing bacteria on mild carbon steel relevant to nuclear waste disposal in the U.K., U.K. DOE Report, DOE/RW/87.065.

Philp, J. C., N. Christofi, and J. M. West. 1984. The geomicrobiology of calcium montmorillonite (Fuller's Earth), U.K. NERC, British Geological Survey Report FLPU 84-4.

Philp, J. C. and K. J. Taylor. 1987. The effect of sulfate-reducing bacteria on the corrosion of mild-steel embedded in a bentonite-granitic groundwater paste, U.K. AERE, R12388.

Philp, J. C., K. J. Taylor, and N. Christofi. 1991. Consequences of sulfate-reducing bacterial growth in a lab-simulated waste disposal regime, *Experientia*, 47: 533.

Rees, J. H. 1989. Gas generation and migration in repositories. Current status, U.K. NIREX Report NSS/G112.

Rees, J. H. and W. R. Rodwell. 1988. Gas evolution and migration in repositories: current status, U.K. NIREX Report NSS/G104.

Rochell, C. A., D. J. Noy, and J. M. West. 1993. Modeling subsurface microbial growth using MGSE (Microbial Growth in Subsurface Environments). Presented at the Int. Symp. Subsurface Microbiology, Bath, U.K. (Abstr.).

Roffey, R. 1990. The Swedish final repository and the possible risk of interactions by microbial activities, *Experientia*, 46: 792.

Roffey, R. and K. Hjalmarsson. 1983. Microbial processes in SFR, the silopart: a theoretical approach and preliminary experiments concerning biodegradation of bitumen. Part 1. FOA Report, C40172-B4 (in Swedish).

Roffey, R. and K. Hjalmarsson. 1984. Microbial processes in the final repository for low and intermediate level wastes. Part 2. FOA Report, C40204-B4 (in Swedish).

Roffey, R. and A. Norquist. 1991. Biodegradation of bitumen used for nuclear waste disposal, *Experientia*, 47: 539.

Rosevear, A. 1986. An assessment of the decomposition of cellulose under alkaline conditions within cement based waste forms, U.K. AERE, R12336.

Rosevear, A. 1991. Review of national research programs on the microbiology of radioactive waste disposal, U.K. Atomic Energy Authority, NIREX Report NSS/R-263.

Rushbrook, P. E. 1985. A survey of possible microbiological effects within shallow land disposal sites designed to accept intermediate level radioactive wastes, U.K. AERE, R11408.

Rushbrook, P. E., Ed. 1987. An assessment of the preliminary microbiological studies on clay cores from Elstow conducted by the Universities of Leicester and Warwick, U.K. AERE Report M3501 (rev.).

Russell, B. F., J. M. Hubbell, and S. C. Minken. 1989. Drilling and sampling procedures to minimize borehole cross-contamination, EG and G Idaho Inc. Report EGG-M-88494.

Sargent, K. A. and C. B. Fliermans. 1989. Geology and Hydrology of the deep subsurface microbiology sampling sites at the Savannah River Plant, South Carolina, *Geomicrobiol. J.*, 7: 3.

Sharland, S. M. 1986. The evolution of the pH in the cementitious components of a model nuclear waste repository. Theoretical Physics Division, Harwell Laboratory, U.K., Report TP1160.

Sharland, S. M. and P. W. Tasker. 1986. The aeration period of a model nuclear waste repository, U.K. DOE Report, DOE/RW/86.120.

Sinclair, J. L. and W. C. Ghiorse. 1989. Distribution of aerobic bacteria, protozoa, algae and fungi in deep subsurface sediments, *Geomicrobiol. J.*, 7: 15.

SKB. 1978. Corrosion resistance of copper canisters for final disposal of spent nuclear fuel, SKB Technical Report, KBS-TR-90.

Snellman, M. 1984. Chemical conditions in a repository for spent fuel, Technical Research Centre for Finland Reactor Laboratory, Report YJY-84-08.

Snellman, M. and H. Uotila. 1984. Chemical conditions in the repository for low- and intermediate-level reactor waste, Technical Research Centre for Finland Reactor Laboratory, Report YJT-84-11.

U.S. DOE. 1987. Subsurface Transport Program: Research summary, Report DOE/ER-0156/4. U.S. Department of Energy, Washington, D.C.

Van Loosdrecht, M. C. M. and A. J. B. Zehnder. 1990. Energetics of bacterial adhesion, *Experientia*, 46: 817.

Walker, S. G., C. A. Fleming, F. G. Ferris, T. J. Beveridge, and G. W. Bailey. 1989. Physicochemical interaction of *Escherichia coli* cell envelopes and *Bacillus subtilis* cell walls with two clays and ability of the composite to immobilize heavy metals from solution, *Appl. Environ. Microbiol.*, 55: 2976.

West, J. M. 1985. The influence of microbial activity on degradation of acetate and EDTA, U.K. DOE Report DOE/RW/89/010.

West, J. M. 1994. Subsurface microbiology — where does the biosphere end?, *Biologist*, 41: 129.

West, J. M. 1995. A review of progress in the geomicrobiology of radioactive waste disposal, *Radioact. Waste Manage. Environ. Restoration*,19: 263-283.

West, J. M., M. A. W. Abbott, and E. J. Rowe. 1986a. Microbial activity tolerances and influences on radionuclide sorption in nutrient depleted natural materials, U.K. NERC, British Geological Survey Report, FLPU 86-8.

West, J. M., M. A. W. Abbott, and E. J. Rowe. 1987a. The influences of sulfate reducing bacteria on ^{137}Cs sorption and their tolerances in nutrient depleted natural materials — A preliminary study, U.K. DOE Report DOE/RW/87.087.

West, J. M. and S. C. Arme. 1985. Tolerance of microorganisms to extreme environmental conditions, U.K. NERC, British Geological Survey Report, FLPU 85-14.

West, J. M., P. J. Chilton, I. N. Gale, S. J. Gardner, and A. R. Lawrence. 1992a. Deep subsurface microbiological research in the United Kingdom, in *Proc. Int. Symp. Environ. Aspects of Pesticide Microbiology,* Swedish University of Agricultural Sciences, Uppsala.

West, J. M., N. Christofi, and S. C. Arme. 1985a. The influence of groundwater and extreme environments on microbial activity and radionuclide transport, NEA Workshop on the Effects of Natural organic Compounds and of Microorganisms on Radionuclide Transport, Paris, June 1985, 19-38.

West, J. M., N. Christofi, and I. G. McKinley. 1985b. An overview of recent microbiological research relevant to the geological disposal of nuclear waste, *Radioact. Waste Manage. Nucl. Fuel Cycle*, 6: 79.

West, J. M., N. Christofi, J. C. Philp, and S. C. Arme. 1985c. Investigations of the populations of introduced and resident microorganisms in deep repositories and their effects on containment of radioactive wastes, U.K. DOE Report, DOE/RW/85/116.

West, J. M., P. Coombs, S. J. Gardner, and C. A. Rochell. 1995. The microbiology of the Maqarin site of Jordan. A natural analogue for cementitious radioactive waste repositories, Materials Research Society Symposium Proceedings Vol. 353. Scientific Basis for Nuclear Waste Management XVIII Kyoto, Japan. pp 181-188. MRS, Pittsburgh.

West, J. M. and H. A. Grogan. 1986. Microbiological research relevant to the Swiss radioactive waste disposal programme. A review of work performed by potential contractors, NAGRA Internal Report.

West, J. M., D. G. Haigh, P. J. Hooker, and E. J. Rowe. 1987b. ^{137}Cs sorption onto Fullers Earth (calcium montmorillonite) — the influence of sulfate-reducing bacteria, U.K. DOE Report, DOE/RW/88.072.

West, J. M., D. G. Haigh, P. J. Hooker, and E. J. Rowe. 1991. Microbial influence on the sorption of radionuclides onto materials relevant to the geological disposal of radioactive waste, *Experientia*, 47: 549.

West, J. M. and I. G. McKinley. 1984. The geomicrobiology of nuclear waste disposal, in *Scientific Basis for Radioactive Waste Management*, McVay, G., Ed., Elsevier, New York, 1984.

West, J. M., I. G. McKinley, and N. A. Chapman. 1982a. Microbes in deep geological systems and their possible influence on radioactive waste disposal, *Radioact. Waste Manage. Nucl. Fuel Cycle*, 3: 1.

West, J. M., I. G. McKinley, and N. Christofi. 1982b. Geomicrobiology and its relevance to nuclear waste disposal. An annotated bibliography, U.K. NERC, Institute of Geological Sciences, Report ENPU 82-7.

West, J. M., I. G. McKinley, H. A. Grogan, and S. C. Arme. 1986b. Laboratory and modeling studies of microbial activity in the near-field of a HLW repository, in *Scientific Basis for Radioactive Waste Management*, Werme, L. O., Ed., Elsevier, New York, 1986.

West, J. M., I. G. McKinley, and A. Vialta. 1988. The influence of microbial activity on the movement of uranium at the Isamu mine, Pocos de Caldas, Brazil, in *Scientific Basis for Radioactive Waste Management*, Proceedings of an MRS meeting held in Berlin, Elsevier Press.

West, J. M., I. G. McKinley, and A. Vialta. 1992. Microbiological analysis at the Pocos de Caldas natural analogue study area, *J. Geochem. Explor.*, 45: 439.

West, J. M. and E. J. Rowe. 1986. Laboratory studies on the production of organic microbial by-products of relevance to radionuclide mobilization and transport. End of Contract Report, NAGRA Internal Report.

West, J. M., E. J. Rowe, G. P. Wealthall, and M. R. Allen. 1989a. The geomicrobiology of the Drigg research site, U.K. DOE Report, DOE/RW/89/093.

West, J. M., E. J. Rowe, G. P. Wealthall, and M. R. Allen. 1989b. The geomicrobiology of the Drigg research site, U.K. NERC, British Geological Survey Technical Report WE 89-8.

West, J. M., A. Vialta, and I. G. McKinley. 1990. Microbiological analysis at the Osamu Utsumi mine and Morro do Ferro analogue study sites, Pocos de Caldas, Brazil, U.K. DOE Report WR 90-050.

Wilkins, J. D. 1988. The influence of organic materials on the near field of an intermediate level waste repository, U.K. DOE Report No. DOE/RW/89/032.

Winograd, I. J. and F. N. Robertson. 1982. Deep oxygenated groundwater. Anomaly or common occurrence, *Science*, 216: 1227.

Wolf, M. 1989. Mikrobieller Abbau von Bitumen (Microbial degradation of bitumen), NAGRA Technical Report, NTB 89-14 (in German).

Wolf, M. and R. Bachofen. 1991. Microbial degradation of bitumen, *Experientia*, 47: 542.

Wyndham, R. C. and J. W. Costerton. 1981. *In vitro* microbial degradation of bituminous hydrocarbons and *in situ* colonisation of bitumen surfaces within the Athabasca Oil Sands deposit, *Appl. Environ. Microbiol.*, 41: 791.

Zhukova, S. V., G. M. Mogilritskii, G. A. Sapozhnikoba, S. A. Eliseenkova, T. B. Tsitovich, T. A. Bykhobskaya, O. I. Volkova, and K. P. Zakharova. 1982. Biostability of compounds based on organic binders formed during solidification of liquid medium activity wastes, *Sov. At. Energy*, 52: 319.

16

Subsurface Microbiology: Effects on the Transport of Radioactive Wastes in the Vadose Zone

Larry E. Hersman

CONTENTS

KEY WORDS: *microorganisms, transport, radionuclides, transuranics, high-level wastes, repository, redox, chelation, sorption, mobilization.*

The processes involved in radionuclide migration from a repository are complex even without the possible influence of microorganisms — **J. M. West et al. (1991)**

16.1   Introduction

The purpose of below-ground nuclear waste disposal is to protect the safety and health of the public by limiting or preventing radioactive wastes (radionuclides) from reaching the accessible environment (the surface or groundwater). To insure this protection, and in agreement with Federal Statute 10 CFR 60, the U.S. Department of Energy (DOE) is using a multi-barrier strategy that combines both engineered and natural containment. In fact, the decision to characterize the Yucca Mountain, Nevada site was due mainly to the presence of the Calico Hills formation, a natural geochemical barrier which lies beneath the proposed location of the high-level repository.

Microbiological investigations pertaining to nuclear repositories in other parts of the world are reviewed in Chapter 15. Microbial investigations are also being conducted for a Canadian repository (Stroes-Gascoyne, 1996, 1994, 1989; Stroes-Gascoyne et al., 1994).

Because of the difficulty in predicting the long-term performance (>10,000 years) of only an engineered barrier system in the altered (near-field) system, it is essential to utilize the multibarrier concept. Inherent in this concept is the necessity of understanding the performance of the natural unaltered system (far-field) with respect to its ability to contain the movement of radionuclides.

Subsurface microbiological work associated with Yucca Mountain is not yet completed. Topics of investigation conducted within the Microbiological Sorption and Tranport Study (Hersman, 1992) include anaerobic microbiology (Clarkson et al., 1996), autotrophic and heterotrophic diversity (Khalil et al., 1996), distribution and limiting nutrients of heterotrophic microbiota (Haldeman et al., 1996; Kieft et al., 1996), *in situ* characterization of microbiota utilizing lipid analyses (Ringelberg et al., 1996), and chelation and transport of metals (Story et al., unpublished data). Other microbiological work associated with Yucca Mountain includes investigation of microbially influenced corrosion (Pitonzo et al., 1996), and other possible impacts that microbiota may have on man-made materials associated with repository construction (Horn et al., 1996; Horn and Meike, 1995).

The geochemistry of both the near-field and far-field barriers significantly influences their performance, and the main performance issue is the transport of radionuclides. Transport consists of the combined effects of flow (hydrology, including diffusion) and retardation (geochemistry). (For more information on microbial transport, see Chapter 15.) Geochemical factors that will influence transport include the mineral content of both the fractures and matrix, groundwater chemistry as it affects solubility of the radionuclide in pore water and ground water, and ion exchange and surface complexation/sorption reactions.

Solubility and sorption are two areas of geochemistry over which microorganisms have a profound influence. The idea that microorganisms are important in the cycles of various elements in nature was first delivered in a speech to the Russian Academy of Sciences in 1897 by S. N. Winogradsky. The ensuing 100 years contains testament contributed by remarkable scientists (Beijerinck, Vernaldskiy, Klyuver, van Niel, Waksman, Zobell, Starkey, Colmer, Temple, Barker, Umbreit, Beck, Baas Becking, Kuznetsov, and others) that has expanded our understanding of the effects of microorganisms on element cycling within the geosphere. Upon their contributions now rest the modern disciplines of geomicrobiology and biogeochemistry. This body of knowledge contains substantial information regarding the potential for microorganisms to affect the solubility and sorption of radionuclides, so much so that recent reviews by Pedersen and Karlsson (1995), and the journal *Experientia* (Issues 46 and 47) discuss microbial involvement with both repository performance and radionuclide transport. The purpose of this chapter is to combine more recent experimental results with our current knowledge of the potential for microbial effects on the transport of radionuclides in an unaltered barrier.

Until recently it was believed that soil microorganisms, both in number and metabolic activity, were largely confined to the "near surface" (within a few meters) environment, due in large part to the relative inaccessibility and remoteness of the subsurface. As stated by Ghiorse and Wilson (1988), the "out-of-sight, out-of-mind" aphorism "aptly described the traditional view of terrestrial subsurface held by generations of human beings." However, the volume of data produced by the DOE Deep Microbiology Subsurface Program, and subsurface programs in Europe, has provided convincing evidence that significant numbers of microorganisms exist, independent of depth, to hundreds of meters below the surface (Fliermans and Hazen, 1990). These microorganisms are present not only in significant numbers, but have significant diversity as well (Fredrickson et al., 1993; Haldeman and Amy, 1993; Kieft et al., 1993; Russell et al., 1994). (For more on this topic, see Chapter 4.) Microorganisms are now considered to be ubiquitous in subsurface environments: hence, there is an increasing interest in the effect(s) that microorganisms may have on the performance of subsurface nuclear repositories.

Microbial studies of the near-field zone (engineered barrier and altered zone) are concerned primarily with the effects that indigenous and introduced microorganisms have on:

1. The corrosion of the primary metal containers (Horn and Meike, 1995; Pedersen and Karlsson, 1995)
2. The degradation of concrete support structures (Perfettini et al., 1991)
3. The production of gas within a repository (Bachofen, 1991)
4. The microbial activity in backfill material (Mayfield and Barker, 1982)
5. The pH and redox (Eh) conditions within a repository (Hersman, 1992)
6. The radio- and thermoresistance of microorganisms in the near-field environment (Horn and Meike, 1995; Stroes-Gascoyne et al., 1994)

The interactions between an indigenous microbial population and radioactive wastes are more subtle in the far-field. Here the total numbers of microorganisms are expected to be lower than found in the near-field, as a result of fewer energy and carbon sources and the limited availability of electron acceptors. Although the potential for interactions is reduced, the performance of the unaltered barrier will be as important as near-field performance because of the time and distance constraints placed on repository performance. For example, in order for Yucca Mountain to be licensed it must be demonstrated that the repository system will retard radionuclide transport for 10,000 years over a distance of approximately 1000 m (to ground water); roughly equal to an average transport rate of 0.1 m (approximately 4 in.) per year. Obviously, a small error made in calculating transport rates could have a profound effect on predicting performance. Therefore, it is essential that all aspects of geochemistry affecting transport be considered carefully.

16.2 Subsurface Vadose Zone Microbiology

The unsaturated (vadose) zone is important because some repositories, such as Yucca Mountain, will be located within a vadose zone environment, and therefore represents the path that radionuclides and other contaminants must pass through to the vadose zone to reach groundwater. Several investigators have demonstrated that microorganisms are common in subsurface vadose zone environments at the Hanford Site, WA; the Idaho National Engineering Laboratory, ID; the Nevada Test Site,

NV; Los Alamos National Laboratory, NM; the Cerro Negro Site, NM; and in unsaturated sediments at various locations below the eastern coastal plain of the U.S. (Amy et al., 1992; Brockman et al., 1992; Colwell, 1989; Fredrickson et al., 1993; Haldeman et al., 1994; Hersman et al., 1988; Kieft et al., 1995; Konopka and Turco, 1991; Russell et al., 1994).

The results of studies on vadose zone environments vary greatly. For example, some of the investigators have reported lower total numbers of vadose zone bacteria than found in (1) saturated sediments (Brockman et al., 1992; Colwell, 1989; Haldeman and Amy, 1993; Johnson and Wood, 1992; Konopka and Turco, 1991), (2) in parent material (Fredrickson et al., 1993), or (3) in paleosols (buried soils) (Brockman et al., 1992). Conversely, other studies have shown that numbers and microbial activities were not higher in the saturated zones (Kieft et al., 1995), numbers did not correspond to depth (Haldeman et al., 1994), and diversity and numbers were not higher in paleosols (Haldeman et al., 1994; Kieft et al., 1995). Furthermore, some studies have indicated that higher microbial populations could be correlated to a coarser texture of the matrix (Konopka and Turco, 1991), while others found the opposite (Kieft et al., 1995).

Vadose zones are, by definition, unsaturated. However, it is believed that water is not limiting to microorganisms. Matric water potentials in vadose zones are usually between 0 and –0.5 mPa (Long and Rawson, 1991; McElroy and Hubbell, 1989). Furthermore, the solute diffusion rate does not become limiting until the matric water potential approaches –0.1 mPa (Griffin, 1981). In light of this information, Kieft et al. (1993) stated that "matric water potentials in vadose zones are, therefore, not sufficiently low to cause a direct desiccation stress", and even if desiccation were a factor "it appears that the majority of culturable cells …[are] desiccation resistant." However, it should be mentioned that low water potential can limit the mobility of microorganisms. Generally, microbial movement becomes limited when the water potential drops below –0.1 mPa. At that point the average water film thickness (<1.0 μm) restricts the movement of microorganisms (Griffin, 1981; Pappendick and Cambell, 1981).

Although potentially limiting, energy sources could be available in the vadose zone. The energy required for minimal microbial metabolism in the vadose zone can be derived from a variety of sources, such as endogenous metabolism (Boylen and Ensign, 1970; Dawes, 1985; Kieft et al., 1993; Rosacker and Kieft, 1990), metabolism of exogenous organic material (e.g., humic material remaining in paleosols) (Kieft et al., 1993), or metabolism of diffused, gaseous substrates (volatile organics, H_2, CH_4, NH_3). Concerning the metabolism of humic materials, Fredrickson et al. (1993) concluded that "the ability of microorganisms in vadose zone sediments to respond to artificial recharge and/or contaminants suggests that indigenous microorganisms have maintained cellular integrity over long time periods by metabolizing, at an extremely low rate, organic carbon that accumulated during pedogenesis." Although their metabolism is not completely understood, Morita (1990) emphasized that

> the starvation state of microorganisms in ecosystems is real, and should be considered as the normal state of microorganisms in nature. Unfortunately, microbiologists think in terms of cells grown in rich medium where ideal conditions exist for growth and reproduction. This idealized situation generally is not the real world.

The vadose zone is certainly not an idealized environment, yet it is inhabited by microorganisms, many microorganisms.

Many microorganisms in a subsurface vadose zone environment may be attached to the matrix (Bar-Or, 1990). Dawson et al. (1981) proposed that microorganisms use adhesion as a tactic in starvation survival, and later Kjelleberg and Hermansson (1984) demonstrated an increased degree of irreversible binding upon starvation. This topic will be discussed in greater detail in Section 16.3.7.

Regardless of their origin, microorganisms appear to permanently inhabit the vadose zone. One can conclude that the microbial communities in a vadose zone's far-field environments may consist of small cells (Lappin-Scott and Costerton, 1990) that are attached to the matrix. The cells may have a high proportion of extracellular polysaccharides (Pitonzo et al., 1995) (to aid in attachment and water retention). Some cells may be old and metabolizing slowly. The community may be heterogeneous, with many metabolic groups represented (i.e., heterotrophs, autotrophs, aerobes, facultative anaerobes, anaerobes, etc.). The distribution of microbiota may exhibit spatial heterogeneity (Haldeman and Amy, 1993; Kieft et al., 1993). For more information regarding spatial heterogeneity, see Chapter 6.

It may be tempting, therefore, to dismiss the potential effects of microorganisms on the transport of radionuclides. After all, the cells are small, and generally relatively inactive. Pedersen and Karlsson (1995) wisely advise us otherwise, pointing out that 0.1 g of catalase (a microbial enzyme produced to reduce H_2O_2 to H_2O and O_2) has the same reducing power as a metric ton of Fe^{2+}. Fe^{2+} is a strong reducing agent, able to reduce other metals such as uranium. (See Section 16.3.4 for more about redox reactions and solubility.) So, although the numbers of microorganisms in a subsurface environment may be small, they should not be ignored until their catalytic power has been evaluated (Pedersen and Karlsson, 1995). It would be prudent to assume that subsurface microbiota will affect transport of radionuclides in a manner similar to other, more abundant and diverse microbial communities, but perhaps at a slower rate unless environmental conditions change.

16.3 Microbial Effects on Transport

Francis (1990) has outlined the various types of wastes generated by the nuclear industry, including organics and heavy metals, as well as radionuclides. Although all of these wastes are significant, the present discussion will be limited to those wastes most common to a high-level nuclear repository. High-level wastes (HLW) are either discarded spent fuel rods, or processing wastes that contain most of the fission products and actinides common to reprocessing. Because HLW require extensive shielding, they also require long-term isolation from the biosphere; hence, deep geologic repositories are now being considered by several European countries, the U.S., and Canada as a safe means of HLW disposal (see Chapter 15). By definition, transuranic (TRU) wastes contain alpha-emitting nuclides with half-lives that exceed 20 years and concentrations greater than 100 nCi per gram of waste. Spent fuel reprocessing, fabrication of nuclear weapons, and the recycling of plutonium fuel to nuclear weapons all generate TRU wastes. TRU wastes include the radionuclides ^{232}Th, ^{233}U, ^{235}U, ^{238}U, ^{237}Np, ^{238}Pu, ^{239}Pu, ^{240}Pu, ^{241}Pu, ^{242}Pu, ^{241}Am, ^{244}Cm, and ^{252}Cf.

How will the vadose zone microbial community affect the transport of radionuclides in a far-field environment? In order to answer this question one must examine the various types of interactions that microorganisms have with radionuclides, and heavy metals in general, because each of these interactions will potentially be performed by the vadose zone subsurface microbiota. As presented in Figure 16.1, a radionuclide could be transported by unsaturated flow to a region in the far-field in one of several chemical forms: as a dissolved species (Nitsche et al., 1993), chelated by

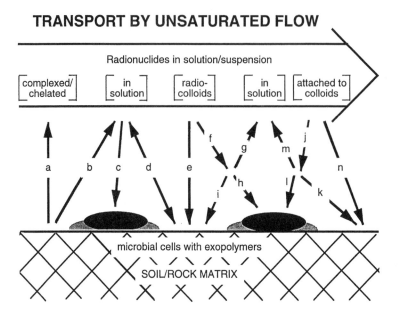

FIGURE 16.1
Interactions between radionuclides in solution/suspension and indigenous microorganisms, where: (a) adsorbed radionuclides are complexed/chelated; (b) dissolution occurs; (c) radionuclides are sorbed by microorganisms; (d) various redox reactions result in solubilization or precipitation; (e and n) agglomeration of colloids; dissolution of (f) radiocolloids or (j) radionuclides attached to colloids, leading to (g and m) solubilization, (h and l) sorption, or (i and k) adsorption/precipitation.

either naturally occurring organics (Francis and Dodge,1993) or microbially produced ligands (Barnhart et al., 1979), sorbed to natural colloids (Buddemeier and Hunt, 1988) and microorganisms (Stroes-Gascoyne and West, 1994), as radiocolloids which are formed by the aggregation or polymerization of hydrolyzed transuranium ions (Kim, 1993), or complexed to naturally occurring anions like CO_3^{-2}, SO_4^{-2}, PO_4^{-3}, and Cl^-. As stated earlier, it is believed that in a vadose zone environment most microorganisms will be attached to the matrix, therefore the transport of radionuclides as sorbed onto mobile microorganisms will not be discussed in this chapter. Certainly, however, in some environments (e.g., those that are saturated, or exhibit fracture flow) transport by mobile microorganisms could be very significant.

Microorganisms are capable of altering the geochemistry of the surrounding environment. (Certainly the ability of the sulfur-oxidizing thiobacilli to create acid mine drainage has caused public concern.) Again, it is important to reemphasize that the indigenous community will probably have a low metabolic activity, and that it will be in relative equilibrium with the geochemistry of the surrounding environment.

Although the indigenous populations may not be able to change the macro environment, they certainly do make changes in their immediate environment. Revsbech and Jorgensen (1986) reviewed the use of various microelectrodes to measure gradients of oxygen, hydrogen sulfide, and pH within microbial mats, biofilms, and around individual cells. Their studies have contributed to the universal acceptance that such gradients exist in and around microorganisms, and that diffusional constraints cause these gradients, thereby establishing the great diversity of microorganisms that would otherwise not occur in such close proximity (Focht, 1992). In a subsurface environment, therefore, exists a continuum of individual microorganisms and consortia intimately surrounded by nutrient, ionic, gaseous, redox, and pH gradients; none of which are reflected in the overall bulk chemistry of the groundwater.

TABLE 16.1

Solubility of Neptunium, Plutonium, and Americium in J-13 Groundwater at pH 5.9, 7.0, and 8.5 and at 25°, 60°, and 90°C

pH	25°C	60°C	90°C
Np Steady-State Concentration (M)			
5.9	5.3×10^{-3}	6.4×10^{-3}	1.2×10^{-3}
7.0	1.3×10^{-4}	9.8×10^{-4}	1.5×10^{-4}
8.5	4.4×10^{-5}	1.0×10^{-4}	8.9×10^{-5}
Pu Steady–State Concentration (M)			
5.9	1.1×10^{-6}	2.7×10^{-8}	6.2×10^{-9}
7.0	2.3×10^{-7}	3.7×10^{-8}	8.8×10^{-9}
8.5	2.9×10^{-7}	1.2×10^{-7}	7.3×10^{-9}
Am Steady–State Concentration (M)			
5.9	1.8×10^{-9}	2.5×10^{-6}	1.7×10^{-9}
7.0	1.2×10^{-9}	9.9×10^{-9}	3.1×10^{-10}
8.5	2.4×10^{-9}	1.2×10^{-8}	3.4×10^{-10}

It is along, into, and out of this microbial community and concomitant gradients that radionuclides may be transported.

16.3.1 Dissolved Species

Because the groundwater flow rate through the waste is expected to be sufficiently low to allow saturation of water, solubility of a radionuclide is a function of saturation limits at a given pH, redox, temperature, osmotic potential, concentration of complexing species, and the oxidation state of the radionuclide. The solubility of radionuclides in groundwater constitutes the first limiting factor in radionuclide transport. Presented in Table 16.1 are summaries of solubility experiments for neptunium, plutonium, and americium at three temperatures (25°, 60°, and 90°C) and three pH values (5.9, 7.0, and 8.5). To summarize, Nitsche et al. (1993) found the following:

1. The neptunium solubility decreased with increasing temperature and pH
2. Plutonium concentrations decreased with increasing temperature and showed no trend with pH
3. The americium solutions showed no clear solubility trend with increasing temperature and increasing pH.

One would anticipate that the chemical and physical conditions within the near-field of a repository would differ from those of the far-field environment. As a radionuclide is being carried away from the near-field to the far-field, the temperature would decrease and the pH may rise (as anticipated at Yucca Mountain). Nitsche et al. (1993) argue that these changing conditions would affect the radionuclides differently, varying from profound effects on neptunium to little effect on americium. It should be noted that in a later study using different well water, Nitsche et al. (1994) found similar results for neptunium and plutonium, but americium solubility decreased significantly with increasing temperature and increased somewhat with increasing pH. Unfortunately, neither study addressed redox or osmotic potential effects on solubility, both of which further complicate the determination of radionu-clide solubility in a natural environment. Obviously, the solubility of a radionuclide can become very complicated and difficult to model.

A dissolved radionuclide would then be involved in a series of sorption/desorp-tion reactions as described by Triay and co-workers (Triay et al., 1993, 1992). Like sol-ubility, the sorption of actinides onto the solid phase has many determinants, such as the species of the actinide and the mineralogy of the matrix. For example, Pabalan et al. (1993) have suggested that the primary sorbing species by surface complexation are neutral hydroxides, where the sorption of uranium is proportional to the amount of $UO_2(OH)_2$ present in solution.

Second, the types of minerals present in the surface of the matrix have a profound effect on sorption. Meijer (1994) summarized data for many Yucca Mountain Project (YMP) investigators, detailing the sorption data for the 20 elements of interest. Briefly, for Am, Pu, and Np, the following were points that relate to the YMP proposed repos-itory site:

1. For most minerals tested, Am likely is sorbed by at least two different mechanisms. At lower pH values (<6.0) ion exchange reactions appeared to dominate. At intermediate values (6 to 8) inner sphere surface complex-ation reactions dominated, while at higher pH ranges (8 to 10) carbonate complexation reactions appeared to compete with surface complexation.

2. The presence of calcite and clay promoted the sorption/coprecipitation of Pu, while zeolitic samples typically had lower sorption values than vitric or devitrified samples.

3. Np did not appear to have high affinity for ion exchange reactions on clays and zeolites; but it did display high sorption values for ferric oxides and oxyhydroxides, apatite, and attapulgite (a magnesium-rich clay). Np had somewhat lower affinity for carbonates (e.g., calcite) and sulfates (e.g., anhydrite), and lowest affinity for silicates.

16.3.2   Microbial Sorption

The soluble radionuclides will participate in sorption/desorption reactions with the minerals on the surface of the matrix, and microorganisms can affect this interaction in several ways (Figure 16.1). Although metal sorption by microorganisms has been discussed extensively in the literature (see Beveridge and Doyle, 1989), it is important to emphasize those aspects of microbial/metal interactions relevant to radionuclide transport. The attached microbial population would occupy a percentage of the total matrix surface area with both cells and exopolymers. This combination of cells and exopolymers would have different sorption characteristics than the bare surface of the matrix. The outer cellular membrane of microorganisms has a net negative charge and generally is regarded as being active in binding heavy metal cations. For Gram-positive bacteria, carboxyl groups of the glutamic acid of peptidoglycan is a major site of metal deposition, as are teichoic and teichuronic acids. Gram-negative bacteria tend to collect metals at the polar head groups of the constituent membrane bilayers of the envelope, or along the peptidoglycan layer between the bilayers in the periplas-mic space (Gadd, 1990). Likewise, the carboxyl residues in exopolysaccharides and the uronic acids in extracellular capsules serve as electron donor groups that form bonds with metals (Geesey and Jang, 1981), are very efficient metal binding agents, and are of major importance in metal removal from solution.

Microorganisms sorb a variety of metals as a means of protection against the toxic effects of the metals. Generally, metals are only toxic when they have activity as ions. By removing metals from solution, microorganisms reduce the activity of a metal, thereby reducing its toxicity. The simplest mechanism used by microorganisms is to deposit a metal on its outer surface. The binding of the metal to the outer surface is specific; the charge, ionic radius, and coordination geometry are the predominant factors. The deposition of the metal is facilitated by the activities of membrane-associated sulfate-reductases, or through the biosynthesis of oxidizing agents such as oxygen or hydrogen peroxide. In this way, metals such as iron (an analog to some radionuclides) are deposited on the surfaces of many species of bacteria; uranium is deposited on the surface of fungi; and iron, nickel, copper, aluminum, and chromium are deposited on the surface of algae (Wood and Wang, 1983). A second means of surface accumulation occurs through the production of extracellular ligands that complex metals outside the cell and prevent their uptake. Both cyanobacteria and green algae have been found to concentrate nickel on their surface to 3000 times greater than the concentration of the culture medium (Wood and Wang, 1983).

Another measure adopted by microorganisms to prevent metals from reaching toxic levels is the biosynthesis of intracellular traps for the removal of metals from solution. One example is the biosynthesis of sulfhydryl-containing metallothionein protein to bind cadmium and copper (Wood and Wang, 1983). Similarly, poly(γ, glutamylcysteinyl)-glycines produced by *Schizosaccharomyces pombe* (a fisson yeast) bind cadmium (Robinson and Jackson, 1986). Most notably, Strandberg and coworkers (1981) reported the intracellular accumulation of uranium by *Pseudomonas aeruginosa*. Although they did not elaborate on a specific mechanism, the uptake of uranium was quite fast, as if mediated by active transport.

By depositing metals, internally or externally, microorganisms are not only protecting themselves from the toxic effects of the metal ions but are also, in effect, concentrating the metal on the matrix surface to an extent greater than would be done by the matrix alone. Hersman (1986) demonstrated that a *Pseudomonas* sp., on a per-gram-dry-weight basis, sorbed Pu(IV) nearly 10,000 times greater than did tuff from a Yucca Mountain repository formation (Table 16.2). Many other researchers have also detailed the sorption of radioactive elements by microorganisms (Faison et al., 1990; Treen-Sears et al., 1984; Shumate et al., 1978; Wildung and Garland, 1982).

Overall, one could predict that with respect to sorption, microorganisms may decrease the amount of radionuclides in solution, thereby reducing the potential for transport.

TABLE 16.2

Optical Density, Dry Weights, Colony Forming Units (CFU), and Sorption Ratio (Pu Solid Phase/Pu Aqueous Phase) for a *Pseudomonas* sp. Over a 48-h Incubation at 22°C

Incubation Time (h)	Optical Density (Klett Unit)	Dry Weight (g)	CFU (per ml)	Sorption Ratio (g/ml)
3	7.53	0.00080	1.0×10^7	$9,280 \pm 707$
6	126.8	0.00179	8.8×10^7	$9,840 \pm 4417$
12	174	0.01411	2.1×10^9	$2,156 \pm 161$
24	173	0.01474	2.7×10^9	$2,080 \pm 230$
48	155	0.01547	3.4×10^9	$4,800 \pm 342$
Control (1 g crushed tuff)	—	1.000	—	5.37 ± 0.56

TABLE 16.3

Functional Groups, and Complexing Compounds Produced
by Microorganisms

Basic		Acidic	

Functional Groups

$-NH_2$	(amino)	$-CO_2H$	(carboxylic)
$=NH$	(imino)	$-SO_3H$	(sulfonic)
$-N=$	(tert. acyclic or	$-PO(OH)_2$	(phosphonic)
	heterocyclic nitrogen)	$-OH$	(enolic, phenolic)
$=CO$	(carbonyl)	$=N-OH$	(oxime)
$-O-$	(ether)	$-SH$	(thioenolic and
$-OH$	(alcohol)		thiophenolic)
$-S-$	(thioether)		
$-PR_2$	(substituted phosphine)		
$-AsR_2$	(substituted arsine)		

Complexing Compounds Produced by Microorganisms

Tricarboxylic acids	Humic acids
Catachol	Fulvic acids
Hydroxamate	
Organic acids (e.g., oxalate, salicylate, acetate, lactate, pyruvate, citrate, and polypeptides)	
Uncharacterized, low molecular weight anions, ions, and organic ligands	

Adapted from Birch, L. and Bachofen, R., *Experientia*, 46, 827, 1990.
With permission.

16.3.3   Complexation/Chelation

A second, and very significant, interaction between soluble radionuclides and vadose zone microorganisms is complexation/chelation. A radionuclide complexed or chelated to an organic ligand would not participate in sorption/desorption reactions as readily as an uncomplexed metal, and would therefore have a potentially higher transport rate. Microorganisms produce a myriad of extracellular compounds, containing a variety of functional groups that complex strongly with metals (Table 16.3). When a compound attaches to a metal with two or more functional groups, forming a ring structure, then this complexation is called chelation (from the Greek "chele" meaning lobster claws) (Birch and Bachofen, 1990; Dwyer and Mellor, 1964). In soil systems Birch and Bachofen (1990) described two groups of microbial compounds which are considered complexing agents:

1. By-products of microbial metabolism and degradation, such as simple organic compounds (low molecular weight organic acids and alcohols), and macromolecular humic and fulvic acids.

2. Microbial exudates induced by metals ions. These include iron-binding siderophores produced in response to low iron concentrations, and toxic-metal binding proteins.

By-products of microbial metabolism — Francis (1982) found low molecular organic acids and alcohols in the leachate from shallow land burial sites of radionuclides. As a rule, the original organic materials present will dictate the types of organic compounds which will be produced by microorganisms. Although various organic acids released by microorganisms have been investigated for metal complexation ability, it is difficult to evaluate the complexation capabilities of the metabolic products as their production is dependent on environmental conditions. The degree of relative metabolic activity in the vadose zone will determine the amount of small molecular weight complexing agents. The metal binding effects of large macromolecular humates (collectively humic and fulvic acids) is well documented in the literature. The following is a brief example of their interactions with metals:

1. Humates alter the adsorption onto particulates (Davis and Leckie, 1978) bioavailability, and toxicity of some heavy metals (Giesy et al., 1977).

2. Humates also reduce the radionuclides Np(V) (Nash et al., 1981), Pu(V), and Pu(VI) (Bondietti et al., 1976).

3. Humates can solubilize [241]Am and Th (Miekeley and Kuchler, 1987; Nash and Choppin, 1980; Sibley et al., 1984).

One could anticipate that the effects of humates will be limited in the vadose zone, because their concentration is minimal in that environment. Nonetheless, even small amounts will affect the concentration of soluble radionuclides.

Microbial exudates: siderophores — Microorganisms have an essential requirement for Fe (a small group of homolactic fermenting bacteria, the lactic streptococci, are the only exception), for example, Fe is used for DNA synthesis, electron transfer proteins, and nitrogen fixation. However, iron suffers from extreme insolubility in aerobic environments at physiological pH; consequently, aerobic and facultatively anaerobic microorganisms have developed a high affinity system for Fe(III) acquisition. In 1981 Neilands (Neilands, 1981) stated that "Siderophores are viewed as the evolutionary response to the appearance of O_2 in the atmosphere, the concomitant oxidation of Fe(II) to Fe(III), and the precipitation of the latter as ferric hydroxide, Ks = $<10^{-38}$."

Siderophores are low-molecular-mass, ferric-specific ligands that are induced at low iron concentrations for the purpose of biological assimilation of Fe(III). Over 200 siderophores have been isolated, of which several score have been fully characterized, with most being classified chemically as either catechols or hydroxamates. Siderophores have a very strong binding affinity for Fe(III), with complex formation constants in the range of 10^{30} or higher. An effective ligand of Fe(III) chelation has a molecular mass of 500 to 1000 Da; however, such a molecular size exceeds the free diffusion limit of the outer membrane of microorganisms. In order to compensate, microorganisms have specialized surface proteins (receptors) and associated internal enzymes and proteins that transport the siderophore-Fe(III) complex to the sites of physiological utilization (Inouye, 1979).

The most common siderophores contain the hydroxamic acid functional group, –R–CO–N(OH)–R′, which forms a five-membered chelate ring with Fe^{3+} through the two oxygen atoms, with the hydroxyl proton being displaced. Frequently three hydroxamate groups are found on a single siderophore molecule and thus supply six oxygen ligands to satisfy the preferred hexacoordinate octahedral geometry of Fe(III). Other bidentate ligand moieties (e.g., catechol, α-hydroxy acid, 2-(2-hydroxyphenyl)-oxazoline, and a fluorescent chromophore) also are found in siderophores (Bossier et al., 1988). Like the hydroxamates, catechols can occur in triplicate as the only unit in a siderophore molecule (e.g., the siderophore enterobactin). It is also common to find siderophores that contain a mixture of functional moieties, such as pseudobactin which contains a hydroxamate, an α-hydroxy acid, and a fluorescent chromophore.

Although the common characteristic of all siderophores is that they are ferric specific, with some reported to have 1:1 complex formation constants with Fe(III) that exceed 10^{50}, Neilands (1981) and Raymond et al. (1984) believe that because Fe(III) and Pu(IV) are similar in their charge/ionic radius ratio (4.6 and 4.2, respectively), Pu(IV) and other (IV) actinides may serve as analogs to Fe(III) and could therefore be chelated by siderophores. Indeed, Hersman et al. (1993) reported that a siderophore purified from a *Pseudomonas* sp. formed complexes with ^{239}Pu(IV). Additionally, siderophore was found to significantly increase the transport of Fe(III) through an unsaturated column of crushed tuff (Story et al., 1996). It is possible that radionuclides could be transported in the environment via the siderophore/iron transport system. Therefore, complexation/chelation would increase the amount of radionuclides in solution, thereby increasing their availability for transport.

16.3.4   Redox Reactions

Radionuclide speciation, which includes (1) the identification of the radionuclide, (2) its oxidation state, and (3) the formula and molecular structure of the ionic and/or solid complex (i.e., the stoichiometry and structure of the metal ion as complexed by ions and/or ligands), is fundamentally important to the determination of radionuclide solubility (Clark et al., 1994). Hence, the solubility of a radionuclide is related to its oxidation state. For example, Pu(IV) is the least soluble oxidation state for all the Pu valences, and U(III) is less soluble than the more oxidized states of uranium. If microorganisms affect the redox chemistry of radionuclides, they would also affect the solubility of the radionuclides.

There is a large body of literature on metal redox reactions that are mediated by microorganisms. In the geosphere, cycling of metals by microorganisms is the foundation for the disciplines of biogeochemistry and geomicrobiology. Microorganisms are able to both directly and indirectly affect redox changes in metals (Erlich, 1990; Nealson 1983a,b). Examples of direct effects are:

1. The oxidation of a metal as a source of energy and electrons
2. The use of a metal as a terminal electron acceptor for respiration, thereby reducing the metal

Examples of indirect effects are:

1. The nonspecific oxidation of metals at the outer surface of the bacterial envelope
2. Reductions of secondary metals caused by the presence of a microbially reduced primary metal, such as the reduction of Fe(III) to Fe(II) via respiration, and the subsequent reduction of surrounding metals by Fe(II)

Pedersen and Karlsson (1995) presented an excellent review of specific microbially mediated metal redox reactions, where they discuss the oxidation of iron, manganese, hydrogen, ammonia, sulfur, and carbon; and the reduction of iron, manganese, nitrate, sulfate, and carbon. Although these discussions and others common in the literature provide mainly circumstantial evidence for the possibility of radionuclide transformations, there is some evidence of microbial involvement. For example, uranium reduction has been reported by several investigators. Uranium is reduced by a variety of bacteria, and dissimilatory iron-reducing bacteria were also shown to be able to obtain energy for growth with the concomitant transfer of electrons to U(VI)

(Lovley et al., 1991). Sulfate reducing bacteria have also been shown to reduce U(VI) with rates comparable to iron reducing bacteria (Lovley and Phillips, 1992). In a later study, the importance of cytochrome C_3 in the reduction was determined, thereby demonstrating that the reduction of uranium was enzymatic and not indirect (Lovley et al., 1993). Francis et al. (1994) have also confirmed enzymatic reduction of uranium by a *Clostridium* sp.; they suggested that direct uranium reduction should occur in the sequence Mn(IV) > U(VI) > Fe(IV) based upon the free energies of reduction: –83.4, –63.3, and –27.2 kcal·mol^{-1} CH_2O for Mn, U, and Fe, respectively. The solubilization of PuO_2(s) by Fe(III)-reducing strains of *Bacillus polymyxa* and *B. circulans* has been reported by Rusin et al. (1994). Although direct reduction was not confirmed, their results implicate the involvement of metal-reducing bacteria in the reduction of a radionuclide.

It is less likely that microorganisms could derive energy directly from the oxidation of radionuclides, particularly at neutral pH. However, indirect oxidation of radionuclides by microorganisms is very probable given the reactive nature of the cell-wall surface. Metal oxides containing Fe are readily deposited on the surfaces of several microbial species, most notably sheathed bacteria. Although the evidence for enzymatic oxidation is equivocal (Ehrlich, 1990), it is recognized that these bacteria do accumulate large amounts of Fe oxides.

Perhaps the indirect mechanism with the most potential for affecting the redox state of radionuclides is the effects of microbially produced oxidized or reduced metals. For example, Mn(III), an intermediate in the microbial reduction of Mn(IV) to Mn(II) or in the microbial oxidation of Mn(II), is a strong oxidant (having a reduction potential close to molecular oxygen). Several studies have discussed the potential importance of dissolved Mn(III) as an environmental oxidant in marine and freshwater sediments (Burdige, 1993; Davidson, 1993; Kostka et al., 1995; Luther et al., 1994). The interrelatedness of Mn and Fe as environmental redox mediators has been discussed in detail by Nealson and Myers (1992). The oxidized and reduced state of these metals, as a function of biological activity, affect the oxidation state of each other and other metals in the surrounding environment.

Because the effect of redox on solubility varies with each radionuclide, the overall effect of microbially mediated redox reactions would have varying effects on radionuclide solubility.

16.3.5 Biodegradation of Organic/Radionuclide Complexes

The microbial metabolites and naturally occurring organic molecules that can complex/chelate radionuclides were discussed earlier. Complexed/chelated radionuclides may therefore be transported to a new region of the vadose zone, and would become subject to microbial degradation. All of the microbial metabolites that complex/chelate radionuclides are subject to microbial degradation, particularly in an environment of limited nutrients. Low molecular weight organic acids, alcohols, and siderophores could all serve as nutrients for indigenous microbial populations. Unfortunately, there have been relatively few studies on the degradation of ligands complexed to radionuclides. Francis and Dodge (1993) have investigated the biodegradation of iron-citrate complexes. They found that the rates of degradation varied, depending on the structure of the metal and citric acid complex formed. For example, when ferric iron formed a bidentate complex with citric acid, the complex was degraded rapidly. However, when ferrous iron formed a tridentate complex with citrate, biodegradation did not occur until the complex underwent oxidation and hydrolysis conversion to a ferric citrate tridentate complex.

Generally, the biodegradation of metal-complexing agents may result in precipitation of the released ion as water-insoluble hydroxides, oxides, or salts, and thus retard transport (Figure 16.1). The stability of naturally occurring and microbially synthesized radionuclide-complexing agents is critical for understanding the mobility of radionuclides in a natural environment (Francis, 1990).

Further research is needed to determine what effects the biodegradation of complexed/chelated radionuclides will have on transport.

16.3.6 Dissolution

Microbially enhanced dissolution reactions will affect the stability of radionuclides in radiocolloids, and radionuclides attached (either sorbed or precipitated) to natural colloidal particles, to the matrix surface, or to the surfaces of microorganisms (Figure 16.1).

As discussed earlier, there is an atomic similarity between Fe(III) and the (IV) valence actinide metals. To better understand how microorganisms can affect the solubility of radionuclides, it is useful to review the dissolution of Fe oxides. There are three general categories of dissolution: (1) proton promoted, (2) ligand promoted, and (3) reductive dissolution. For each of these, the first step involves the formation of a surface complex by fast adsorption of protons, ligands, electron donors, or a combination of these. This leads to a polarizing and thereby a weakening of the Fe-O bond, and then to the detachment of the Fe atom into solution. In dissolving a Fe (III) (hydr) oxide complex the coordinative environment of Fe (III) changes. The Fe^{3+} in the crystalline structure replaces its O^{2-} or OH^- by water or another ligand. In reductive dissolution, it also changes its oxidation state to Fe^{2+} (Stumm and Sulzberger, 1992).

In proton-promoted dissolution, protons are adsorbed to the surface and eventually polarize neighboring Fe-OH groups, followed by the detachment of the Fe (III) cation. The original surface structure with neutral OH or OH^+ groups is then restored by the adsorption of aqueous protons and the process starts again. As discussed earlier, microorganisms are capable of creating acidic conditions, most notably though the production of sulfuric acid, or by excreting any one of numerous organic acids. Therefore, proton-promoted dissolution is certainly one mechanism of dissolution that microorganisms could utilize in a subsurface environment.

Through surface adsorption, organic ligands may either accelerate or retard dissolution, depending upon whether they weaken the Fe-O bond or block adsorption sites. The weakening effect depends on the structure of the surface complex. Generally a bidentate complex is more effective at bond weakening than a monodentate one.

Due to the high formation constants of siderophore/Fe(III) complexes, it has been assumed that the thermodynamic pressure to trap ferric Fe is so great that Fe oxides will undergo ligand-promoted dissolution in the presence of a siderophore ligand. While this may be true, until recently there was no quantitative evidence on this subject in the literature. Hersman et al. (1995) demonstrated that the dissolution of hematite by the siderophore compared favorably to the dissolution of hematite by oxalic and ascorbic acids, and to proton-promoted dissolution. Brainard et al. (1992) were also able to solubilize hydrous plutonium (IV) oxyhydroxide using siderophores.

Reductive dissolution of Fe (III) oxides operates by a mechanism similar to that of ligand-promoted dissolution. Electron transfer occurs following the adsorption of the electron donor species onto the oxide surface. The reduction of Fe (III) will destabilize the coordination sphere and induce the detachment of the Fe as Fe^{2+}.

There is an increasing body of evidence in the literature supporting the concept that microorganisms and plants can assimilate Fe through extracellular reductive dissolution (it is important to remember that intercellular iron reduction is well understood and believed to be the means by which iron is removed from a siderophore within a cell). For plants, Grusak et al. (1990), Sijmons et al. (1984), Romheld and Marschner (1986), and Bienfait (1988) have discussed the uptake of iron by reduction. Meanwhile, Lesuisse and co-workers (1990, 1991, 1987) have demonstrated that *Saccharomyces cerevisiae* can reduce extracellular ferric chelates by a plasma membrane-bound redox system that is induced under Fe-deficient conditions.

In a series of publications dating back to 1986, Lovley and co-workers (Lovley, 1991; Lovley and Phillips, 1986a,b, 1987, 1988a,b, 1992) have discussed dissimilatory iron reduction. In this process, ferric iron is used as a terminal electron acceptor (instead of oxygen) by microorganisms during respiration. This metabolic process is commonly called anaerobic respiration. Although the reduction does not provide nutritional iron, Lovely's group demonstrated that iron reduction by bacteria does occur. More importantly, Lovley's group has coupled the biodegradation of toxic wastes to iron reduction in anoxic sediments, and has heightened our awareness of not only a greater potential for iron cycling, but also an increased rate of biodegradation of toxic wastes in anoxic environments.

In related studies, Arnold and co-workers (1986a,b, 1988, 1990) also have examined dissimilatory Fe(III) reduction by bacteria. In 1986, they reported that, under certain conditions, "dissimilatory iron reduction appeared to be uncoupled from oxidative phosphorylation", meaning that iron reduction was occurring independently of respiration. They went on to suggest that there may be a branching in the electron transport chain, one being constitutive and used for oxidative phosphorylation, while a second (shorter) branch serves as a sink for cellular reducing power (i.e., electrons). In other words, it may be possible for aerobic microorganisms to produce reducing power for use in iron uptake.

Like iron, the radionuclides are redox active metals; therefore their dissolution should be enhanced by redox reactions mediated by microorganisms. Listed in Table 16.4 are the reduction potentials for several metal oxyhydroxides. Because the reduction potentials of hydrous PuO_2 and geothite (α-FeOOH) are similar, one would suspect that microorganisms capable of reducing goethite may be able to reduce Pu(IV) oxyhydroxides (K_{sp} approximately = 10^{-57}) to Pu(III) hydroxide (K_{sp} = $10^{-22.6}$), thus increasing its solubility significantly (Kim and Kanellakopulos, 1989; Morse, 1986). Unfortunately there are few quantitative discussions in the literature regarding microbially mediated dissolution of radionuclides. One notable exception is the recent work of Rusin et al. (1994). In this experiment, several iron-reducing *Bacillus* species were observed to promote the dissolution of hydrous PuO_2. Furthermore, this effect was enhanced by the presence of nitrilotriacetic acid (NTA), which was added to the reaction vessels to stabilize the soluble Pu(III). Although circumstantial, the authors suggested that the bacilli reduced Pu(IV) to Pu(III), which in turn was complexed and spontaneously oxidized by NTA. Fortunately, the Pu(IV)-NTA complex remained in solution, and could be quantified using liquid scintillation techniques.

In conclusion, microorganisms are known to participate in dissolution processes. Due to the similarities between Fe and Mn and radionuclides, there is mounting evidence that microorganisms will affect the dissolution rates of radionuclides. The long-term adsorption/precipitation of radionuclides on the surface of the rock matrix, or the stability of radiocolloids, would be affected by microorganisms. Therefore, the overall result of enhanced dissolution is difficult to predict. On one hand, a portion of the radionuclides removed from the surface of the rock matrix would be in

TABLE 16.4

Reduction Potentials (pH 7, $Mn^+(Aq) = 10^{-6}$ M)
for Selected Metal Oxyhydroxides

Reaction	E'(v)
$UO_2^{2+} + 2e^- = UO_2(s)$	0.94
$\frac{1}{2}\text{-}MnO_2(s) + 2H^+ + e^- = \frac{1}{2}Mn^{2+} + H_2O$	0.64
$\alpha\text{-}FeOOH(s) + 3H^+ + e^- = Fe^{2+} + 2H_2O$	−0.22
$PuO_2\gamma(H_2O)(s) + 4H^+ + e^- = Pu^{3+} + (\dot{O} + 2)H_2O$	−0.27

Adapted from Rusin, P.A., Quintana, J.R., Brainard, B.A.,
et al. *Environ. Sci. Technol.*, 28, 1686, 1994. With permission.

solution and therefore available to be complexed or chelated by microbial metabolites, thereby increasing the migration rate of radionuclides. On the other hand, another portion of the radionuclides would precipitate onto or be readsorbed by the matrix. Alternatively, the removal of adsorbed/precipitated radionuclides from the surface of natural colloids or the dissolution of radiocolloids, both of which have the potential to be very mobile, would have a net negative effect on the transport rate; because before dissolution all of the colloidal radionuclides were potentially mobile, whereas after dissolution only the complexed/chelated portion would remain in solution.

16.3.7   Colloidal Agglomeration

Colloidal dispersion has long been implicated as a means of transporting radionuclides through soil and rock systems (Buddemeier and Hunt, 1988). When irreversibly attached to a colloid, these radionuclides are unable to participate in sorption/desorption reactions with the soil or rock matrix. As a consequence, radionuclides can potentially move at an accelerated rate with colloids through the soil or rock matrix. If, however, colloids become attached to one another to form agglomerates, then the colloids would no longer be available to participate in colloidal dispersion processes because their increased size would exclude them from movement through the small pores, or because their increase in size would result in an accelerated sedimentation rate. Obviously, the net result would be an overall decrease in the transport of metals and wastes.

Within the literature there is a substantial body of information regarding the interactions between microorganisms and solid surfaces (including colloidal particles), and the affects of microbial adhesion on colloids. These interactions include attraction processes: adhesion, adsorption, and flocculation.

16.3.7.1   Attraction of Bacteria to Solid Surfaces

Many natural habitats have a low nutrient status, therefore solid surfaces are potential sites for concentrating nutrients (as ions and macromolecules) and, consequently, for promoting intensified microbial activity. The movement of water across a surface provides increased opportunities for microorganisms to approach solid-liquid interfaces. In addition, there are many physico-chemical and biological attraction mechanisms operative in the immediate vicinity of interfaces. Though not described here, these attraction mechanisms include chemotaxis, Brownian motion, electrostatic attraction, van der Waals interaction, electrical double layer effects, and cell surface hydrophobicity.

Adsorption of organic substrates on soils depends on the nature of the particulate matter, the organization of the fabric, the clay types, the cationic status of the soils, and on the concentration and molecular structure of the substrate. Hence, the availability of substrates to soil microorganisms may be enhanced or reduced by the presence of particulates. If substrates are concentrated at the surface of clay minerals, then these minerals would become populated with microorganisms utilizing those substrates, thereby increasing the potential for interactions between microorganisms and clay particles. Such interactions often lead to formation of multiple bacterial-colloidal particle agglomerates, which then result in the removal of the colloids from suspension.

16.3.7.2 Adhesion

The adhesion of bacteria to inanimate surfaces is widely recognized as having enormous ecological significance. Adhesion of microorganisms is involved in certain diseases of humans and animals, in dental plaque formation, in industrial processes, in fouling of man-made surfaces, in microbial influenced corrosion, and in syntrophic and other community interactions between microorganisms in natural habitats. Most aquatic bacteria appear to adhere to surfaces by means of surface polymers, including cellular lipopolysaccharides, extracellular polymers and capsules, pili, fimbriae, flagella, and more specialized structures such as appendages and prosthecae. Even though these surface components play a role in the initial, reversible adhesion, they often serve to anchor the bacterium at an interface by polymeric bridging.

The composition and quantity of bacterial cell surface polymers vary considerably and are strongly influenced by growth and environmental conditions. Although extracellular polysaccharides have been reported as being responsible for irreversible adhesion, this is not always true. For example, Brown and co-workers (1977) demonstrated adhesion in mixed, carbon-limited populations despite no evidence of extracellular polymer production. Also, a nitrogen-limited culture resulted in poor adhesion, although large extracellular polymer production was observed. It should always be kept in mind that polymers present between the cell and the substratum, but not observed in light or scanning electron microscopy, could be responsible for the adhesion. It appears that, with respect to inanimate surfaces, there is a subtle balance between cell surface components (extracellular polysaccharides or lipopolysaccharides [LPS]) that may reduce or promote adhesion to inanimate surfaces. Jonsson and Wadstrom (1983) observed that an encapsulated *Staphylococcus aureus* strain did not bind to hydrophobic octyl-Sepharose® gel, whereas a noncapsulated variant showed binding capabilities. In fact, polyanionic extracellular carbohydrate material may not be of primary concern with the initial adhesion processes, but rather with development of subsequent bacterial film (Pringle et al., 1983). The presence of LPS reduces cell surface hydrophobicity and decreases adhesion to the air-water interface of *Salmonella typhimurium* (Hermansson et al., 1983).

16.3.7.3 Adsorption to Colloidal Particles

The adsorption of colloidal clays to bacterial surfaces has been studied by Lahav (1962) and Marshall (1968, 1969a,b). Lahav (1962) suggested that clay platelets may be oriented in a number of ways at the bacterial surface, as a consequence of the net negative charge on clay platelets and the existence of some positive charges on broken edges of platelets. Marshall (1968, 1969b) reported that a species of *Rhizobium* with a carboxyl type ionogenic surface sorb more Na^+-illite per cell than do species with a carboxyl-amino ionogenic surface. Using sodium hexametaphosphate (HMP) to suppress positive charges on platelet edges, Marshall (1969a) found that the HMP-clay did not sorb to a carboxyl type bacteria, whereas a limited amount of this clay sorbed to carboxyl-amino type bacteria. He interpreted these results in terms of the electrostatic attraction between the platelets and the bacterial cell surfaces. Normal sodium montmorillonite particles sorb in an edge-to-face manner to carboxyl type bacterial surfaces, with positive-charged edges of the clay attracted to the negative-charged bacterial surface. Sorption in this manner is prevented by neutralization of positive edge charges of the clay by HMP.

16.3.7.4 Microbially Enhanced Flocculation

In recent years, several processes have been patented for the flocculation of clays, particularly clays derived from phosphate beneficiation. Microbial polysaccharides from such organisms as *Pullularia, Xanthomonas, Arthrobacter, Cryptococcus, Hansenula,* and *Plectania* were found to flocculate finely divided inorganic solids in an aqueous medium (Goren, 1968). In another application, the use of alkaline-treated microbial nucleoprotein is described for flocculating organic and inorganic wastes (Bomstein, 1972). Used in this process were nucleoproteins from the microbes *Polangium, Myxococcus, Sorangium, Flavobacterium, Leuconostoc, Micrococcus,* and *Alcaligenes.* Nucleoprotein derived from these organisms was treated with any one of a variety of alkaline compounds, including $Ca(OH)_2$, KOH, NaOH, NH_3, Na_3PO_4, and quaternary ammonium compounds, which would raise the pH to the point where the microbial material would lyse and form a sol. Flocculation of suspended waste resulted when the concentration of alkaline-treated microbial material was present in concentrations of 1 to 500 ppm (Bomstein, 1972). Floc deterioration can result from biological factors as well as physical factors. Synthetic and natural polymers used for flocculation of colloids may be subject to degradation by microorganisms, which could result in floc destabilization (Brown and Lester, 1979; Obayashi and Gaudy, 1973). Finally, Hersman (1995) has demonstrated that both microbial cells and their metabolites enhanced the agglomeration rates of bentonite particles.

Although the overall results of colloidal/radionuclide transport is not completely understood, in general, microbial-mediated colloidal agglomeration will retard the transport of radionuclides as colloids.

16.4 Conclusions

Microorganisms inhabit subsurface environments. Evidence suggests that many subsurface microbial communities have been in place for thousands of years (for more information, see Chapters 9 and 11). Therefore, the subsurface microbial communities should be regarded as being stable, and this stability becomes an important parameter when developing performance evaluations for the migration of radionuclides in the far-field environment. Because of its stability, the subsurface community will be in place and metabolically active, however diminutively, for the entire time frame required for repository performance. Given the time frame required for far-field performance (e.g., 10,000 years for Yucca Mountain), the potential for significant microbial effects on transport are increased, and therefore the metabolic process performed by microorganisms when interacting with radionuclides must become part of performance assessment. As discussed, microbial metabolic processes affecting radionuclide solubility are significant and varied. These processes include, but are not limited to, sorption/precipitation, complexation/chelation, dissolution, oxidation/reduction reactions, and colloidal agglomeration. Additionally, microorganisms create microenvironments of nutrient and chemical gradients, capable of altering radionuclide solubilities.

As West et al. (1991) reminded us, the migration of radionuclides is very complicated, without the involvement of microorganisms. Unfortunately, the subsurface locations of proposed high-level repositories are not sterile. The overall effects of microorganisms on transport must be integrated into transport calculations — they cannot be ignored.

Acknowledgments

This work was supported by the Yucca Mountain Site Characterization Office as part of the Civilian Radioactive Waste Management Program. This project is managed by the U.S. Department of Energy, Yucca Mountain Site Characterization Project.

References

Amy, P. S., D. L. Haldeman, D. Ringelberg, D. H. Hall, and C. Russell. 1992. Comparison of identification systems for classification of bacteria isolated from water and endolithic habitats within the deep subsurface. *Appl. Environ. Microbiol.,* 58:67.

Arnold, R. G., T. J. DiChristina, and M. R. Hoffmann. 1986a. Inhibitor studies of dissimilative Fe(III) reduction by *Pseudomonas* sp. strain 200 *("Pseudomonas ferrireductans")*. *Appl. Environ. Microbiol.,* 52:281.

Arnold, R. G., T. M. Olsen, and M. R. Hoffmann. 1986b. Kinetics and mechanism of dissimilative Fe(III) reduction by *Pseudomonas* sp. 200. *Biotechnol. Bioeng.,* 28:1657.

Arnold, R. G., T. J. DiChristina, and M. R. Hoffmann. 1988. Reductive dissolution of Fe(III) oxides by *Pseudomonas* sp. 200. *Biotechnol. Bioeng.,* 32:1081.

Arnold, R. G., M. R. Hoffman, T. J. DiChristina, and F. W. Picardal. 1990. Regulation of dissimilatory Fe(III) reduction activity in *Shewanella putrifaciens. Appl. Environ. Microbiol.,* 56:2811.

Bachofen, R. 1991. Gas metabolism of microorganisms. *Experientia* 47:508.

Bar-Or, Y. 1990. The effect of adhesion on survival and growth of microorganisms. *Experientia,* 46:823.

Barnhart, B. J., E. W. Campbell, J. M. Hardin, E. Martinez, D. E. Caldwell, and R. Hallett. 1979. Potential microbial impact on transuranic wastes under conditions expected in the Waste Isolation Pilot Plant, LA-7788-PR, Los Alamos National Laboratory, Los Alamos, NM.

Beveridge, T. J. and Doyle, R. J., 1989. *Metal Ions and Bacteria,* John Wiley & Sons, New York.

Bienfait, H. F. 1988. Proteins under the control of the gene for Fe efficiency in tomato. *Plant Physiol.,* 88:785.

Birch, L. and R. Bachofen. 1990. Complexing agents from microorganisms. *Experientia,* 46:827.

Bomstein, R. A. 1972. Waste Treatment With Microbial Nucleoprotein Flocculating Agent. U.S. Patent No. 3,684,706.

Bondietti, E. A., S. A. Reynolds, and M. H. Shanks. 1976. Interaction of plutonium with complexing organics in soils and natural waters, in *Transuranium Nuclides in the Environment,* I.A.E.A, Vienna, pp. 273.

Bossier, P., M. Hofte, and W. Verstraete. 1988. Ecological significance of siderophores in soil. *Adv. Microbiol. Ecol.,* 10:385.

Boylen, C. W. and J. C. Ensign. 1970. Intracellular substrates for endogenous metabolism during long-term starvation of rod and spherical cells of *Arthrobacter crystallopietes. J. Bacteriol.*, 103:578.

Brainard, J. R., B. A. Strietelmeier, P. A. Smith, P. J. Langston-Unkefer, R. R. Barr, and R. R. Ryan. 1992. Actinide binding and solubilization by microbial siderophores, *Radiochim. Acta*, 58/59:357.

Brockman, F. J., T. L. Kieft, J. K. Fredrickson, B. N. Bjornstad, S. W. Li, W. Spangenburg, and P. E. Long. 1992. Microbiology of vadose zone paleosols in south-central Washington State. *Microb. Ecol.*, 23:279.

Brown, C. M., D. C. Ellwood, and J. R. Hunter. 1977. Growth of bacteria at surfaces. Influence of nutrient limitations. *FEMS Microbiol. Lett.*, 1:163.

Brown, M. J. and J. N. Lester. 1979. Metal removal in activated sludge: The role of extra-cellular polymers. *Water Res.*, 13:817.

Buddemeier, R. W. and J. R. Hunt. 1988. Transport of colloidal contaminants in groundwater. Radionuclide migration at the Nevada Test Site. *Appl. Geochem.*, 3:535.

Burdige, D. J. 1993. The biogeochemistry of manganese and iron reduction in marine sediments. *Earth Sci. Rev.*, 35:249.

Clark, D. L., J. G. Watkin, D. E. Morris, and J. M. Berg. 1994. Molecular models for actinide speciation, Report, LA-12780-MS, UC-802. Los Alamos National Laboratory, Los Alamos, NM.

Clarkson, W. W., L. R. Krumholz, and J. M. Suflita. 1996. Enumeration of anaerobic bacteria in Yucca Mountain, Nevada rock samples. Proc. 7th Annu. Int. High-Level Radioactive Waste Manage. Conf., Las Vegas, NV, April 29-May 3.

Colwell, F. S. 1989. Microbiological comparison of surface soil and unsaturated subsurface soil from a semiarid high desert. *Appl. Environ. Microbiol.*, 55:2420.

Davidson, W. 1993. Iron and manganese in lakes. *Earth Sci. Rev.*, 34:119.

Davis, J. A. and J. Leckie. 1978. Effects of adsorbed complexing ligands on trace metal uptake by hydrous oxides. *Environ. Sci. Technol.*, 12:1309.

Dawes, E. A. 1985. Starvation, survival, and energy reserves, in *Bacteria in Their Natural Environments*, M. Fletcher and G. Floodgate, Eds., Academic Press, New York, pp. 43.

Dawson, M. P., B. A. Humphrey, and K. C. Marshall. 1981. Adhesion: a tactic in the survival strategy of a marine vibrio during starvation. *Curr. Microbiol.*, 6:195.

Dwyer, F. P. and D. P. Mellor. 1964. *Chelating Agents and Metal Chelates*, Academic Press, New York.

Ehrlich, H. L. 1990. *Geomicrobiology*, Marcel Dekker, New York.

Faison, B. D., C. A. Cancel, S. N. Lewis, and H. I. Adler. 1990. Binding of dissolved strontium by *Micrococcus luteus. Appl. Environ. Microbiol.*, 56:3649.

Fliermans, C. B. and T. C. Hazen. 1990. Proc. First Int. Symp. Microbiol. of the Deep Subsurface, OSR 25-82-W(4-89), 1. Westinghouse Savannah River Co., Aiken, SC, 1.

Focht, D. D. 1992. Diffusional constraints on microbial processes in soil. *Soil Sci.*, 154:300.

Francis, A. J. 1982. Microbial transformation of low-level radioactive waste, in Environmental Migration of Long-Lived Radionuclides, Report SM-257/72. I.A.E.A. Vienna.

Francis, A. J. 1990. Characteristics of nuclear and fossil energy wastes. *Experientia*, 46:794.

Francis, A. J. and C. J. Dodge. 1993. Influence of complex structures on the biodegradation of iron-citrate complexes. *Appl. Environ. Microbiol.*, 59:109.

Francis, A. J., C. J. Dodge, F. Lu, G. P. Halada, and C. R. Clayton. 1994. XPS and XANES studies of uranium reduction by *Clostridium* sp. *Environ. Sci. Technol.*, 28:636.

Fredrickson, J. K., F. J. Brockman, B. N. Bjornstad, P. E. Long, S. W. Li, J. P. McKinley, J. V. Wright, J. L. Conca, T. L. Kieft, and D. L. Balkwell. 1993. Microbiological characteriztics of pristine and contaminated deep vadose sediments from and arid region. *Geomicrobiol. J.*, 11:95.

Gadd, G. M. 1990. Heavy metal accumulation by bacteria and other microorganisms. *Experientia*, 46:834.

Geesey, G. G. and L. Jang. 1981. Interactions between metal ions and capsular polymers, in *Metal Ions and Bacteria*, T. J. Beveridge and R. J. Doyle, Eds., John Wiley & Sons, New York, pp. 325.

Ghiorse, W. C. and J. T. Wilson. 1988. Microbial ecology of the terrestrial subsurface. *Adv. Appl. Microbiol.*, 33:107.

Giesy, J. P., G. T. Leversee, and D. R. Williams. 1977. Effects of naturally occurring aquatic fractions of cadmium toxicity to *Simomcephalus serrulatus* (Daphnidae) and *Gambusia affinis* (Poecilliidae). *Water Res.*, 11:1013.

Goren, M. B. 1968. Process for Flocculating Finely Divided Solids Suspended in an Aqueous Medium with a Microbial Polysaccharide, U.S. Patent No. 3,406,114.

Griffin, D. M. 1981. Water potential as a selective factor in the microbial ecology of soils, in *Water Potential Relations in Soil Microbiology*, Parr, J. F., Gardner, W. R., and Elliott, L. F., Eds., Soil Science Society of America, Madison, WI., pp. 141.

Grusak, M. A., R. M. Welch, and L. V. Kochian. 1990. Physiological characterization of a single-gene mutant of *Pisum sativum* exhibiting excess iron accumulation. *Plant Physiol.*, 93:976.

Haldeman, D. L. and P. S. Amy. 1993. Bacterial heterogeneity in deep subsurface tunnels at Rainier Mesa, Nevada Test Site. *Microb. Ecol.*, 25:183.

Haldeman, D. L., B. J. Pitonzo, S. P. Story, and P. S. Amy. 1994. Comparison of the microbiota recovered from surface and deep subsurface rock, water, and soil along an elevational gradient. *Geomicrobiol. J.*, 12:99.

Haldeman, D. L., L. Ragatz, and P. S. Amy. 1996. Distribution and nutrient limitations of heterotrophic bacteria from Yucca Mountain. Proc. 7th Annu. Int. High-Level Radioactive Waste Manage. Conf., Las Vegas, NV, April 29-May 3.

Hermansson, M., S. Kjelleberg, T. K. Korhonen, and T. A. Stenstrom. 1983. Hydrophobic and electrostatic characterization of surface structures of bacteria and its relationship to adhesion at an air/water surface. *Arch. Microbiol.*, 131:308.

Hersman, L. E. 1986. Sorption of [239]Pu by a *Pseudomonas* species. Abstracts of the Annual Meeting, American Society for Microbiology, Washington, D.C., pp. 258.

Hersman, L. E. 1992. Study plan for biological sorption and transport, Yucca Mountain Project Report YMP-LANL-8.3.1.3.4.2, RO. Los Alamos National Laboratory, Los Alamos, NM.

Hersman, L. E. 1995. Microbial effects on colloidal agglomeration, Los Alamos National Laboratory Report, LA-12972-MS, Los Alamos, NM.

Hersman, L. E. 1996. Microbiological sorption and transport: Field and laboratory experiments. Proc. 7th Annu. Int. High-Level Radioactive Waste Manage. Conf., Las Vegas, NV, April 29-May 3.

Hersman, L., T. Lloyd, and G. Sposito. 1995. Siderophore-promoted dissolution of hematite. *Geochim. Cosmochim. Acta*, 59:3327.

Hersman, L. E., P. D. Palmer, and D. E. Hobart. 1993. The Role of Siderophores in the Transport of Radionuclides. Materials Research Society Proceedings, Materials Research Society, Pittsburgh, PA, pp. 294, 765.

Hersman, L., W. Purtymun, and J. Sinclair. 1988. Preliminary microbiological analysis of the vadose zone, Pajarito Plateau, NM., Abstr. Ann. Meet. Am. Soc. Microbiol., American Society for Microbiology, Washington, D.C., pp. 252.

Horn, J. M., B. Economides, A. Meike, and R. D. McRight. 1996. Initial studies to assess microbial impacts on nuclear waste deposition. Proc. 7th Annu. Int. High-Level Radioactive Waste Manage. Conf., Las Vegas, NV, April 29-May 3.

Horn, J. M. and A. Meike. 1995. Microbial activity at Yucca Mountain, Report UCRL-ID-122256, Lawrence Livermore National Laboratory, Livermore, CA, pp. 17.

Inouye, M. 1979. *Bacterial Outer Membranes*, John Wiley & Sons, New York.

Johnson, A. C. and M. Wood. 1992. Microbial potential of sandy aquifer material in the London Basin. *Geomicrobiol. J.*, 10:1.

Jonsson, P. and T. Wadstorm. 1983. High surface hydrophobicity of *Staphylococcus aureus* as revealed by hydrophobic interaction chromatography. *Curr. Microbiol.*, 8:347.

Khalil, M., D. Haldeman, A. Igbinovia, P. Castro, L. Ragatz, and P. Amy. 1996. Autotrophic and heterotrophic bacterial diversity from Yucca Mountain. Proc. 7th Annu. Int. High-Level Radioactive Waste Manage. Conf., Las Vegas, NV, April 29-May 3.

Kieft, T. L., P. S. Amy, F. J. Brockman, J. K. Fredrickson, B. N. Bjornstad, and L. L. Rosacker. 1993. Microbial abundance and activities in relation to water potential in the vadose zones of arid and semiarid sites. *Microb. Ecol.*, 26:59.

Kieft, T. L., J. K. Fredrickson, J. P. McKinley, B. N. Bjornstad, S. A. Rawson, T. J. Philps, F. J. Brockman, and S. M. Pfiffner. 1995. Microbiological comparisons within and across contiguous lacustrine, paleosol, and fluvial subsurface sediments. *Appl. Environ. Microbiol.*, 61:749.

Kieft, T. L., W. P. Kovacik, and J. Taylor. 1996. Factors limiting microbial activity in volcanic tuff at Yucca Mountain. Proc. 7th Annu. Int. High-Level Radioactive Waste Manage. Conf., Las Vegas, NV, April 29-May 3.

Kim, J. I. 1993. The chemical behavior of transuranium elements and barrier functions in natural aquifer systems, *Materials Research Society Symposium Proceedings*, Materials Research Society, Pittsburgh, PA, 294, 3.

Kim, J. I. and B. Kanellakopulos. 1989. Solubility products of plutonium(IV) oxide and hydroxide. *Radiochim. Acta*, 79:26.

Kjelleberg, S. and M. Hermansson. 1984. Starvation induced effects on bacterial surface characteriztics. *Appl. Environ. Microbiol.*, 48: 497.

Konopka, A. and R. Turco. 1991. Biodegradation of organic compounds in vadose zone and aquifer sediments. *Appl. Environ. Microbiol.*, 57: 2260.

Kostka, J. E., G. W. Luther, and K. H. Nealson. 1995. Chemical and biological reduction of Mn(III)-pyrophosphate complexes. Potential importance of dissolved Mn(III) and an environmental oxidant. *Geochim. Cosmochim. Acta*, 59:885.

Lahav, N. 1962. Adsorption of sodium bentonite particles on *Bacillus subtilis*. *Plant Soil*, 17: 191.

Lappin-Scott, H. M. and J. W. Costerton. 1990. Starvation and penetration of bacteria in soils and rocks. *Experientia*, 46:807.

Lesuisse, E., R. R. Crichton, and P. Labbe. 1990. Iron-reductases in the yeast *Saccharomyces cerevisiae*. *Biochem. Biophys. Acta*, 1038: 253.

Lesuisse, E., B. Horion, P. Labbe, and F. Hilger. 1991. The plasma membrane ferrireductase activity of *Saccharomyces cerevisiae* is partially controlled by cyclic AMP. *Biochem. J.*, 280: 545.

Lesuisse, E., F. Raguzzi, and R. R. Crichton. 1987. Iron uptake by the yeast *Saccharomyces cerevisiae:* involvement of a reduction step. *J. Gen. Microbiol.*, 133: 3229.

Long, P. E. and S. A. Rawson. 1991. Hydrologic and geochemical controls on microorganisms in subsurface formations, in Pacific Northwest Laboratory Annual Report for 1990 to the DOE Office of Energy Research, Part 2. Environmental Sciences (PNL-7600 pt. 2), Department of Energy, Washington, D.C.

Lovley, D.R. 1991. Dissimilatory Fe(III) and Mn(IV) reduction. *Microbiol. Rev.*, 55:259.

Lovley, D. R. and E. J. P. Phillips. 1986a. Availability of ferric iron for microbial reduction in bottom sediments of the freshwater tidal Potomac River. *Appl. Environ. Microbiol.*, 52:751.

Lovley, D. R. and E. J. P. Phillips. 1986b. Organic matter mineralization with reduction of ferric iron in anaerobic sediments. *Appl. Environ. Microbiol.*, 51:683.

Lovley, D. R. and E. J. P. Phillips. 1987. Rapid assay for microbially reducible ferric iron in aquatic sediments. *Appl. Environ. Microbiol.*, 53:1536.

Lovley, D. R. and E. J. P. Phillips. 1988a. Manganese inhibition of microbial iron reduction in anaerobic sediments. *Geomicrobiol. J.*, 6:145.

Lovley, D. R. and E. J. P. Phillips. 1988b. Novel mode of microbial energy metabolism: organic carbon oxidation coupled to dissimilatory reduction of iron or manganese. *Appl. Environ. Microbiol.*, 54:1472.

Lovley, D. R. and E. J. P. Phillips. 1992. Reduction of uranium by *Desulfovibrio desulfuricans*. *Appl. Environ. Microbiol.*, 58:850.

Lovley, D. R., E. J. P. Phillips, Y. A. Gorby, and E. R. Landa. 1991. Microbial reduction of uranium. *Nature*, 350:413.

Lovley, D. R., P. K. Widman, J. C. Woodward, and E. J. P. Phillips. 1993. Reduction of uranium by cytochrome C3 of *Desulfovibrio vulgaris*. *Appl. Environ. Microbiol.*, 59:3572.

Luther, G. W., D. B. Nuzzio, and J. Wu. 1994. Speciation of manganese in Chesapeake Bay waters by voltammetric methods. *Anal. Chem. Acta*, 284:473.

Marshall, K. C. 1968. Interaction between colloidal montmorillonite and cells of Rhizobium species with difference ionogenic surfaces. *Biochem. Biophys. Acta*, 156:179.

Marshall, K. C. 1969a. Orientation of clay particles sorbed on bacteria possessing different ionogenic surfaces. *Biochem. Biophys. Acta*, 193:472.

Marshall, K. C. 1969b. Studies by microelectrophoretic and microscopic techniques of the sorption of illite and montmorillonite to *Rhizobia*. *J. Gen. Microbiol.*, 56:301.

Mayfield, C. I. and J. F. Barker. 1982. Biogeochemsitry of the backfill/buffer environment, Technical Record TR-186, Atomic Energy of Canada, Whitshell Nuclear Research Establishment, Pinawa, Manitoba.

McElroy, D. L. and J. M. Hubbell. 1989. Vadose zone monitoring at the radioactive waste management complex, Idaho National Engineering Laboratory, in *Proc. Nuclear Waste Isolation in the Unsaturated Zone, Focus '89*, P. A. Witherspoon, J. H. Fiore, and D. B. Slemmons, Eds., American Nuclear Society, Inc., LaGrange Park, IL, pp. 359.

Miekeley, N. and I. L. Kuchler. 1987. Interactions between thorium and humic compounds on surface waters. *Inorg. Chem. Acta*, 140:315.

Morita, R. Y. 1990. The starvation-survival state of microorganisms in nature and its relationship to the bioavailable energy. *Experientia*, 46:813.

Morse, L. R. 1986. Thermodynamic properties, in *The Chemistry of the Actinide Elements*, 2nd ed., G. Seaborg and T. Morse, Eds., Chapman and Hall, New York, pp. 1289.

Nash, K. and G. Choppin. 1980. Interaction of humic and fulvic acids with Th(IV). *J. Inorg. Nucl. Chem.*, 42: 1045.

Nash, K., F. Sherman, A. M. Fiedman, and J. C. Sullivan. 1981. Redox behavior, complexing and adsorption of hexavalent actinides by humic acid and selected clays. *Environ. Sci. Technol.*, 15:834.

Neilands, J. B. 1981. Microbial iron compounds. *Ann. Rev. Biochem.*, 50:715.

Nealson, K. H. 1983a. The microbial iron cycle, in *Microbial Geochemistry*, W. E. Krumbein, Ed., Blackwell Scientific, Oxford, pp. 159.

Nealson, K. H. 1983b. The microbial manganese cycle, in *Microbial Geochemistry*, Krumbein, W. E., Ed., Blackwell Scientific, Oxford, pp. 191.

Nealson, K. H. and C. R. Myers. 1992. Microbial reduction of manganese and iron. New approaches to carbon cycling. *Appl. Environ. Microbiol.*, 58:439.

Nitsche, H., R. C. Gatti, E. M. Standifer, S. C. Lee, A. Muller, T. Prussin, R. S. Deinhammer, H. Maurer, K. Becraft, S. Leung, and S. A. Carpenter. 1993. Measured solubilities and speciations of neptunium, plutonium, and americium in a typical groundwater (J-13) from the Yucca Mountain Region, Milestone Report 3010-WBS 1.2.3.4.1.3.1, LA-12562-MS, Los Alamos National Laboratory, Los Alamos, NM.

Nitsche, H., K. Roberts, T. Prussin, A. Muller, K. Becraft, D. Keeny, S. A. Carpenter, and R. C. Gatti. 1994. Measured solubilities and speciations from oversaturation experiments of neptunium, plutonium, and americium in UE-25P well water from the Yucca Mountain region, Milestone Report 3329 — WBS 1.2.3.4.1.3.1, LA-12563-MS,LANL. Los Alamos National Laboratory, Los Alamos, NM.

Obayashi, A. W. and A. F. Gaudy. 1973. Aerobic digestion of extracellular microbial polysaccharides. *J. Water Pollut. Control Fed.*, 45:1584.

Pabalan, R. T., J. D. Prikryl, P. M. Muller, and T. B. Dietrich. 1993. Experimental study of uranium (6+) sorption on the zeolite mineral clinoptilolite, in Scientific Basis for Nuclear Waste Management XVI, C. G. Interrante, and R. T. Pabalan, Eds., *Materials Research Society Symposium Proceedings*, Materials Research Society, Pittsburgh, PA., pp. 777.

Pappendick, R. I. and G. S. Cambell. 1981. Theory and measurement of water potential, in *Water Potential Relations in Soil Microbiology*, J. F. Parr, W. R. Gardner, and L. F. Elliott, Eds., Soil Science Society of America, Madison, WI., pp. 1.

Pedersen, K. and F. Karlsson. 1995. Investigations of subterranean microorganisms. Their importance for performance assessment of radioactive waste disposal, SKB Technical Report 95-10, Swedish Nuclear Fuel and Waste Management Co., Stockholm, pp. 168.

Perfettini, J. V., E. Revertegat, and N. Langomazino. 1991. Evaluation of cement degradation induced by the metabolic products of two fungal strains. *Experientia*, 47:527.

Pitonzo, B., P. Amy, D. Ringelberg, and D. White. 1995. Isolation and characterization of exopolysaccharide-producing organisms from Yucca Mountain, Abstracts of the Annual Meeting, America Society for Microbiology, Washington, D.C., pp. 346.

Pitonzo, B., P. Castro, P. Amy, D. Jones, and D. Bergman. 1996. Microbially influenced corrosion capability of Yucca Mountain bacterial isolates, Proc. 7th Annu. Int. High-Level Radioactive Waste Manage. Conf., Las Vegas, NV, April 29-May 3.

Pringle, J. H., M. Fletcher, and D. C. Ellwood. 1983. Selection of attachment mutants during the continuous culture of *Pseudomonas fluorescens* and relationship between attachment ability and surface composition. *J. Gen. Microbiol.*, 129:2557.

Raymond, K. N., G. Muller, and B. F. Matzanke. 1984. Complexation of iron by siderophores. A review of their solution and structural chemistry and biological function. *Curr. Chem.*, 123:49.

Revsbech, N. P. and B. B. Jorgensen. 1986. Microelectrodes: their use in microbial ecology, in *Advances in Microbial Ecology*, Vol. 9, Marshall, K. C., Ed., Plenum Press, New York, 293.

Ringelberg, D. B., J. O. Stair, D. C. White, and L. H. Hersman. 1996. In situ characterization of the microbiota in Yucca Mountain sediments. Proc. 7th Annu. Int. High-Level Radioactive Waste Manage. Conf., Las Vegas, NV, April 29-May 3.

Robinson N. J. and P. J. Jackson. 1986. Metallothionein-like metal complexes in angiosperms; their structure and function. *Plant Physiol.*, 67:499.

Romheld V. and H. Marschner. 1986. Mobilization of iron in the rhizosphere of different plant species. *Adv. Plant Nutr.*, 2:155.

Rosacker, L. L. and T. L. Kieft. 1990. Biomass and adenylate energy charge of a grassland soil during drying. *Soil Biol. Biochem.*, 22:1121.

Rusin, P. A., L. Quintana, J. R. Brainard, B. A. Strietelmeier, C. D. Tait, S. A. Elkerg, P. D. Palmer, T. W. Newton, and D. L. Clark. 1994. Solubilization of plutonium hydrous oxide by iron-reducing bacteria. *Environ. Sci. Technol.*, 28:1686.

Russell, C. E., R. Jacobson, D. L. Haldeman, and P. S. Amy. 1994. Heterogeneity of deep subsurface microorganisms and correlations to hydrogeological and geochemical parameters. *Geomicrobiol. J.*, 12:37.

Shumate, S. E., G. W. Strandberg, and J. R. Parrot. 1978. Biological removal of metal ions from aqueous process streams, in Biotechnology in Energy Production, Conservation Biotechnology, and Bioengineering Symposium, No. 8, C. D. Scott, Ed., John Wiley & Sons, New York, pp. 13.

Sibley, T. H., J. R. Clayton, E. A. Wurtz, A. L. Sanchez, and J. J. Alberts. 1984. Effects of dissolved organic compounds on the adsorption of transuranic elements. *Dev. Biochem.*, 1: 289.

Sijmons, P. J., F. C. Lanfermeijer, A. H. de Boer, H. B. A. Prins, and H. F. Bienfait. 1984. Depolarization of cell membrane potential during trans-plasma membrane electron transfer to extracellular electron acceptors in iron-deficient roots of *Phaseolus vulgaris* L. *Plant Physiol.*, 76:943.

Story, S., L. Hersman, M. Martinez, and M. Aldrich. 1996. Unpublished data.

Strandberg, G. W., S. E. Shumate, and J. R. Parrot. 1981. Microbial cells as biosorbants for heavy metals: Accumulation of uranium by *Saccharomyces cerevisiae* and *Pseudomonas aeruginosa*. *Appl. Environ. Microbiol.*, 41:237.

Stroes-Gascoyne, S. 1989. The potential for microbial life in a Canadian high-level nuclear waste disposal vault. A nutrient and energy source analysis. Atomic Energy of Canada Limited, Report NO. AECL-9574. Whiteshell Nuclear Research Establishment, Pinawa, Manitoba.

Stroes-Gascoyne, S. 1996. Microbial studies in the Canadian Nuclear Fuel Waste Management Program. Proc. 7th Annu. Int. High-Level Radioactive Waste Manage. Conf., Las Vegas, NV, April 29-May 3.

Stroes-Gascoyne, S., L. M. Lucht, J. Borsa, T. L. Delaney, S. A. Haveman, and C. J. Hamon. 1994. Radiation resistance of the natural microbial population in buffer materials, Mater. Res. Soc. Symp. Scientific Basis for Nuclear Waste Manage. XVIII, Kyoto, Japan, Materials Research Society, Pittsburgh, PA.

Stroes-Gascoyne, S. and J. M. West. 1994. Microbiological issues pertaining to the Canadian concept for the disposal of nuclear fuel waste, AECL-10808, Scientific Document Distribution Office, AECL Research, Chalk River, Ontario, pp. 18.

Stumm, W. and B. Sulzberger. 1992. The cycling of iron in natural environments. Considerations based on laboratory studies of heterogeneous redox processes. *Geochem. Cosmochim. Acta*, 56:3233.

Treen-Sears, M. E., B. Volesky, and R. Neufeld. 1984. Ion exchange/complexation of the uranyl ion by *Rhizopus* biosorbant. *Biotechnol. Bioeng.*, 26:1323.

Triay, I. R., K. H. Birdsell, A. J. Mitchell, and M. A. Ott. 1993. Diffusion of sorbing and non-sorbing radionuclides in tuff, in High Level Waste Management: Proc. Fourth Annu. Int. Conf., Las Vegas, NV, American Nuclear Society, La Grange Park, IL, pp. 1527.

Triay, I. R., A. J. Mitchell, and M. A. Ott. 1992. Radionuclide migration studies for validation of batch sorption date, in Proceedings of the DOE/Yucca Mountain Site Characterization Project Radionuclide Adsorption Workshop at LANL, Canepa, J. A., Compiler, Report LA-12325-C, 2014, Los Alamos National Laboratory, Los Alamos, NM.

West, J. M., D. J. Haigh, P. J. Hooker, and E. J. Rowe. 1991. Microbial influence on the sorption of 137Cs onto materials relevant to the geological disposal of radioactive waste. *Experientia*, 47:549.

Wildung, R. E. and T. R. Garland. 1982. Effects of plutonium on soil microorganisms. *Appl. Environ. Microbiol.*, 43:418.

Wood, J. M. and H.-K. Wang. 1983. Microbial resistance of heavy metals. *Environ. Sci. Technol.*, 17:521A.

17

Biofilm Processes in Porous Media — Practical Applications

Al Cunningham, Bryan Warwood, Paul Sturman, Kevin Horrigan, Garth James, J. William Costerton, and Randy Hiebert

CONTENTS

KEY WORDS: *biobarrier, plugging, porous media, ultramicrobacteria, ground water, bioremediation, biofilm, containment, contaminants.*

17.1 Introduction

There are many microbial functions that are of interest to scientists and engineers who seek to manipulate processes in the subsurface environment. In bioremediation, we consider the capabilities of different groups of bacteria somewhat avidly because we know that these organisms can live in pore spaces in this environment and biodegrade organic pollutants and reduce metallic pollutants to an insoluble immobile form. In oil recovery, we are equally attracted to the well-established bacterial properties of plugging by slime (exopolysaccharide) production and of wax mobilization by surfactant production. There are literally dozens of potential uses for microorganisms in the subsurface and most of these involve bacteria because their small size (1 to 4 µm) makes them potentially more mobile in a porous medium than larger eukaryotic microorganisms.

The first attempts to use bacteria in the subsurface date back to the 1940s, in central Europe, when vegetative cells of a variety of potentially useful organisms were simply pumped down water injection wells with the intent of plugging the high-permeability "thief" zones that cause water breakthrough to the producing wells, and signal the end of effective secondary oil recovery. Success was claimed in some field experiments, but these well treatments were generally ineffective and the idea was largely abandoned. In 1984, the biofilm group at the University of Calgary undertook a controlled study of the transport of bacteria through porous media in cooperation with a geologist — Dr. Norman Wardlaw (Shaw et al., 1985). Pure monospecies and mixed natural populations of vegetative bacteria were passed into a porous medium composed of scintered glass beads whose pore throats averaged 27 µm. Direct examination by scanning electron microscopy (SEM) clearly showed that these vegetative cells adhered so avidly to the first available surfaces offered by the porous medium that they quickly plugged the inlet surface to produce a slimly "skin plug" only a few hundreds of microns deep. Direct examination was pivotal in this case because we could actually see the plugging phenomenon, which is very similar to biofilm formation in linear systems, and this process was not apparent to researchers extrapolating from culture data. Workers who had relied on culture methods, and had used bacteria that had been modified by repeated laboratory culture, found that some of these bacteria would pass through various porous media and be found and cultured at the effluent end of the matrix. When we used "wild-type" bacteria that had not lost their adherent properties during repeated lab culture, and when we used microscopy to examine the cells *in situ*, we proved that vegetative bacteria adhere avidly to available surfaces at the inlet of the matrix. This avid adhesion produces a *de facto* biofilm on the inlet surface which is heavily colonized, and even plugged, by metabolically active vegetative bacteria. This active population also may release a number of small starved bacteria that traverse the porous medium and emerge at the effluent end of the matrix. These small starved bacterial cells are well dispersed and they produce relatively large "counts" by traditional plating methods when the effluent fluids are cultured.

When metabolically active mixed populations of bacteria from the subsurface, in our case from "produced" water from a oil recovery operation, were introduced into a porous medium they adhered avidly to the first available surfaces and formed a very shallow "skin plug". Bacteria that had been cultured repeatedly were much more mobile in porous media because they had lost the tendency of wild-type bacteria to adhere to surfaces. It is clear that vegetative bacteria vary enormously in their adhesion capabilities and we must expect to find a spectrum of vegetative cells that span between the stickiness of wild-type biofilm bacteria, and the relative mobility of wild-type motile bacteria, to the higher mobility of cells that have lost their adhesive tendencies upon culture or have been genetically manipulated to delete enzymes involved in exopolysaccharide production. Mixed wild-type populations may be expected to contain cells in all of these states and slime formation by wild-type biofilm bacteria tends to trap cells of all types in the slimy "skin plug". We must also remember that wells that have received injection water containing live wild-type bacteria may already be extensively plugged by an existing skin plug (McKinley et al.,

1988) that will trap subsequently introduced organisms. We conclude that careful manipulation is necessary to deliver vegetative bacteria into the subsurface and that their penetration is inherently limited by their size (about 1.0 μm) and by at least some tendency to adhere to available surfaces.

Having concluded that "wild-type" metabolically active bacterial cells penetrate only very poorly through porous media, even though they are much smaller than the pore throats in those media, we turned to the study of the transport of ultramicrobacteria (UMB) through these same media. UMB were first described in the deep-sea environment by Morita and colleagues (Novitsky and Morita, 1976). They are simply bacterial cells that have undergone radical size reduction and adopted metabolic dormancy in response to starvation (see Chapter 11). In the very oligotrophic deep-sea environment, a wide variety of bacterial species respond to low levels of carbon and energy sources by a sigma factor-directed phenotypic change analogous to a bacterial stress response, in which they reduce their size (0.3 to 0.5 μm) and slow their metabolic processes until they approach complete dormancy (Kjelleberg, 1993). UMB are stable for many years, in the absence of nutrients, but they are capable of rapid resuscitation when suitable nutrients are supplied. These very small, dormant, starved bacterial cells have now been discovered in the deep subsurface (>8,000 ft) in formations where their presence clearly indicates that they are capable of free movement in porous media >150 milliDarcies (mD) in permeability. Because of their small size, their modified adhesion properties, and their capacity to form metabolically active vegetative cells through resuscitation, UMB appeared to constitute an ideal means of transporting bacteria with useful properties deep into the subsurface. A large number of aquatic and subsurface bacteria have been observed to form UMB, and a majority show this response to starvation (Kjelleberg, 1993). A minority of these bacteria form UMB that retain large cell envelopes with shrunken protoplasts, or form irregular-shaped UMB, and these UMB have no inherent advantages over most vegetative cells in terms of their penetration of porous media.

In the sections that follow we will describe experiments in which UMB can be seen to penetrate porous media (>110 mD in permeability) and respond to a suitable nutrient "chaser" by returning to the vegetative state, reproducing, and then producing exopolysaccharide to an extent that the pore spaces of the porous medium are completely occluded by bacterial biomass (McLeod et al., 1988). UMB derived from a variety of bacterial species have been shown to penetrate Berea sandstone cores and sand packs of various permeabilities, ranging from 1.4 Darcies (D) to >18 D. These UMB appear at the effluent end of the test system in small numbers at first, and then in numbers approaching those being injected at the inlet. Mass balance calculations and direct observations of fractured cores and sand samples have shown that some of the UMB that are introduced into porous media are retained in the pores, probably by simple trapping and by adsorption to the matrix material. Experiments conducted by the Alberta Research Council, using medium scale (45 cm diameter) oilfield simulators composed of sandpacks (1.4 D), indicated by extrapolation that UMB can penetrate at least 200 m through these consolidated sands (Cusack et al., 1992). The ability of UMB to readily penetrate porous media >110 mD, and the tendency of these small, dormant, starved cells to be retained within pore spaces as they are transported, sets the stage for resuscitation by nutrient "chasers".

UMB are not motile, and therefore they follow the path of injected water in the subsurface environment. They penetrate wherever pore sizes are permissive and a significant number are retained in each pore structure by trapping and/or adsorption. When a solution containing a suitable nutrient mixture for UMB resuscitation is subsequently injected into the porous medium, these nutrients "catch up " with the 0.3 to 0.5 μm UMB and initiate the relatively rapid process of resuscitation. The UMB return to their vegetative 1.0 to 1.4 μm form and begin to reproduce by binary fission. Immediately following their resuscitation, vegetative bacteria derived from these UMB begin to produce exopolysaccharides (EPS) and the simultaneous nutrient-driven processes of reproduction and EPS production rapidly fill the available pore space if the nutrients provided are sufficient in concentration and readily assimilated (Lappin-Scott et al., 1988). The processes of resuscitation and plugging can be controlled by the judicious choice of nutrients that are fast-acting or slow-acting, by virtue of the extent to which they are assimilable by the resuscitating organisms, and by the rate at which these "wake up" nutrients are pumped into the porous medium in question. Rapid resuscitation of UMB, and slow rates of nutrient injection, can cause plugging very close to the injection point and limit the depth of the biobarrier that is put in place. We have been able to use modeling and other modern engineering tools to design UMB injection and resuscitation protocols that enable us to place biobarriers with considerable accuracy. The barriers to groundwater flow are attractive for environmental purposes because they form preferentially in zones of high permeability, and because they form in any part of the porous medium matrix that exceeds 110 mD in permeability. This latter characteristic is valuable because biobarriers formed by this method would actually be "anchored" in strata such as highly permeable sandstone. Biobarriers would avoid the problems of physical barriers, such as sheet piling and grout, which are readily bypassed by groundwater that flows between them and the underlying bedrock.

Because biobarriers can be constructed from the virtually limitless armamentarium of bacterial species with different physiological capabilities, an almost infinite number of different types of barriers can be constructed. We will show that bacterial strains that reproduce readily and form large amounts of EPS can be used to construct "tight" biobarriers that plug 99.9% of groundwater flow and require subsequent nutrient addition less than every 110 years. On the other hand, biobarriers constructed using strains that produce very little EPS would plug the porous media of the subsurface to a much lesser extent and bacterial cells within these "loose" biobarriers would be capable of pollutant degradation and/or metal reduction as groundwater passed through them. Our present understanding of the microbiology of the subsurface environment has enabled us to devise a UMB-based technology to deliver bacteria of interest into specific regions of this environment for a wide variety of purposes ranging from virtually complete plugging to bioremediation and metal reduction. Perhaps the most reassuring aspect of this new technology is that it uses natural strains of subsurface bacteria in UMB and vegetative forms.

17.2 Biofilm Accumulation

In porous media, as in other aqueous environments, microbial cells may exist in suspension or may adsorb firmly to solid surfaces comprising the porous medium matrix. If favorable environmental conditions persist, adsorbed cells will grow and reproduce at the surface, increasing the amount of attached biomass. If rates of cell adsorption and growth exceed the rate of desorption, a net accumulation of biomass will result on the surface. As the accumulation process continues, additional cells may attach (and detach) directly to (or from) the existing biomass surface. Attachment and detachment are probably the least understood processes affecting the accumulation of biofilm. In this chapter we are concerned with the condition where the substrate supply to the biofilm is sufficient to permit the formation of thick, continuous biofilms, as opposed to patchy, isolated cell colonies. The relationship between the substrate loading rate and biofilm morphology is discussed in detail by Rittmann (1993). Under high substrate loading conditions, the average biofilm thickness on individual media particles will increase, resulting in a corresponding decrease in effective pore space. In systems where the flow rate through the porous medium remains constant, the average pore velocity will increase with the increasing biofilm thickness, while in systems where the piezometric gradient remains constant, pore velocity will decrease. Increased thickness may result in depletion of nutrients within the biofilm structure. It is probable that the net rate of biofilm detachment increases with increasing biofilm thickness, but additional research is needed to fully confirm that assumption. However, it is clear that as the temporal progression of biofilm thickness reaches a quasi steady-state condition, the average specific growth rate for the biofilm must be balanced by the net detachment rate. Particles of organic and inorganic material flowing in suspension may be removed by the attached biomass through filtration processes including diffusion, interception, and sedimentation (Cunningham et al., 1990). The entire deposit of cells and polymers, together with captured organic and inorganic particles, is termed the "biofilm". The amount of biofilm accumulation occurring in a porous medium flow system is therefore the net result of the biomass added through adsorption, growth, attachment, and filtration, less the amount removed by desorption and detachment (Figure 17.1).

Individual biofilm processes are considerably more difficult to examine in porous media than in other common reactor geometries such as flasks, tanks, reservoirs, and pipelines. Biofilm growth, for example, is complicated by the nature of fluid and nutrient transport which, in porous media, occurs along tortuous flow paths of variable geometry. Similarly, the wide distribution of pore velocities introduces considerable variation in the processes of adsorption, desorption, attachment, and detachment.

17.3 Microscale Observations

The accumulation and activity of biofilms varies from point to point along individual pore channels, and thus are considered to be microscale phenomena. Observations of biofilm processes have been reported by several investigators. Cunningham et al. (1991) conducted microscopic observations of porous media biofilm reactors under high substrate loading conditions (25 mg l^{-1}). Media tested consisted of 1 mm glass spheres, 0.70 mm sand, 0.54 mm sand, and 0.12 mm glass and sand. *Pseudomonas aeruginosa* was used as the inoculum and 25 mg l^{-1} glucose substrate was continuously supplied to the reactor. Reactors were operated under constant piezometric head conditions resulting in a flow rate decrease as a biofilm developed. Biofilm accumulation was measured by determining the average thickness on the exposed edges of individual media particles (Figure 17.2). The progression of biofilm thickness followed a sigmoidal curve, reaching a maximum thickness after about 5 days, and maximum thickness was proportional to medium diameter (Cunningham et al., 1991).

Wanner et al. (1995) investigated microscale biofilm accumulation in a packed-bed biofilm reactor inoculated with a pure culture of *P. aeruginosa* that was run under high substrate loading and constant flow rate conditions. The 3.1-cm-diameter cylindrical reactor was 5 cm in length and packed with 1 mm glass beads. Daily observations of

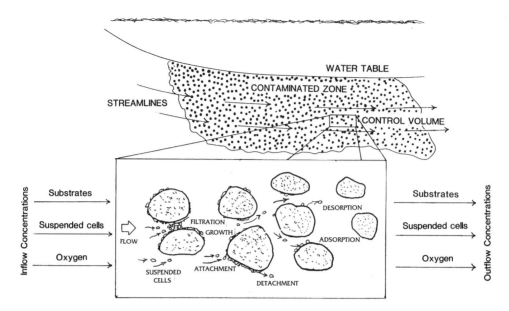

FIGURE 17.1
Schematic showing individual processes contributing to biofilm accumulation and detachment in porous media.

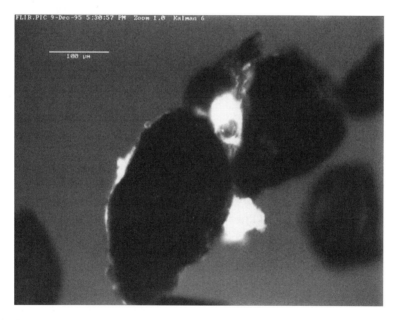

FIGURE 17.2
Pseudomonas aeruginosa biofilm growing on the exposed edges of 150 μm sand grains. The biofilm is the light-colored material attached to the dark-colored sand grains.

biofilm thickness, influent and effluent glucose substrate concentration, and effluent dissolved and total organic carbon were made during the 13-day experiment. Biofilm thickness appeared to reach a quasisteady-state condition after 10 days. AQUASIM, a published biofilm process simulation program (Reichert, 1994), was used to analyze the experimental data. Comparison of observed and simulated variables revealed three distinct phases of biofilm accumulation during the experiment: (1) an initial

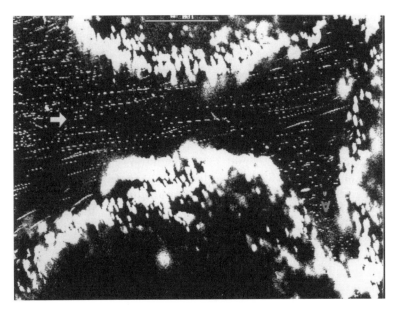

FIGURE 17.3
Streak lines through a porous medium-biofilm matrix. Local fluid velocity was measured by particle image velocimetry using 0.282 μm latex beads and scanning confocal laser microscopy. The arrow indicates the direction of flow. Beads that were transported rapidly were captured by time-lapse photography as strings of beads, while in areas of virtually no flow, where beads demonstrated little or no movement, they were captured as a single bead, or slightly elongated bead.

phase in which substrate removal is determined by cell mass and mass transfer resistance between biofilm and bulk fluid, (2) a growth phase in which cell mass, mass transfer resistance, and area of the biofilm-bulk fluid interface are important, and (3) a mature biofilm phase in which the area of the interface is the important parameter. In this last phase mass transfer resistance between the biofilm and the bulk fluid is insignificant and the biofilm thickness has little effect on the substrate removal rate. It was also found that, in a porous medium under high substrate loading, biofilm detachment is correlated with shear stress at the biofilm surface.

Development of the confocal scanning laser microscope has allowed more detailed examinations of live biofilms under flowing conditions. Stoodley et al. (1994) investigated liquid flow in a model biofilm system consisting of *P. aeruginosa*, *P. fluorescens*, and *Klebsiella pneumoniae*. Local fluid velocity was measured by particle image velocimetry (neutral density 0.282 μm latex spheres) and scanning confocal laser microscopy. Figure 17.3 shows the streaklines defined by the latex spheres flowing between two 250-μm-diameter glass beads in the porous medium flow cell reactor. A biofilm approximately 30 μm in thickness had accumulated on the exposed edges of the beads. Analysis of particle velocity data indicates that as biofilm accumulates in a porous medium it can have a significant effect on local flow velocities and shear stresses. Some channels may become completely blocked while some are only partially restricted.

17.4 Influence on Hydrodynamics

The foregoing experimental investigations indicate that as biofilm thickness increases, the diffusional path length within the biofilm increases, thereby decreasing nutrient concentrations in the base film which will subsequently reduce growth rate. Accumulation of biofilm will continue until the specific growth rate is balanced by the detachment rate. These interactions give rise to the sigmoidal shape exhibited by the accumulation progressions, which indicate that a quasisteady-state thickness is eventually reached (Cunningham et al., 1991).

If sufficient biomass accumulates so as to reduce the effective pore space, a corresponding decrease in medium porosity and permeability as well as an increase in friction factor will ensue. Cunningham et al. (1991) observed that as thickness approached steady-state, medium porosity decreased between 50 and 96% while permeability decreased between 92 and 98%. The porous medium friction factor also increased substantially for all media tested. In all experimental results reported to date, observations of permeability in the biofilm-medium matrix indicate that a minimum permeability (3 to 7×10^{-8} cm^2) persisted after biofilm thickness had reached a maximum value. These observations suggest that, for the experimental conditions used herein, the biofilm accumulation process stabilizes so as to preserve a minimum permeability within the medium biofilm matrix. Such results indicate substantial interaction between mass transport, hydrodynamics, and biofilm accumulation at the fluid/biofilm interface in porous media.

17.5 Mesoscale Biobarrier Evaluation

Mesoscale evaluation of biobarrier formation and persistence was conducted in a series of column and small lysimeter reactors. Up-flow column reactors were used to study biobarrier development and long-term persistence in a one-dimensional flow configuration (nutrients and inocula provided in the hydraulic flow direction), and resistance to heavy metals and organic solvents. Lysimeter reactors were used to study biobarrier formation and persistence in two-dimensional flow systems (nutrients and inocula provided perpendicular to and independent from the hydraulic flow) and their resistance to heavy metals.

17.5.1 One-Dimensional Biobarrier Formation in Column Test Chambers

Test chambers consisted of 91.4-cm lengths of clear polyvinylchloride (PVC) pipe 15.2 cm in diameter. The chamber walls were coated with PVC adhesive and sand to limit bypass flow along the chamber walls. A column test chamber manufactured entirely from 316 stainless steel in the same dimensions as the PVC columns was used for organic solvent challenge work. The columns were packed with varying grades of foundry sand resulting in initial hydraulic conductivities of 0.26 to 2.97 cm/min. The chamber inlet and outlet were layered with large gravel, pea gravel, and course sand to prevent settling of sand into the inlet tubing and wash-out of sand from the outlet tubing. The chamber was fitted with 4 sample ports, spaced exponentially along the column at distances of 8.5, 18.2, 38.0, and 76.6 cm from the column inlet, to allow pressure measurements and withdrawal of fluid samples (Figure 17.4). Pressure was measured using piezometers located at each sample port. Flow of solutions through the chambers was provided by maintaining an 8-l constant-head tank approximately 2 m above the column inlet.

The hydraulic conductivity (K) of the column between each sample port was calculated using Darcy's law,

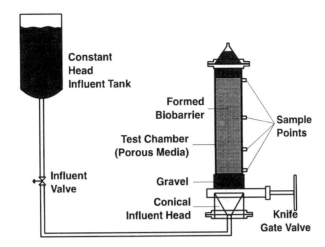

FIGURE 17.4
Sand-packed column for evaluation of biobarrier formation and performance. Nutrient and challenge solutions up-flow through the column from a constant-head reservoir.

$$K = Q/A \; (dh/dl) \tag{17.1}$$

where Q is the flow rate, A is the cross-sectional area of the column, and *(dh/dl)* is the piezometric gradient.

17.5.2 Inoculum Preparation and Column Inoculation

The inoculum for the nonsterile columns consisted of starved cells of a streptomycin-resistant environmental isolate of *Klebsiella oxytoca* suspended in a phosphate-buffered saline (PBS) solution (MacLeod et al., 1988). The bacterial culture was diluted in Sodium Citrate Medium (SCM; Lappin-Scott et al., 1988) to a volume equal to 2 column pore volumes and gravity fed to the column (final cell inoculum of 2.5×10^6 CFU/ml). The inoculated columns were then allowed to incubate under no-flow conditions for 2 days. Thereafter, the columns were perfused daily with SCM until the permeability became constant, after which the columns were continuously supplied with SCM.

Recoverable heterotrophic cells and streptomycin-resistant cells were enumerated in the column effluent during inoculation. Prior to the inoculation, the column effluent contained 1.4×10^5 CFU/ml recoverable heterotrophic cells and $<1 \times 10^3$ streptomycin-resistant cells. The columns were incoculated with a streptomycin-resistant *K. oxytoca* cell population at a concentration of approximately tenfold higher than the initial effluent cell population. Following inoculation, the effluent heterotrophic cell populations were not significantly higher than the initial effluent cell density; however, the number of streptomycin-resistant cells in the column effluent increased more than 100-fold. This result indicates that the streptomycin-resistant inoculum comprised a significant proportion of the culturable microbial community in the column, and in the column effluent. Following the inoculation and during nutrient resuscitation, the distribution of bacteria and nutrient concentrations were determined along the column length. These distributions indicated a homogeneous microbial recovery and effective biobarrier formation (Figures 17.5A and B).

Citrate utilization increased in proportion to the cell number increase, resulting in no detectable citrate throughout the column after 5 days. Phosphate concentrations

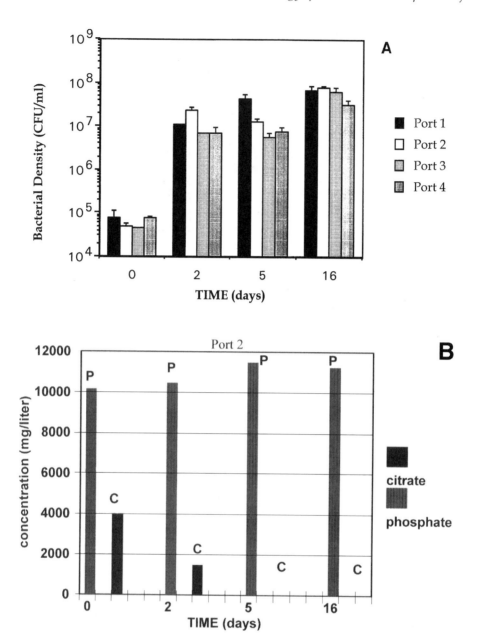

FIGURE 17.5

The column was inoculated with a starved cell suspension of *Klebsiella oxytoca* and incubated under no-flow conditions for 2 days, followed by 2 pore volumes of sodium citrate medium (SCM) and 2 more days of incubation at no flow. At day 5, continuous flow of SCM was initiated. Cells were recovered along the length of the column (ports 1, 2, 3, 4) after inoculation (time 0) and at 2, 5, and 16 days after inoculation (A). Typical citrate and phosphate concentrations were measured throughout a packed-sand column during nutrient resuscitation with SCM (B).

did not change appreciably over the 16-day-biobarrier-formation monitoring period, and provided evidence of homogeneous nutrient medium distribution (Figure 17.5B).

Bacteria were enumerated in sand samples taken from the columns after biobarrier formation (Figure 17.6). The distribution of bacteria was relatively uniform throughout the columns with a mean bacterial density of 108 CFU/g (wet-weight) of sand.

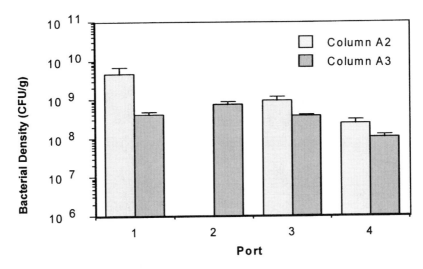

FIGURE 17.6
Bacterial densities in sand samples taken throughout columns A2 (control) and A3 (strontium challenge) after 96 days of biobarrier formation. Densities are reported as CFU per gram of wet sand. Note the even distribution of bacteria throughout the columns.

17.5.3   Reduction in Hydraulic Conductivity

Nutrient resuscitation of the starved bacterial inoculum resulted in a rapid reduction of hydraulic conductivity throughout the column length during the first 30 days. The hydraulic conductivity continued to decrease steadily at a much slower rate until reaching a relatively constant value after about 150 days. The net reduction in column hydraulic conductivity is approximately one order of magnitude (Figure 17.7). Stable hydraulic conductivity reductions were accompanied by a relatively constant number of culturable cells in the column effluent (107 CFU ml^{-1}).

Initial biobarrier development (0 to 50 days) indicates the column inlet sections reduce the hydraulic conductivity to a very low level. During continued barrier formation, the hydraulic conductivity at the effluent end of the column slowly decreases over an extended period (approximately 100 days) until a common hydraulic conductivity exists for the entire column. Biobarrier operation beyond 150 days indicates the column has a homogeneous biobarrier distribution and a common hydraulic conductivity throughout the column at approximately 0.1 cm/min.

Biobarrier persistence was evaluated by challenging the formed barrier with nutrient-free (DI-H_2O; challenge 1) and carbon-free (Mg/Fe solution; challenge 2) hydraulic flow (Figure 17.7). The hydraulic conductivity remained relatively constant for the first 5 days of challenge 1 and the first 10 days of challenge 2, and then rapidly increased in column sections farthest from the inlet. The effluent culturable cell density decreased from 10^7 to 10^6 CFU ml^{-1} during these starvation challenges. Following the barrier challenges, reintroduction of nutrient SCM medium resulted in a rapid reformation of the biobarrier throughout the column.

Resistance of the biobarrier to heavy metals was tested by addition of SCM containing 1 mg l^{-1} strontium (1 ppm) as strontium chloride, and SCM containing 1 mg l^{-1} cesium as cesium chloride (Figure 17.8). A carbon tetrachloride challenge was conducted by adding the solvent to the reservoir to provide a final target concentration of 200 mg l^{-1} carbon tetrachloride in SCM to the column. Neither of the two heavy metals nor the organic solvent had any measurable effect on biobarrier stability as measured by changes in hydraulic conductivity (data not shown).

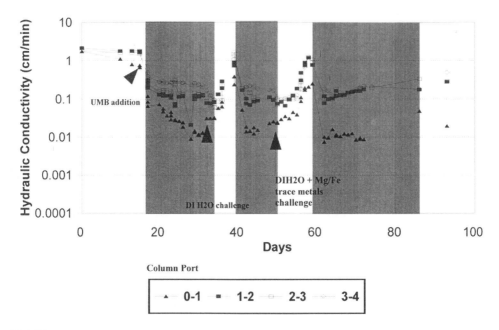

FIGURE 17.7

Biobarrier formation in a packed sand column from *Klebsiella oxytoca* fed sodium citrate medium (SCM). Nutrient-free (DI-water challenge) and carbon-free (Mg/Fe solution challenge) hydraulic flow resulted in a decrease in biobarrier performance. Biobarrier formation and performance returned quickly when SCM was provided (shaded regions).

17.5.4 Biobarrier Formation by Indigenous Microorganisms

An inoculated, nonsterile column packed with sand was fed SCM medium in a similar manner as the previously described column experiments. Another nonsterile column was flooded with the SCM medium, without inoculation, in order to observe the plugging pattern caused by indigenous organisms present in the water and in the sand itself. In the indigenous population column experiment, a reduction in hydraulic conductivity was achieved in the first section of the column compared to the remainder of the column (approximately two orders of magnitude greater; Figure 17.9). In the uninoculated column, the bacterial population was significantly higher near the column inlet and contained a relatively low number of streptomycin-resistant cells (Figure 17.10). In contrast, the bacterial population in the inoculated column was relatively uniform throughout the column length and contained a high proportion of streptomycin-resistant cells. We conclude that column flooding with UMB and nutrients results in a homogeneous plug while the simple resuscitation of indigenous bacteria lends to the formation of a shallow "skin" plug at the column influent.

17.6 Two-Dimensional Biobarrier Formation in Pilot-Scale Lysimeters

Two small pilot-scale lysimeters were designed and manufactured to facilitate evaluation of biobarrier performance in a two-dimensional configuration. This configuration

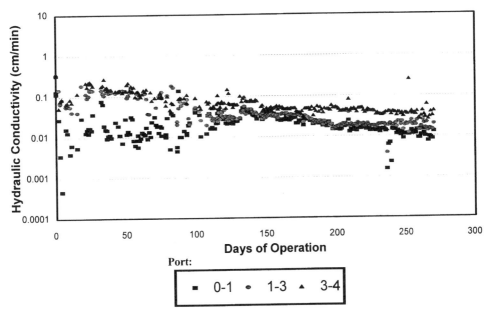

FIGURE 17.8
Established biobarrier resistance to 1 ppm cesium in sodium citrate medium. No deleterious effect of the heavy metal was noted during the 60 day challenge period (days 145 to 207).

provided hydraulic flow perpendicular to biobarrier formation, as compared to column design which consisted of hydraulic flow parallel and in conjunction with biobarrier formation. Evaluation of biobarrier performance in a two-dimensional configuration approximates field conditions more closely. As will be discussed later, the lysimeter provided a three-dimensional test of biobarrier performance. Two lysimeters were manufactured and used during these experiments to provide a side-by-side comparison of parameters affecting biobarrier formation and performance, such as heavy metal challenges and nutrient pulsing, and to provide destructive sampling opportunities (one reactor could be sampled while the other was maintained).

17.6.1 Lysimeter Design

The pilot-scale test lysimeters were manufactured from 0.95 cm and 0.48 cm 316 stainless steel (Figure 17.11). The inlet section was 91.4 cm wide by 61.0 cm deep by 15.2 cm long; the remainder of the reactor was 91.4 cm by 30.4 cm deep by 106.7 cm long, for an overall length of 122 cm. Four 7.6-cm flow distribution baffles are located on the top plate perpendicular to the flow direction at approximately 12.7, 41.9, 61.0, and 88.9 cm and 88.9 cm from the influent end of the top plate to limit sheet flow between the top of the sand and the top plate. A rigid support ran around the circumference of the reactor perpendicular to flow and approximately midway along the top plate, to provide support to the sides, base, and top plate. The support was approximately 1.3 cm high. The support also acted to disrupt flow along the walls, top, and base and redirect it into the sand fill.

Ten sampling ports were installed in two rows with logarithmically increased spacing from the inlet to the effluent end. Piezometer ports were installed in three evenly spaced rows across the reactor that had logarithmically increased spacing from inlet to effluent. Piezometer ports were also located on the hydraulic head tank and the effluent line, providing pressure-drop measurements between the head tank and

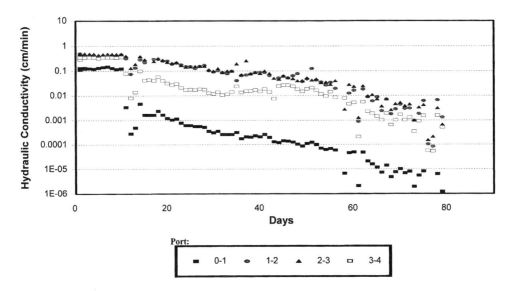

FIGURE 17.9
Skin-plug formation by indigenous bacteria in a packed sand column continuously fed with sodium citrate medium is indicated by the relatively high hydraulic conductivity drop between ports 0-1 and 1-2.

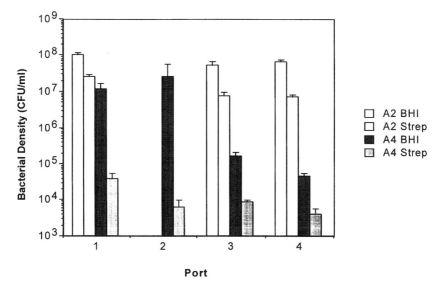

FIGURE 17.10
Total viable bacterial counts (BHI) and streptomycin-resistant bacterial counts (Strep) throughout biobarriers formed in a column (A2) inoculated with streptomycin-resistant *K. oxytoca* UMB and in a column (A4) that was not inoculated.

lysimeter effluent. Each reactor piezometer port was plumbed with a perforated 0.95-cm stainless steel tube to provide pressure-drop information across the full depth of the reactor.

A head tank (constant fluid level) provided reverse osmosis purified water at a constant head to the inlet of the reactor, simulating a constant-head hydraulic flow. This hydraulic gradient flow entered the reactor at the influent section of the reactor (approximately the first 15.2 cm of length), which was filled with coarse gravel to a depth of 45.7 cm. A total of 11 nutrient/inoculation injection ports were located in

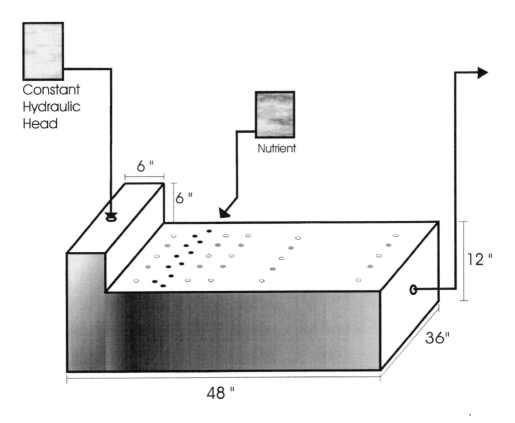

○ Piezometer Ports
● Injection Ports
◒ Sample Ports

FIGURE 17.11
Pilot-scale lysimeter schematic. A constant-head hydraulic head tank provides reverse osmosis water as simulated groundwater flow. Nutrients and UMB are pumped into the injection ports downstream from the hydraulic head inlet. Biobarrier formation is monitored by flow rate measurements an piezometric monitoring of the hydraulic head throughout the lysimeter.

two rows (5 and 6 ports, respectively) near the inlet section of the reactor. These injection ports were also plumbed with perforated 0.95-cm stainless steel tube and are 2-way valved to a stainless T-fitting to provide sampling capabilities through a septum from the top and nutrient/inoculation injection into the "T".

Effluent from the reactor was collected on three levels by perforated 1.3-cm stainless steel tubing buried in coarse gravel in the last 15.2 cm of the reactor. The effluent head was maintained at a constant level approximately 3.8 cm below the head reservoir level.

The reactor was filled (except for the influent and effluent sections described above) with Ottawa Sand (F110) to its full depth. The hydraulic conductivity was calculated based on Darcy's Law (Equation 17.1) and the pore volume of the sand was determined through liquid displacement measurements. A pore volume of approximately 35% was measured for the F-110 sand used to fill the lysimeter beds. The initial overall hydraulic conductivity in the lysimeters was approximately 4.0 to 4.8 cm/min, respectively (calculation based on effluent flow rate and pressure drop across the entire reactor).

Tracer studies were conducted to evaluate the hydraulic flow characteriztics of the reactor prior to inoculation. A pulse of chloride as NaCl was introduced at varying

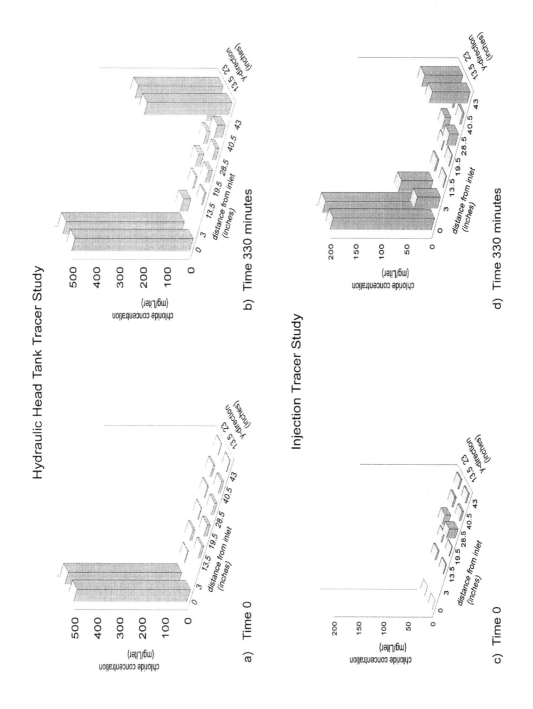

concentrations into the hydraulic head tank and injection ports. The results of these tracer studies indicated preferential flow channels within the reactor beds (lysimeter II tracer results shown in Figure 17.12). These channels are indicated by zones of higher and lower salt concentration within the reactor, independent of their location. In addition, the effluent concentration increased to near-influent concentration much quicker than the sample ports along the length of the reactor. These results are indicative of heterogeneous flow in preferential flow channels compared to a uniform concentration gradient that would be expected under a homogeneous flow pattern.

Inoculation of the lysimeters was conducted by suspending the UMB culture in sodium citrate medium amended with a small amount of nitrate (mSCM) and pumping the suspension into the nutrient/inoculum injection ports using a multihead peristaltic pump. The injection flow rate was set to provide the inoculum flow rate at 10% of the total flow. The lysimeters were inoculated with a total of 0.2 pore volumes (PV) of UMB suspension [total of 2 PVs (inocula + hydraulic flow)] through the reactor. Following inoculum injection, mSCM injection commenced at the same flow rate. Nutrient injection continued until the reactor effluent flow rate equaled the injection flow rate, at which time nutrient injection was terminated. Nutrient injection at 10% of the pre-inoculation effluent flow rate was initiated as needed. The flow rate reduction achieved by the formation of a biobarrier in one lysimeter is plotted in Figure 17.13. The final flow reduction due to biobarrier formation was in excess of 99.99% of the initial flow. This flow reduction was maintained in excess of 99.9% for 4 months without further nutrient addition. The biobarrier challenge was conducted with 1 mg l^{-1} strontium (as strontium chloride) added to the hydraulic head tank. Column flow reduction continued during the strontium challenge (data not shown).

17.6.2 Column vs. Lysimeter Studies

Biobarrier persistence in the columns and lysimeters was significantly different. Column barriers appear to require continuous nutrient addition for barrier maintenance, while the lysimeter barriers were able to persist for a considerable period without nutrient addition. The persistence of the biobarriers in the lysimeters compared to the column experiments may be due to the higher head pressure continually present in the column reactor design. The difference is approximately 122 cm of head above the top of the 91.4-cm column compared to approximately 3.6 cm across the 122-cm length (91.4 cm of sand bed) of the lysimeters. The higher head provides a higher flow rate and pore velocity through the columns, which may increase the wash-out rate.

The results of the tracer studies and piezometric gradient data in the lysimeters indicate the presence of channels or wall flow. The presence of heterogeneous flow through the reactor cross section implies a three-dimensional reactor model rather than a two-dimensional model. The three-dimensional reactor model is therefore useful for field simulation of biobarrier formation and performance.

FIGURE 17.12

Lysimeter II chloride tracer study. Hydraulic head tank tracer study: 500 mg/l chloride as NaCl was added to the head reservoir. Sampling results are plotted for time 0 (a) and 1 pore volume (b). Injection port tractor study: 200 mg/l chloride was pumped into the injection ports at 10% of the combined flow rate (hydraulic + injection). The sampling results for time 0 (c) and 1 pore volume (d) are plotted. Reactor flow is characterized as primarily channel flow.

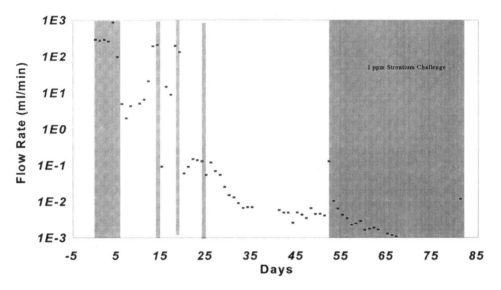

FIGURE 17.13
Lysimeter II effluent flow rate vs. time. UMB/nutrient addition occurred at time 0. Continuous nutrient pumping through the injection ports at 10% of the combined flow rate (hydraulic + injection) during the first 7 days and pulse injection at 3 time periods thereafter (shaded regions) reduced the flow rate 99.999%. A strontium challenge at 1 mg/l in the hydraulic flow was initiated on day 52 (cross-hatch region) and this toxic metal did not affect the biobarriers.

17.7   Conclusions

17.7.1   Biobarrier Development

Initial biobarrier development is a rapid and effective method for reducing the hydraulic conductivity throughout the length of the 3-ft packed sand columns and small lysimeters. An initial reduction in the hydraulic conductivity to less than 10% of the original value was observed in all columns and reactors, and reduction to less than 0.0001% of the initial hydraulic conductivity was obtained in the lysimeter studies.

17.7.2   Biobarrier Maintenance

Reduced reactor flow rates during biobarrier persistence leads to low nutrient addition flow rates and, therefore, minimal nutrient requirements for maintenance (less than 10% of the original liquid flow through the column). However, starvation conditions had a deleterious effect on biobarrier effectiveness in one-dimensional column models, suggesting that nutrient supply is required for biobarrier maintenance under high hydraulic head situations. The results of the column and lysimeter experiments suggest wash-out of the barrier is a function of the hydraulic head and flow rate through the formed barrier. After formation, biobarrier persistence under low flow conditions may require only limited nutrient addition to maintain extremely low hydraulic conductivities. In addition, extended operation of an engineered biobarrier resulted in a homogeneous biobarrier distribution throughout the 91.4-cm packed sand columns and 122-cm lysimeters.

17.7.3   Biobarrier Heavy Metal and Organic Solvent Challenge

Biobarrier performance and formation was not altered by the presence of heavy metal contaminants (strontium and cesium) at 1 ppm levels for periods up to 120 days, or influenced by the presence of carbon tetrachloride at concentrations of 100 mg/l for extended periods and up to 300 mg/l for short periods.

17.7.4   Three-Dimensional Flow Simulation

Documentation of the presence of flow heterogeneities in the pilot-scale lysimeters indicates that a three-dimensional flow pattern was developed, and was effectively plugged, using the biobarrier technology. Preferential wall or flow channeling was significantly reduced in addition to low flow zones within the sand.

Overall, these results indicate the hydrodynamic properties of porous media can be manipulated using microbial biobarriers, and these barriers may be an effective technology for the containment of groundwater contaminants. The UMB-based biobarrier technology is being considered for commercial use in the selective plugging of high-permeability water breakthrough zones in secondary oil recovery by water flooding. A very similar use of the technology is being considered for the construction of barriers to protect natural ecosystems (e.g., rivers and lakes) from pollutant plumes in approaching groundwater. Plugging biobarriers are being considered as reinforcements to the linings of landfills, especially those containing toxic organic molecules and/or heavy metals. UMB introduction, with subsequent nutrient infusion, has been suggested as a means of plugging water "fingering" channels in earthen dams and berms (Lappin-Scott and Costerton, 1992). UMB have been suggested as a delivery mechanism for the introduction of pollutant degrading/reducing organisms into subsurface plumes of organic and metallic pollutants. Perhaps the most intriguing possibility is using UMB to reduce indigenous metals (e.g., iron) in the path of groundwater containing chlorinated organic molecules and/or oxidized heavy metals. These reduced indigenous metals (e.g., ferrous iron) would then reductively dechlorinate the organic pollutant (Caccavo et al., 1996; Gorby et al., 1995) and/or reduce the mobile heavy metals to produce immobile crystals of insoluble metal salts. A mixture of UMB and vegetative bacterial cells could readily be used to produce a zone of nutrient-stimulated metabolic activity at the surfaces of mine tailings that would serve as an oxygen-consuming "cap" that would prevent the penetration of oxygen and the consequent development of acid mine drainage (Blenkinsopp et al., 1991).

Acknowledgments

This research was funded by DOE through MSE, Inc. under agreement 95-C213-CR, and by the U.S Environmental Protection Agency under assistance agreement R-815709 to Montana State University through the Hazardous Substance Research Center for U.S. EPA Regions 7 and 8, headquartered at Kansas State University.

The authors wish to acknowledge the contributions of personnel at MSE, Inc. and at the Center for Biofilm Engineering (an NSF-supported Engineering Research Center under Cooperative Agreement EEC-8907039) located at Montana State University.

References

Blenkinsopp, S.A., D.C. Herman, and J.W. Costerton, The Use of Biofilm Bacteria to Exclude Oxygen from Acidogenic Mine Tailings, in Proc. *Second Int. Conf. Abatement of Acidic Drainage*, Montreal, pp. 369-377, 1991.

Caccavo, F., Jr., N.B. Ramsing, and J.W. Costerton, Starvation and Resuscitation of the Dissimilatory Metal-Reducing Bacterium Shewanella Alga Strain BrY., Manuscript Submitted.

Cunningham, A.B., E.J. Bouwer, and W.G. Characklis. Biofilms in porous media, in *Biofilms*, W.G. Characklis and K.C. Marshall, Eds., John Wiley & Sons, New York, 1990, pp. 697-732.

Cunningham, A.B., W.G. Characklis, F. Abedeen, and D. Crawford. Influence of biofilm accumulation on porous media hydrodynamics, *Environ. Sci. Technol.*, 25:1305-1310, 1991.

Cusack, F., S. Surindar, C. McCarthy, J. Grieco, M. DeRocco, D. Nguyen, H. Lappin-Scott, and J. W. Costerton, Enhanced oil recovery — three-dimensional sandpack simulation of ultramicrobacteria resuscitation in reservoir formation, *J. Gen. Microbiol.*, 138:647-655, 1992.

Gorby, Y.A., D. J. Workman, S.E. Amonette, and J.S. Fruchter, Transformation of carbon tetrachloride by biogenic Fe(II), in *Emerging Technologies in Hazardous Waste Management, VII*, Division of Industrial & Engineering Chemistry, American Chemical Society, Washington, D.C., 1995, p. 593.

Kjelleberg, S., *Starvation in Bacteria*, Plenum Press, New York, 1993.

Lappin-Scott, H.M., F. Cusack, and J.W. Costerton, Nutrient resuscitation and growth of starved cells in sandstone cores: a novel approach to enhanced oil recovery, Appl. *Environ. Microbiol.*, 54(6):1373-1382, 1988.

Lappin-Scott, H.M. and J.W. Costerton, Ultramicrobacteria and Their Biotechnological Applications, *Curr Opinion Biotechnol.*, 3:283-285, 1992.

MacLeod, F.A., H.M. Lappin-Scott, and J.W. Costerton, Plugging of a rock model system by using starved bacteria. Appl. *Environ. Microbiol.*, 54(6):1365-1372, 1988.

McKinley, V.L., J.W. Costerton and D.C. White, Microbial biomass, activity, and community structure of water and particulates retrieved by backflow from a waterflood injection well, *Appl. Environ. Microbiol.*, 54(6):1383-1393, 1988.

Novitsky, J.A. and R.Y. Morita, Morphological characterization of small cells resulting from nutrient starvation of a psychrophilic marine vibrio, *Appl. Environ. Microbiol.*, 32:617-622, 1976.

Reichert, P., AQUASIM — a tool for simulation and data analysis of aquatic systems, *Water Sci. Technol.*, 30(2):21-30, 1994.

Rittmann, B. E., The significance of biofilms in porous media, *Water Resour. Res.*, 29(7):2195-2202, 1993.

Shaw, J.C., B. Bramhill, N.C. Wardlaw, and J.W. Costerton, Bacterial fouling in a model core system, Appl. *Environ. Microbiol.*, 49(3):693-701, 1985.

Stoodley, P., D. DeBeer, and Z. Lewandowski, Liquid flow in biofilm systems, Appl. *Environ. Microbiol.*, 60(8):2711-2716, 1994.

Wanner, O., A.B. Cunningham, and R. Lundman, Modeling biofilm accumulation and mass transport in a porous medium under high substrate loading, *Biotechnology and Bioengineering*, 47:703-712, 1995.

Index

Index